PLANETARY SCIENCE AND EXOPLANETS IN THE ERA OF JWST

IAU SYMPOSIUM 393

COVER ILLUSTRATION:

Abdelkrim Boskri from Oukaimeden Observatory

IAU SYMPOSIUM PROCEEDINGS SERIES

INTERNATIONAL ASTRONOMICAL UNION
UNION ASTRONOMIQUE INTERNATIONALE

International Astronomical Union

IAU

PLANETARY SCIENCE AND EXOPLANETS IN THE ERA OF JWST

PROCEEDINGS OF THE 393rd SYMPOSIUM OF THE INTERNATIONAL ASTRONOMICAL UNION CAPE TOWN, SOUTH AFRICA 13–15 AUGUST 2024

Edited by

ZOUHAIR BENKHALDOUN
Sharjah University, United Arab Emirates and University Cadi Ayyad, Morocco

YOUSSEF MOULANE
Mohammed VI Polytechnic University, Morocco

and

ANDREW SZENTGYORGYI
Harvard-Smithsonian Center for Astrophysics, USA

CAMBRIDGE
UNIVERSITY PRESS

CAMBRIDGE UNIVERSITY PRESS
University Printing House, Cambridge CB2 8BS, United Kingdom
1 Liberty Plaza, Floor 20, New York, NY 10006, USA
10 Stamford Road, Oakleigh, Melbourne 3166, Australia

First published 2026

Printed in Great Britain by Henry Ling Limited, The Dorset Press, Dorchester, DT1 1HD

Typeset in System LaTeX 2ε

A catalogue record for this book is available from the British Library Library of Congress Cataloguing in Publication data

ISBN 9781009546126 hardback
ISSN 1743-9213

For EU product safety concerns, contact us at Calle de José Abascal, 56, 1°, 28003 Madrid, Spain, or email eugpsr@cambridge.org

Table of Contents

Preface viii

Editors x

List of Participants xii

Magnetism of Saturn in the origin of the visible dense rings and in their peculiarities recorded by the Cassini probe 1
Vladimir V. Tchernyi and Sergey V. Kapranov

Migration of bodies ejected from the Earth and the Moon 9
Sergei I. Ipatov

Unveiling New Parent Bodies in Meteor Showers: A Robust Methodological Approach 14
M. Guennoun, J. Vaubaillon, D. Čapek, P. Koten and Z. Benkhaldoun

Host stars and their effects on exoplanet atmospheres: Reconstructing the missing extreme-ultraviolet and Lyman-alpha radiation that exoplanets see 20
Jeffrey L. Linsky and Seth Redfield

Migration of bodies in the Proxima Centauri and Trappist 1 planetary systems . 29
Sergei I. Ipatov

The magnetic field and space environment of known exoplanet hosts: GJ 436 and HD 63433 34
S. Bellotti, S. Boro Saikia, V. Bourrier, C. Danielski, J. F. Donati, D. Evensberget, R. Fares, É. M. Hébrard, G. A. J. Hussain, A. Lavail, T. Lüftinger, G. Micela, J. Morin, C. Moutou, P. Petit and A. A. Vidotto

Observing Exoplanet Biosignatures outside the Conservative Habitable Zone . . 45
Amri Wandel

Impact of Surface Pressure on Nitrogen-Dominated USP Super-Earths Atmospheres 53
J. Chouqar, J. Lustig-Yaeger, Z. Benkhaldoun, A. Szentgyorgyi, A. Jabiri and A. Soubkiou

Water at the earliest stages of star and planet formation 60
Itziar de Gregorio-Monsalvo, Santiago Gil, Alejandro Santamaría-Miranda, José Francisco Gómez, Andrés Pérez-Sánchez, Adele Plunkett, Elizabeth Artur de la Villarmois, Viviana Guzmán and Matthias R. Schreiber

Probing the gas that builds planets: Results from the JWST MINDS program 65
Ewine F. van Dishoeck and the MINDS team

Implementing an AstroLab Project in Africa 72
Youssef Moulane and Abdelouahed El Fatimy

Exoplanets in the Classroom: A Bilingual K12 Educational Suite for Exploring Exoplanet Science 77
Marie-Eve Naud, Julie Bolduc-Duval, Heidi White, Frédérique Baron and Nathalie Nguyen-Quoc Ouellette

Bringing the sky to the living room: Reflections on more than three years of live streaming astronomical events 87
Claudia Mignone, Federica Duras, Giulia Mantovani, Federico Di Giacomo, Gianluigi Filippelli, Livia Giacomini, Sandro Bardelli and Emanuele Scalise

Approach of the NGC1977 star cluster to the TOI-2796 host star. II 97
H. Y. Wakjira, S. H. Negu, M. D. Sizova, E. S. Postnikova, D. A. Mosunova, N. V. Chupina, Anna M. Fateeva, Alexander M. Fateev and Sergei V. Vereshchagin

Stellar variability monitoring and vistas of its support of JWST programs 100
Alisher S. Hojaev

Instability of In-plane Equilibrium Points of the Restricted Three-Body Problem with P-R Drag and Variable Masses 105
Tajudeen Oluwafemi Amuda and Leke Oni

Solar Hydrologic Cycle Determines Earth and Venus Atmospheres Mass and Composition 114
Xuguang Leng

Numerical Geodynamic Simulations of the Thermal Evolution within the Outer Ice Shell of Ganymede 121
Katherine Villavicencio Valero

Observational and Chemical Modelling of Protoplanetary Disks after FU Ori outbursts 126
Lis Zwicky, Tamara Molyarova, Vitaly Akimkin, Dmitry Semenov, Ágnes Kóspál and Péter Ábrahám

Exploring the collinear Lagrangian points of exoplanet systems with P-R drag and oblateness 129
S. N. Hasan, Priya Hasan and Ibtesam Shaikh

Search for new members and age estimation of the young Emilkowalski asteroid family 133
E. D. Kuznetsov, A. S. Perminov, V. S. Safronova and M. A. Vasileva

Advances in Coronagraphy for Future Space Missions: Interferometric Apodization through Homothety (IAH) and its Application in APLC Coronagraphs 138
J. Chafi, Y. El Azhari, O. Azagrouze, A. Jabiri, Z. Benkhaldoun, A. Boskri and A. Qazbour

Spectral Characterization of a Complete Equatorial Sample of 615 K Dwarfs . . 144
Hodari-Sadiki Hubbard-James

Detecting and characterising exoplanets with HARPS-N 150
K. Rice, A. Mortier, L. Malavolta, F. Pepe, A. Cameron, A. Ghedina, D. Latham, S. Udry, L. Affer, S. Aigrain, A. S. Bonomo, V. Bourrier, L. A. Buchhave, H. Cegla, P. Cortes-Zuleta, R. Cosentino, M. Cretignier, M. Damasso, X. Dumusque, D. Ehrenreich, A. A. John, B. Klein, A. Leleu, M. Lopez-Morales, C. Lovis, J. Maldonado, A. Martínez Fiorenzano, G. Micela, E. Molinari, B. Nicholson, N. O'Sullivan, M. Pinamonti, G. Piotto, D. Sasselov, A. M. Silva, A. Sozzetti, M. Stalport, A. Vanderburg, S. Vissapragada, C. Watson and T. Wilson

Anticipating additional exoplanets in multi-planetary systems 156
M. Mousavi-Sadr, G. Gozaliasl and D. M. Jassur

Refining the exoplanet mass-radius relation using machine learning 161
M. Mousavi-Sadr, G. Gozaliasl and D. M. Jassur

Dynamic evolution of the compact planetary system K2-72 167
E. D. Kuznetsov and A. S. Perminov

Advancing Apodization Techniques: Optimizing Interferometric Apodization by Homothety for Exoplanet Detection 171
A. Qazbour, Y. El Azhari, J. Chafi, Z. Benkhaldoun, O. Azagrouze and A. Boskri

Interferometric Apodization using Deformable Mirror in a Michelson Interferometer to Improve Exoplanet Detection – I. Theoretical Foundations 176
A. Qazbour, Y. El Azhari, J. Chafi, Z. Benkhaldoun, O. Azagrouze and A. Boskri

Apatite Formation in Shergottite NWA 7397: Implications for the Mapping of Martian Meteorites 182
Amber Bonsall-Pisano, Geoffrey H. Howarth, Chad Peel and Gelu Costin

Discovery of Three Sub-Neptune Exoplanets Orbiting M Dwarfs with TESS . . . 187
A. Soubkiou, K. Barkaoui, T. Gan and Z. Benkhaldoun

VLT imaging survey of giant planets around nearby young white dwarfs 196
Hans Zinnecker, Wolfgang Brandner and Matthias Schreiber

Author Index 199

Preface

The IAU Symposium S393 brought together planetary scientists, exoplanet researchers, astrobiologists, and educators from across the world to discuss groundbreaking advances in planetary science, exoplanet detection, and characterization, and the potential for life beyond Earth. With a central focus on the capabilities of the James Webb Space Telescope (JWST), the meeting covered a wide range of topics related to the solar system, exoplanets, and outreach initiatives. We summarize the key scientific themes and highlights from the symposium below:

Exoplanet Detection and Characterization: A major theme of the symposium was the detection and characterization of exoplanets, particularly using ground- and space-based telescopes like JWST. Several talks explored how JWST's high-resolution imaging, spectroscopy, and mid-infrared sensitivity are offering new insights into exoplanet atmospheres, climate dynamics, and composition. The discussions emphasized the challenge of observing temperate, Earth-sized exoplanets and the need for continued innovation in detection techniques. JWST's ability to measure chemical signatures of life (e.g., oxygen, methane) in exoplanet atmospheres was heralded as a key advancement in the search for biosignatures. The symposium also focused heavily on the concept of planetary habitability, the potential for life beyond Earth, and the growing field of astrobiology. JWST is poised to make significant contributions to these fields, as it is designed to detect and analyze the atmospheres of potentially habitable planets. The concept of dynamic habitability was discussed at length, with researchers emphasizing that the potential for life is not static but changes over time due to planetary processes, stellar activity, and orbital evolution. Understanding the geochemical and biochemical interactions in early planetary environments is key to astrobiology.

Solar System and Protoplanetary Disks: The JWST is not only revolutionizing exoplanet studies but is also opening up new opportunities for solar system exploration. Participants shared insights into the study of the planets, moons, and small bodies in our solar system, as well as the dynamics of protoplanetary disks, which are key to understanding planet formation. The study of small bodies like asteroids, comets, and Kuiper Belt objects was central to many talks. These objects are thought to hold the building blocks of life and have preserved the early conditions of the solar system, offering a window into the early solar system's chemical makeup.

Education, Outreach, and Public Engagement: A significant part of the symposium focused on education and outreach in planetary science, with an emphasis on making the latest discoveries accessible to the general public and fostering global collaboration in the field of exobiology. The symposium recognized the importance of making planetary science a truly global endeavor, where access to educational resources, data, and telescopes is shared widely to inspire future generations of scientists.

The IAU-S393 meeting in Cape Town provided a comprehensive overview of the current state of planetary science and exoplanet research, particularly in the JWST era. The telescope's ability to study both solar system bodies and exoplanets is expected to revolutionize our understanding of planet formation, atmospheric chemistry, and the potential for life beyond Earth. The meeting also underscored the importance of collaboration between observational astronomers and theorists, as well as the need for global outreach to inspire and educate the public about the wonders of the universe.

Looking forward, JWST's high-precision spectroscopy and imaging capabilities will continue to drive major discoveries, from the analysis of biosignatures on distant exoplanets to the dynamics of planetary systems. With this remarkable new tool in hand, planetary science stands on the brink of a new golden age of discovery, one that promises to answer some of humanity's oldest questions about the origins of life and the nature of worlds beyond our own.

On behalf of the organizing committee, we extend our deepest gratitude to all contributors, participants, and supporters of this symposium held within the framework of the International Astronomical Union (IAU) General Assembly in South Africa. This gathering has provided a unique platform to advance discussions in planetary science and exoplanetary research in the era of the James Webb Space Telescope (JWST), bringing together experts from around the world to share their insights and discoveries.

We express our sincere appreciation to the Coordinating IAU Division, **Division F**: Planetary Systems and Bioastronomy, for its leadership in making this symposium a success. Our gratitude also extends to the Supporting IAU Divisions, **Division C**: Education, Outreach and Heritage, whose invaluable support has helped bridge scientific advancements with broader public engagement and historical perspectives.

Finally, we thank all researchers, educators, students, and attendees whose enthusiasm and contributions continue to drive the exploration of planetary systems and the search for life beyond Earth. Your dedication fuels the progress of our field, and we look forward to the continued impact of this symposium on future discoveries.

Zouhair Benkhaldoun
Youssef Moulane
Andrew Szentgyorgyi

Editors

Chair: Zouhair Benkhaldoun, Sharjah University, United Arab Emirates and University Cadi Ayyad, Morocco
Co-Chair: Youssef Moulane, Mohammed VI Polytechnic University, Morocco
Co-Chair: Andrew Szentgyorgyi, Harvard-Smithsonian Center for Astrophysics, USA

Organizing Committee

Scientific Organizing Committee

Zouhair Benkhaldoun, Sharjah University, United Arab Emirates and University Cadi Ayyad, Morocco
Andrew Szentgyorgyi, Harvard-Smithsonian Center for Astrophysics, USA
Youssef Moulane, Mohammed VI Polytechnic University, Morocco
Aline Vidotto, Leiden Observatory, Leiden University, Netherlands
Dennis Bodewits, Auburn University, USA
Stefanie Milam, NASA Goddard Space Flight Center, USA
Itziar de Gregorio-Monsalvo, ESO, Chile
Xian Shi, Shanghai Astronomical Observatory, China
Marie-Eve Naud, Institut Trottier de recherche sur les Exoplanets, Canada
Antonella Barucci, Observatoire Paris-Site de Meudon LESIA, France
Richard De Gris, Macquarie University, Australia

Local Organizing Committee

Chafi Jamal, Cadi Ayyad University, Morocco
Benkhaldoun Zouhair, Sharjah University, United Arab Emirates and University Cadi Ayyad, Morocco
Moulane Youssef, Mohammed Vi Polytechnic University, Morocco
Boskri Abdelkarim, Cadi Ayyad University, Morocco
Fares Rim, United Arab Emirates University, UAE
Elyajouri Meriem, Université Paris Saclay, France
Guennoun Meryem, Cadi Ayyad University, Morocco
Hmiddouch Said, Cadi Ayyad University, Morocco
Chouqar Jamila, Cadi Ayyad University, Morocco
Jabiri Abdelhadi, Cadi Ayyad University, Morocco
Benhida Abdelmajid, Cadi Ayyad University, Morocco
El Azhari Youssef, Ministry Of Education, Morocco
Takalana Charles, African Astronomical Society, South Africa

List of Participants

Saad, Somaya
Schreiber, Matthias
Vereshchagin, Sergei
Tchernyi, Vladimir
Linsky, Jeffrey
Ipatov, Sergei
Hojaev, Alisher
Ipatov, Sergei
Amuda, Tajudeen
Leng, Xuguang
Villavicencio Valero, Katherine
Bellotti, Stefano
Moulane, Youssef
Guennoun, Meryem
Zwicky, Lis
Shah, Priya
Kuznetsov, Eduard
Wandel, Amri
Chafi, Jamal
Hubbard-James, Hodari-Sadiki
Rice, Ken
Naud, Marie-Eve
Mousavi-Sadr, Mahdiyar
Mousavi-Sadr, Mahdiyar
Kuznetsov, Eduard
Mignone, Claudia
Qazbour, Aziz
Chouqar, Jamila
Bonsall-Pisano, Amber
Soubkiou, Abderahmane
de Gregorio-Monsalvo, Itziar
Brandner, Wolfgang
Villavicencio Valero, Katherine
Van Dishoeck, Ewine

TRAPPIST-North
https://www.trappist.uliege.be
- Exoplanet Program: 50%
- Comet & Asteroid Program: 50%
- Model: Gambato
- Diameter: 4.2 meters
- Automatization: TauTec
XXXII IAU GENERAL ASSEMBLY

UM6P
University Mohammed VI Polytechnic
Institute of Applied Physics
Implementing an AstroLab Project in Africa
Mohammed VI Polytechnic University (UM6P), Morocco
XXXII IAU GENERAL ASSEMBLY
CAPE TOWN, SOUTH AFRICA, 2024

science & innovation
Department:
Science and Innovation
REPUBLIC OF SOUTH AFRICA
XXXII IAU GENERAL ASSEMBLY
CAPE TOWN, SOUTH AFRICA, 2024

Planetary Science and Exoplanets in the Era of JWST
Proceedings IAU Symposium No. 393, 2026
Z. Benkhaldoun, Y. Moulane & A. Szentgyorgyi, eds.
doi:10.1017/S1743921324002746

Magnetism of Saturn in the origin of the visible dense rings and in their peculiarities recorded by the Cassini probe

Vladimir V. Tchernyi[1] and Sergey V. Kapranov[2]

[1]Modern Science Institute, SAIBR, Moscow, Russia
email: chernyv@bk.ru

[2]A.O. Kovalevsky Institute of Biology of the Southern Seas of RAS, Sevastopol, Russia
email: sergey.v.kapranov@yandex.ru

Abstract. Saturn's magnetism and diamagnetism of ice can help explain the emergence of visible dense rings by transformation of the protoplanetary cloud containing ice bodies into a disk-shaped system of stable visible dense rings with extreme flatness at Saturn's equator. It can also explain the thin structure of the rings as a whole, the sharp edges of dense rings, the existence and specific features of *B*-rings, and the stability of the entire dense ring system around Saturn. Additionally, it allows calculation of equilibrium separation of ice bodies, as predicted by J.C. Maxwell, with the magnetic repulsion of the ice bodies compensating for their gravitational attraction.

Keywords. Planetary ring divisions, planetary rings, Saturn, Solar system, astrophysical magnetism

---

1. Introduction

Gravity and magnetism are the main forces of nature in the Universe. The problem of the origin of Saturn's dense rings still remains unresolved. Recently, it was demonstrated how Saturn's magnetism makes an important contribution to the origin of visible dense rings at the planet's equator and can explain the equilibrium separation of ice bodies in the rings as predicted by J.C. Maxwell in the middle of the 19th century (Maxwell 1859; AAS 2020; Tchernyi & Kapranov 2020; Tchernyi et al. 2021; Tchernyi & Kapranov 2024; Tchernyi et al. 2024). It also explains sharp edges of rings, thin structure in the group of separated rings, existence of spokes and specific features in the *B*-ring, and how the 'hanging' disk of dense rings with individual ice bodies around Saturn can remain stable in the outer space under various gravitational influences.

In fact, the gravitational model explains only how and from where ice could get into the protoplanetary cloud and rings. Different extensions of the gravitational model with various degrees of success explaining genesis and peculiarities of Saturn's rings system were proposed (Safronov 1972; Fridman & Gorkavyi 1999; Charnoz et al. 2009; Canup 2010; Hyodo et al. 2017; Wisdom et al. 2022; Teodoro et al. 2023). The rings may be a relic of a proto-satellite disk, which may also have been filled with ice debris from a Titan-sized body tidally disrupted by Saturn, or with ice debris from various types of collisions in a protoplanetary cloud with a radius less than the Roche limit. James Webb Space Telescope recently detected water in the protoplanetary cloud of a forming exoplanet (Perotti et al. 2023), and this is an argument in favour of Safronov's theory.

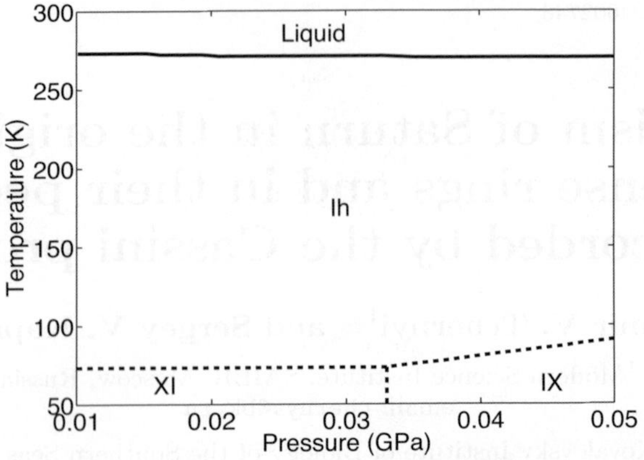

Figure 1. Low-pressure and low-temperature part of the phase diagram of ice encompassing allotropic modifications: ice Ih, XI and IX (Hemley 2000)

Unfortunately, there is no single universal gravitational theory that would provide a consistent explanation for all the observed features of the above-mentioned visible dense rings of Saturn (Crida & Charnoz 2010; Cuzzi et al. 2010; Esposito 2010; Estrada et al. 2017; Porco 2017; Tiscareno & Murray 2018; Spilker 2019). Without a convincing model that would explain the observed features of rings, there is no point to discuss their age (Crida et al. 2019).

The Cassini probe determined that rings consist mostly of frozen water ice, 90-95%; their physical properties and chemical composition are presented in (Poulet & Cuzzi 2002; Cuzzi et al. 2010; Esposito 2010; Miller et al. 2024). The probe detected the ratio of deuterium and hydrogen isotopes of the ice of Saturn's rings to be close to the terrestrial bulk-Earth value (Clark et al. 2019). This fact emphasizes the similarity of ice in the rings and Earth's ice. Ice XI is more likely a material that makes up Saturn's dense rings as it has stable thermodynamic parameters below 73 K (Hemley 2000) (Fig. 1) and is diamagnetic (Yen et al. 2016; Tchernyi & Kapranov 2021). The Cassini discovered that Saturn's magnetic equator almost perfectly coincides with the geographic one (Dougherty et al. 2018). In the inner region of Saturn's magnetosphere, where dense rings are located, the own magnetic field of Saturn with the largest contribution of the dipolar component of B_θ (Andre et al. 2008) predominates and appears to be very stable. This is also confirmed by the fact that Saturn's strongly dipolar magnetic field is capable of trapping high-energy charged particles and forming radiation belts outside the dense rings (Roussos et al. 2018). The Cassini probe recorded that under any gravitational influence of satellites in the gap of rings or oscillations of Saturn's gravity on the rings, their icy constituents in the visible dense rings remain separate; otherwise, they would have stuck together due to their own gravity and formed other satellites.

2. Magnetic anisotropic accretion in the origin of Saturn's dense rings

Based on the above observations, we proposed a new concept of Magnetic Anisotropic Accretion, which is associated with the influence of Saturn's magnetic field on diamagnetic ice bodies in the protoplanetary cloud. In the presence of Saturn's gravitational field, its magnetic field exerts a force of diamagnetic expulsion on ice bodies. As a consequence, all the orbits of ice bodies within the Roche limit-sized cloud started tilting into the plane of magnetic equator where their magnetic energy has a minimal value. At the end of this process, a system of stable visible dense rings formed in Saturn's equator plane under the combined effect of the gravitational and magnetic fields exerted to ice

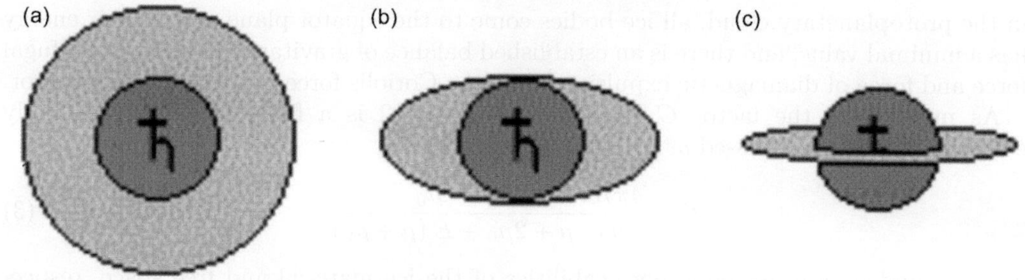

Figure 2. Evolution of Saturn's protoplanetary cloud of Roche size radius into a disk-shaped dense ring structure under the influence of gravitational, centrifugal and magnetic forces on diamagnetic bodies of ice: from (a) → (b) → to (c).

bodies (Tchernyi & Kapranov 2021, Tchernyi & Kapranov 2022, Tchernyi & Kapranov 2023). This scenario is shown in Fig. 2.

Thus, in addition to the gravitational force acting on an ice body in a protoplanetary cloud, a diamagnetic expulsion force emerges. As result of the combined action of both forces and random collisions of ice bodies, the protoplanetary cloud collapsed into a stable disk of dense rings. The force of diamagnetic expulsion has a stabilizing action on the ice bodies in the rings. The action of magnetic field on ice bodies ultimately leading to the emergence of a stable structure of dense rings is referred to as the magnetic anisotropic accretion. All icy constituents of the dense rings are eventually trapped inside a three-dimensional magnetic well in the plane of Saturn's magnetic equator coinciding with the geographic one.

It is worthwhile to note that the gravitational field of Saturn is spherically symmetric and its magnetic field is axisymmetric. Given this, first, a problem of magnetization of an ice sphere in the external magnetic field was solved, and then, by analogy, the solution was extended to a system of identical uniformly magnetized ice spheres with constant number density in an infinite disk-like structure (AAS 2020; Tchernyi & Kapranov 2020). The equations of collisionless motion of a magnetized ice sphere orbiting at a constant radius to the center in the gravitational and magnetic fields of the planet are reduced to the equation for the polar angle θ_{Sp} of the orbiting ice body:

$$\ddot{\theta}_{Sp} + \dot{\theta}_{Sp}^2 \cot\theta_{Sp} = \frac{GM_S}{r_{Sp}^3}\cot\theta_{Sp} + \frac{3C\mu_0^2 m_S^2}{2\pi^2 r_{Sp}^8 M_p}\cot\theta_{Sp}\cos^2\theta_{Sp} \tag{1}$$

where $G = 6.67430 \cdot 10^{-11}$ N·m^2·kg^{-2} is the gravitational constant, $M_S = 5.683 \cdot 10^{26}$ kg is the mass of Saturn, r_{Sp} is the radial distance between the centers of Saturn and the ice body, $m_S = 4.6 \cdot 10^{18}$ T·m^{-3} is the magnitude of magnetic moment of Saturn (Russell & Dougherty 2010), M_p is the mass of body with magnetic moment m_p, and C is the Clausius-Mossotti factor for magnetic spheres, which depends on the ice body environment.

The only solution of equation 1 is singular: $\theta_{Sp} = \pi/2$. It accounts for the essentially planar structure of Saturn's dense rings and their location in the magnetic equator plane. The relationship for the azimuthal velocity of the ice body is

$$\dot{\varphi}_{Sp} = \sqrt{GM_S/r_{Sp}^3 + 3C\mu_0^2 m_S^2/\left(8\pi^2 r_{Sp}^8 M_p\right)}. \tag{2}$$

It is seen from equation 2 that both the centrifugal force and the force of diamagnetic expulsion counterbalance the gravitational force acting on the orbiting ice body. As it follows from the consideration, eventually, after certain amount of collisions of ice bodies in the protoplanetary cloud, all ice bodies come to the equator plane where their energy has a minimal value, and there is an established balance of gravitational force, centrifugal force and force of diamagnetic expulsion while the Coriolis force is not active at equator.

As mentioned, the factor C in equations 1 and 2 is a function of the ice body environment. It is expressed as

$$C = \frac{4\pi R^3}{\mu_0} \frac{\mu - \mu_0}{\mu + 2\mu_0 + E\left(\mu + \mu_0\right)} \tag{3}$$

where μ and μ_0 are magnetic permeabilities of the ice material and free space, respectively; R is the ice sphere radius; and $E = -\pi R^3 \sigma / r_0$ is the environment factor for the ice sphere in a disk-like structure with the constant planar hexagonal lattice density $\sigma = 2/r_0^2\sqrt{3}$ of identical spheres at a distance of r_0 from each other.

In general, for a variable density $\sigma\left(r_{Sp}\right)$, C can be found by solving an integral equation (see equation 4 below and equations (26), (31) and (32) in Tchernyi & Kapranov 2020). The magnetized spheres, each having magnetic dipole moment $\mathbf{m}_p$, will then collectively generate the induced global magnetic field, which is convenient to use as an integral characteristics instead of the summed magnetic field intensities of discrete spheres. If the angle between the disk and magnetic equator plane is ϑ_t and the distance from the center of the sphere to the nearest edge of the infinite annular disk representing a host of ice bodies in the rings is r_0, a magnetized sphere will experience the magnetic field intensity of the surrounding magnetized spheres $\mathbf{H}_d$ in the disk:

$$\mathbf{H}_d = \int_S \mathbf{H}_{out} \sigma \mathrm{d}s = \tfrac{\sigma m_p}{4\pi} \int_0^{2\pi} \mathrm{d}\varphi \int_{r_0}^{\infty} \sin\varphi \cos\vartheta_t \tfrac{2\mathbf{e}_r \sqrt{1-\sin^2\varphi \cos^2\vartheta_t} + \mathbf{e}_\theta \sin\varphi \cos\vartheta_t}{r'^2} \mathrm{d}r' \\ = -\mathbf{e}_\theta \sigma m_p \cos^2\vartheta_t / (4r_0). \tag{4}$$

Then magnetization of the disk is:

$$\mathbf{M} = \frac{3\mathbf{B}_0}{\mu_0} \frac{\mu - \mu_0}{\mu + 2\mu_0 - (\mu + \mu_0)\pi R^3 \sigma \cos^2\vartheta_t / r_0}, \tag{5}$$

and its magnetic moment is:

$$\mathbf{m}_p = \frac{4\pi R^3 \mathbf{B}_0}{\mu_0} \frac{\mu - \mu_0}{\mu + 2\mu_0 - (\mu + \mu_0)\pi R^3 \sigma \cos^2\vartheta_t / r_0}. \tag{6}$$

In equation 4, the integration is performed over the disk area and the disk is assumed to be continuous. Note that for a sole ice sphere $E = 0$, and for the greatest possible R of the ice spheres in the magnetic equator plane ($R_{\lim} = r_0/2$), it can reach the limit of $-\pi R_{\lim}^3 \sigma / r_0 = -0.45$. Thus, for diamagnetic spheres, C is always negative, and the diamagnetic expulsion force is stronger for ice bodies accreted in a disk-like structure than for sole ice spheres.

It is possible to demonstrate that for a spherical cloud with radius r consisting of identical spherical ice bodies with the 3D hexagonal lattice arrangement and volumetric density $\rho = \sqrt{2}/r_0^3$, the environment factor is

$$E = \pi R^3 \rho \ln\left(r/2r_0\right) = \pi\sqrt{2}(R/r_0)^3 \ln\left(r/2r_0\right). \tag{7}$$

Consequently, the diamagnetic expulsion of an ice sphere in the cloud is weaker than that of an ice body in a disk-like structure and that of a sole ice sphere. Because there are no constraints imposed on the nebula radius, the environment factor can be infinitely great and the diamagnetic force vanishingly small. Thus, the magnetic force acting on the ice bodies, in the evolution of the planetary nebula to the disk-shaped system of dense rings, increases in magnitude.

3. Magnetic well of the disk of dense rings

The model of Saturn's dense rings can be represented as spatially separated and uniformly magnetized spherical ice bodies in a disk-shaped structure consisting of identical spheres with the same planar density. We find that the magnetization ($\mathrm{M_d}$) and magnetic moment (m_d) of ice sphere in the disk-shaped structure is much higher than that of a sole sphere due to the alignment of many magnetic dipoles with the field. And the ratio of the corresponding parameters for the spherical ice body in the disk and for the sole ice body is

$$\frac{\mathrm{M}_d}{\mathrm{M}_0} = \frac{m_d}{m_0} = \left[1 - \frac{\pi R^3 \sigma\,(\mu + \mu_0)}{r_0\,(\mu + 2\mu_0)}\right]^{-1} \tag{8}$$

where the parameters are the same as in equation 3. The resilience of ice bodies to the vertical shift is due to their minimum energy at the equator, and the horizontal orbit stability is ensured by the inhomogeneity of the magnetic field along the radius and by the centrifugal force.

4. Role of magnetism in the observed features of the *B*-ring

The spokes in the *B*-ring are located almost radially in the spatial anomalies of magnetic field of Saturn. The questions arise as to why the spokes exist at all and why they are located only in the opaque dense *B*-ring system. The Cassini probe found that there is three times less matter in the *B*-ring than previously thought; it consists of small ice bodies and has high density (Hedman & Nicholson 2016).

Following the mathematical analysis in (Tchernyi & Kapranov 2020), it is possible to explain the existence of spokes in the *B*-ring by the fact that when small ice bodies in the *B*-ring get into the anomalies of Saturn's magnetic field, they change their position and mix, trying to obtain a new trajectory. The observer sees this picture as spokes. After having passed through the anomaly, the ice bodies occupy the same orbit as they had before arriving at the anomaly.

The existence of spokes and high density in the *B*-ring becomes understandable, since it was found that the role of the axisymmetric magnetic field increases with decreasing magnetized body size. As it follows from equation 8, the magnetic well of the *B*-ring at the equator plane is deeper, and it makes sharp edges of the *B*-ring. The magnetic pressure on the *B*-ring from the side of both gaps is greater. The separate small ice bodies are packed more densely inside the well, and this is why it looks like a bright solid ring. An interesting feature is that the small ice bodies inside the *B*-ring, though assembled extremely densely, are separated from each other because the gravitational attraction between them is balanced by the magnetic repulsion. This is why the separated bodies do not stick together; otherwise they would have formed a satellite. In addition, diamagnetic repulsion can prevent these bodies from sticking together when they fall into spiral density waves in the *B*-ring described by Hedman & Nicholson (2013).

5. Origin of the thin structure of dense rings due to the influence of Saturn's magnetism

Given the presence of magnetic field of Saturn, orbits of all ice bodies in dense rings acquire stability with respect to the shift in the horizontal and vertical directions. The force of diamagnetic expulsion from the inhomogeneous magnetic field inside the ring structure forms sharp edges and separates the ice bodies. The stability of dense rings, the thin structure and sharp edges of dense rings are the result of the ice body redistribution in the magnetic field gradient in the magnetic equator plane under the action of the radial and polar components of the magnetic force (Tchernyi & Pospelov 2007; Tchernyi et al. 2021; Tchernyi & Kapranov 2022). Lines of Saturn's magnetic field are concentrated

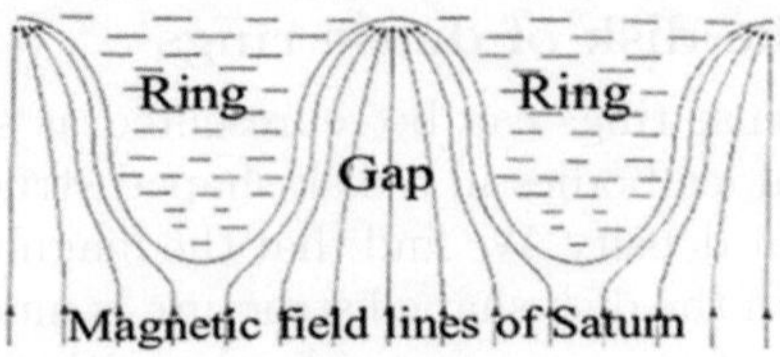

Figure 3. Dense and rarefied areas of magnetic flux density in ice bodies look like the system of dense rings.

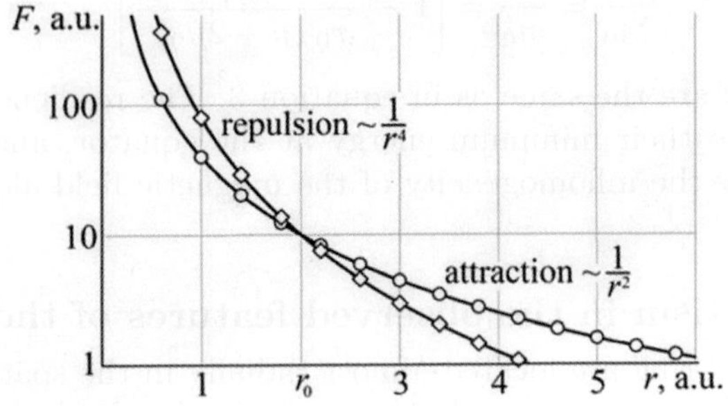

Figure 4. Dependence of the repulsion and attraction forces on distance between magnetized ice bodies.

in areas with the highest (signed) magnetic permeability creating the largest magnetic flux density, and the diamagnetic ice bodies are gathering in areas with low density of magnetic flux (Fig. 3). Additionally, the action of diamagnetic expulsion force forms sharp edges of dense rings.

The stability of dense rings inside the disk-like structure of rings is ensured by the fact that bodies of diamagnetic ice in dense rings are captured in a magnetic well in Saturn's magnetic equator plane (Tchernyi & Kapranov 2020). The resilience in the equatorial plane against the vertical shift (along the meridians) is ensured by the minimum energy of ice bodies at the equator. The radial stability is ensured by the inhomogeneity of the magnetic field in the horizontal direction and by the balance of the gravitational, centrifugal, and magnetic forces.

6. Repulsion force and separation of ice bodies in dense rings

The Cassini probe registered that radial and azimuthal density waves caused by fluctuations in Saturn's gravity, and the dynamics of epicycles of the ring boundaries caused by the movement of the moons between the rings, do not change the entire structure of the rings consisting of individual ice bodies (Shepelyansky et al. 2009; Hedman & Nicholson 2016; Mankovich et al. 2019; see also images of rings from NASA/JPl, Caltech/Space Science Institute). J. Maxwell proved that the rings consisted of individual bodies (Maxwell 1859). In the magnetic field of Saturn, bodies of diamagnetic ice acquire their own magnetization, opposite in sign to the magnetic field of Saturn. The study of separation of the ice bodies due to their magnetization also contributes to Maxwell's analysis (Tchernyi & Kapranov 2024; Tchernyi et al. 2024). For diamagnetic ice bodies, the magnetic force results in their repulsion. The force of magnetic interaction of two bodies represented by point magnetic dipoles varies as r^{-4} with distance (r) between them (Furlani 2001). It is counterbalanced by the gravitational force attracting the bodies to each other, which is known to vary as r^{-2} in the point mass approximation. Schematically, these two relationships are shown in Fig. 4. When the gravitational force and the magnetic repulsion force are equal, the equilibrium distance between the ice bodies in the rings is:

$$r_0 = \left(\frac{R}{r_{Sp}}\right)^3 \frac{m_S}{M_p} \left|\frac{\mu - \mu_0}{\mu + 2\mu_0}\right| \sqrt{\frac{3\mu_0}{4\pi G}} \tag{9}$$

where the notations are the same as used above.

7. Conclusions

We have demonstrated that the magnetization of diamagnetic ice bodies of the protoplanetary cloud and the magnetic anisotropic accretion allow understanding the origin of visible dense rings at Saturn's equator. Saturn's magnetic field has been shown to provide stability and resilience to the entire system of visible dense rings. These effects also account for the thin structure of the dense rings, their sharp edges and the stability of dense rings in the vertical and horizontal directions. Saturn's magnetism can explain the observed features of the B-ring, such as the presence of spokes due to the compactness of the ice bodies, high density and strong brightness, ice body separation, and sharp edges.

The Newtonian dynamics of a single body of ice differs from the dynamics of ice bodies assembled in rings. Ice bodies are repelled from each other under the influence of magnetic force and are attracted to each other under the action of gravity. From the balance of the forces of gravitational attraction and magnetic repulsion, we have derived the expression for the equilibrium distance between ice bodies in the rings. This result complements Maxwell's inference of visible dense rings being not continuous, but rather consisting of individual bodies.

The results of the presented theory are confirmed by measurements and images of visible dense rings obtained by the Cassini probe in 2004–2017.

References

AAS 2020, *AAS Journal Author Series: Vladimir Tchernyi on 2020ApJ...894...62T.* https://youtu.be/La7RmcWGUTQ (accessed on 03 September 2024)

Andre, N., et al. 2008, *Rev. Geophys.*, 46, 4

Canup, R.M. 2010, *Nature*, 468, 943

Charnoz, S., Morbidelli, A., Dones, L., & Salmon, J. 2009, *Icarus*, 199, 413

Crida, A., & Charnoz, S. 2010, *Nature*, 468, 903

Crida, A., Charnoz, S., Hsu, H.-W., & Dones, L. 2019, *Nat. Astron.*, 3, 967

Clark, R., et al. 2019, *Icarus*, 321, 791

Cuzzi, J.N., et al. 2010, *Science*, 327, 1470

Dougherty, M.K., et al. 2018, *Science*, 362, 6410

Esposito, L.W. 2010, *Ann. Rev. Earth Planet. Sci.*, 38, 383

Estrada, P.R., Durisen, R.H., & Cuzzi J.N. 2017, in: *AGU Fall Meeting*, Dec. 11-15, New Orleans, LA, P23B-2734

Fridman, A.M., & Gorkavyi, N.N. 1999, *Physics of Planetary Rings: Celestial Mechanics of Continuous Media* (Berlin: Springer)

Furlani, E.P. 2001, *Permanent Magnet and Electromechanical Devices: Materials, Analysis, and Applications* (San Diego, CA: Academic Press)

Hedman, M.M., & Nicholson, P.D. 2013, *AJ*, 146, 12

Hedman, M.M., & Nicholson, P.D. 2016, *Icarus*, 279, 109

Hemley, R.J. 2000, *Ann. Rev. Phys. Chem.*, 51, 763

Hyodo, R., Charnoz, S., Ohtsuki, K., & Genda, H. 2017, *Icarus*, 282, 195

Mankovich, C., Marley, M.S., Fortney, J.J., & Movshovitz, N. 2019, *ApJ*, 871, 1

Maxwell, J.C. 1859, *MNRAS*, 19, 297

Miller, K.E., Filacchione, G., Cuzzi J.N., et al. 2024, *Space Sci. Revs*, 220, 70

Perotti, G., et al. 2023, *Nature*, 620, 516

Porco, C. 2017, *Sci. Am.*, 317(4), 78

Poulet, F., & Cuzzi, J.N. 2002, *Icarus*, 160, 350
Roussos, E., et al. 2018, *Science*, 362(6410), eaat1962
Russell, C.T., & Dougherty, M.K. 2010, *Space Sci. Revs*, 152, 251
Safronov, V.S. 1972, *Evolution of the Protoplanetary Cloud and Formation of the Earth and Planets* (Washington, D.C.: NASA)
Shepelyansky, D.L., Pikovsky, A.S., Schmidt, J., & Spahn F. 2009, *MNRAS*, 395, 1934
Spilker, L. 2019, *Science*, 364, 1046
Tchernyi, V.V., & Kapranov, S.V. 2020, *ApJ*, 894, 62
Tchernyi, V.V., & Kapranov, S.V. 2021a, *RNAAS*, 5(10), 255
Tchernyi, V.V., & Kapranov, S.V. 2021b, arXiv: 2104.03967
Tchernyi, V.V., & Kapranov, S.V. 2022, arXiv: 2204.03414
Tchernyi, V.V., & Kapranov, S.V. 2023, *Opt. Spectrosc.*, 131, 327
Tchernyi V.V., & Kapranov S.V. 2024, *Space Res. Today*, 219, 80
Tchernyi V.V., Kapranov S.V., & Belodedov, M.V. 2024, *Adv. Theor. Comp. Phys.*, 7(3), 1
Tchernyi V.V., Kapranov S.V., & Pospelov, A.Yu. 2021, *Radio Sci. Lett.*, 3, 69
Tchernyi V.V., & Pospelov, A.Yu. 2007, *Ap&SS*, 307(4), 347
Teodoro, L.F.A., et al. 2023, *ApJ*, 955, 137
Tiscareno, M.S., & Murray, C.D. (eds.) 2018, *Planetary Ring Systems: Properties, Structure, and Evolution* (Cambridge, UK: Cambridge University Press)
Wisdom, J., Dbouk, R., Militzer, B., Hubbard, W.B., Nimmo, F., Downey, B.G., & French, R.G. 2022, *Science*, 377, 1285
Yen, F., Gao, T., Liu, Y., & Berlie, A. 2016, arXiv: 1605.03383

Planetary Science and Exoplanets in the Era of JWST
Proceedings IAU Symposium No. 393, 2026
Z. Benkhaldoun, Y. Moulane & A. Szentgyorgyi, eds.
doi:10.1017/S1743921324001911

Migration of bodies ejected from the Earth and the Moon

Sergei I. Ipatov

Vernadsky Institute of Geochemistry and Analytical Chemistry of RAS,
119991, 19 Kosygin st., Moscow, Russia
email: siipatov@hotmail.com

Abstract. This study analyzes the motion of bodies ejected from the Earth or the Moon. We studied the ejection of bodies from several points on the Earth's surface, as well as from the most far point of the Moon from the Sun. Different velocities and angles of ejection of bodies were considered. The dynamical lifetimes of bodies reached a few hundred million years. Over the entire considered time interval, the values of the probability of a collision of a body ejected from the Earth with the Earth were approximately 0.3, 0.2, and 0.15–0.2 at an ejection velocity v_{ej} equaled to 11.5, 12, and 14 km/s, respectively. At $v_{ej} \leq 11.3$ km/s, most of the ejected bodies fell back onto the Earth. The total number of bodies delivered to the Earth and Venus probably did not differ much. The probabilities of collisions of bodies with Mercury and Mars usually did not exceed 0.1 and 0.02, respectively. At $v_{ej} \geq 11.5$ km/s, the probability of a collision of a body ejected from the Earth with the Moon was about 15–35 times less than that with the Earth, and it was about 0.01. The probability of a collision with the Earth for a body ejected from the Moon moving in its present orbit was about 0.3–0.32, 0.2–0.22, and 0.1–0.14 at v_{ej}=2.5 km/s, v_{ej}=5 km/s, and at $12 \leq v_{ej} \leq 16.4$ km/s, respectively.

Keywords. methods: n-body simulations, Earth, Moon, solar system: general

1. Introduction

The terrestrial planets and the Moon were bombarded by impactors during their formation and at the stage of the Late-Heavy Bombardment. Some material could be ejected at such impacts. Earlier Gladman et al. (2005) and Reyes-Ruiz et al. (2012) studied the motion of bodies ejected from the Earth during time interval equal to 30 Kyr. Considered initial velocities were perpendicular to the surface of the Earth. According to (Shuvalov & Trubetskaya (2011)), the values of an ejection angle i_{ej} are mainly between 20° and 55°, especially between 40° and 50°. More detailed introduction information on the item of the below paper can be found in (Ipatov (2024)).

2. The model and initial data for studies of ejection of bodies from the Earth

In each calculation variant, the motion of 250 ejected bodies was studied for the same ejection point on the Earth's surface and for the fixed values of an ejection angle i_{ej}, a velocity v_{ej} of ejection, and a time step t_s of integration. In different variants, the values of i_{ej} equaled to 15°, 30°, 45°, 60°, 89°, or 90°, and v_{ej} equaled mainly to 11.22, 11.5, 12, 12.7, 14, 16.4, or 20 km/s. In most calculations, bodies started directly from the Earth. In each variant, bodies started from one of six considered opposite points on the Earth's surface (Ipatov (2024)). Bodies ejected from points F and C started their motion from the most far and most close points to the Sun, respectively. For points W and B, bodies started from the points on Earth's surface located on the forward side (in the direction of motion) and on the opposite side, respectively. For points U and D, bodies started from the Earth at maximum and minimum values of z at Oz axis perpendicular to the plane of the orbit of the Earth. The rotation of the Earth around its axis was not taken into account, as the additional velocity of its surface caused by such rotation is smaller than the considered ejection velocity. The velocity of rotation doesn't exceed 0.5 km/s, and it is different for different points on the Earth's surface.

The motion of bodies was studied during the dynamical lifetime T_{end} of all bodies, which equaled to a few hundreds of Myr and could exceed 1 Gyr. During this time interval, all bodies either collided with planets or the Sun, or they were ejected into hyperbolic orbits. The gravitational influence of the Sun and all eight planets was taken into account. Bodies that collided with planets or the Sun or reached 2000 AU from the Sun were excluded from integration. The motion of the Moon was not included in the integration of motion equations. The probabilities of collisions of bodies with the Moon were calculated with the use of some formulas. Such calculations were based on the arrays of orbital elements of the bodies that left the Hill sphere of the Earth (Ipatov (2024)). After a body had left the Hill sphere, it could enter this sphere again several times during its long dynamical lifetime. For such body, the contribution to the probability of its collision with the Moon was relatively small before the body left the Hill sphere for the first time. However, if an ejected body moved for a long time inside the Hill sphere of the Earth, then the probability of its collision with the Moon could be not small. Actually such long motion inside the Hill sphere could be only for a small ejected velocity especially if the Moon embryo moved close to the Earth (see Section 6) or if the ejected material was not in the form of solid bodies (e.g. was in the form of vapor).

The symplectic code from the SWIFT integration package by Levison, Duncan (1994) was used for integration of the motion equations. In different variants the considered time integration step t_s equaled to 1, 2, 5, or 10 days. The results of calculations with different t_s were compared. Most of calculations have been made with an integration time step equal to 5 days. With this step, the obtained probabilities of collisions with planets were about the same as those obtained with a smaller t_s.

3. Probabilities of collisions of bodies ejected from the Earth with the Earth

Exclusive for ejection velocities close to the parabolic velocity, on average, about a half and less than 10% of initial ejected bodies still moved in elliptical orbits after 10 and 100 Myr, respectively. A few ejected bodies collided with planets after 250 Myr. Impactors with a diameter of about 10 km formed Chicxulub and Popigai craters on the Earth with ages 66 and 36 Myr. About one 1-km impactor could collide the Earth during 1 Myr. Some bodies ejected from the Earth during the last 100 Myr can still move in the zone of the terrestrial planets and can collided with them.

The probability p_E of collisions of bodies ejected from the Earth with the Earth is greater for a smaller ejection velocity. Most of the ejected bodies fell back onto the Earth at v_{ej} slightly greater than the parabolic velocity (at $v_{ej}\leq$11.25 km/s). At v_{ej}=11.5 km/s, the values of p_E were about 0.2-0.3 and did not differ much for different starting points on the Earth's surface. At v_{ej}=12 km/s, p_E was about 0.16-0.27. For the point W (when bodies were ejected from the forward point of the Earth's motion - from apex), the probabilities p_E can differ much from those for other points. For this point at $v_{ej}\geq$16.4 km/s and 45°$\leq i_{ej}\leq$60°, the value of p_E could be 0. In this case more than 80% of the bodies were thrown into hyperbolic orbits, and most of other bodies collided with the Sun. At v_{ej}=20 km/s, the values of p_E equaled to 0 for ejection from the forward point W at $i_{ej}\geq$30°. They exceeded 0.2 for ejection from the back point B (from antapex), and were about 0.1 for ejection from other four considered points of ejection. The ratios of the probability of collisions of bodies with the Earth to the probabilities of collisions of bodies with other planets and the Sun usually decreased with time.

Time variations of the probabilities of collisions of bodies ejected from the Earth with the Earth and the Moon were considered by Ipatov (2024) for time interval T between 10 Kyr and 100 Myr. The probabilities (especially at smaller T) depended much on v_{ej} and the point of ejection. At i_{ej}=45° and v_{ej}=12 km/s for all 6 ejection points, the ratio of the values of p_E at T to those at T_{end} was in the ranges 0.02-0.04, 0.2-0.3, and 0.6-0.8 at T equal to 10 Kyr, 1 Myr, and 10 Myr, respectively. The intervals of the values of p_E for the above case were 0.004-0.008, 0.06-0.07, and 0.15-0.16 at T equal to 10 Kyr, 1 Myr, and 10 Myr, respectively. The above intervals can be wider for $v_{ej} \geq 12.7$ km/s.

The average velocities of collisions of bodies ejected from the Earth with the Earth and the Moon are greater for greater ejection velocity. The collision velocities of bodies with the Earth were about 13, 14–15, 14–16, 14–20, 14–25, and 15–35 km/s at v_{ej} equal to 11.3, 11.5, 12, 14, 16.4, and 20 km/s, respectively. The collision velocities of ejected bodies with the Moon were mainly in the range of 7–8, 10–12, 10–16, 11–23, and 12–33 km/s at v_{ej} equal to 11.3, 12, 14, 16.4, and 20 km/s, respectively.

4. Probabilities of collisions of bodies ejected from the Earth with the Moon

The probability $p_M = p_E/k_{EM}$ of collisions of bodies ejected from the Earth with the Moon was calculated to be equal to the ratio of the value of p_E obtained at integration to the ratio k_{EM} of the probabilities of collisions of bodies with the Earth and the Moon. The value of k_{EM} was calculated with the use of some formulas based on the arrays of orbits of bodies during considered time interval T similar to (Ipatov (2019)). At $v_{ej} \geq 11.5$ km/s, the probability of a collision of a body ejected from the Earth with the Moon was about 15–35 times less than that with the Earth. For the Moon moved in its present orbit, the probability p_M was of the order of 0.01. It was about 0.01-0.02 at v_{ej}=11.3 km/s and 0.004-0.008 at $v_{ej} \geq 14$ km/s. Based on this result, Ipatov (2024) concluded that a large Moon embryo should be formed close to the Earth in order to accumulate material not rich in iron ejected from forming Earth. The estimates of p_M have been made for the model of ejection of bodies. In the multi-impact and giant-impact models, some of the ejected matter (e.g. in the form of vapor) went into orbits around the Earth. In this case the growth of the Moon embryo could be greater than for the ejection of bodies considered above. Ipatov (2018) supposed that the embryos of the Earth and the Moon with masses smaller by one or two orders of magnitude than their present masses could be formed as a result of contraction of a common rarefied condensation.

5. Probabilities of collisions of bodies ejected from the Earth with planets and probabilities of their ejection into hyperbolic orbits

The probability p_V of collisions of the bodies ejected from the Earth with Venus during a whole considered time interval of evolution (at T=T_{end}) were often about 0.2-0.35 at $11.5 \leq v_{ej} \leq 16.4$ km/s. The values of p_V were were less than p_E at $v_{ej} \leq 11.4$ km/s, were about p_E at v_{ej}=11.5 km/s, and were greater than p_E at $v_{ej} \geq 12$ km/s. The total number of ejected bodies delivered to the Earth and Venus probably did not differ much, and the upper layers of Earth and Venus can contain similar material.

The probability of collisions of bodies ejected from the Earth with Mercury during a whole considered time interval was about 0.02-0.08 and 0.03-0.05 at $11.3 \leq v_{ej} \leq 11.5$ and $12 \leq v_{ej} \leq 20$ km/s, respectively. The probabilities of collisions of bodies ejected from the Earth with Mars were smaller than those with Mercury. At T=10 Myr and T=T_{end}, the probability of collisions of bodies with Mars did not exceed 0.012 and 0.025, respectively.

Therefore, more material ejected from the Earth was delivered to Mercury than to Mars. The fraction of bodies ejected from the Earth that collided with Mars in 1 Myr was about 0.0002. It is supposed by Mileikowsky et al. (2000) that microbes can survive during the flight between the Earth and Mars with duration of about 1 Myr.

The fraction p_J of bodies ejected from the Earth and then collided with Jupiter was of the order of 0.001. However, at $T{=}T_{end}$ in a few variants, mainly with large v_{ej}, the value of p_J exceeded 0.01. The probability of collisions of bodies ejected from the Earth with the Sun often was between 0.2 and 0.5 at $v_{ej}{\geq}11.3$ km/s.

The fraction of bodies that had been ejected from the Earth and then were ejected into hyperbolic orbits did not exceed 0.1 at $v_{ej}{\leq}12$ km/s and $T{=}T_{end}$. It was mainly greater for a greater ejection velocity. At $v_{ej}{\geq}16.4$ km/s and $i_{ej}{\geq}60°$, all bodies were ejected from the Solar System if they started from the front point W. The fraction of bodies ejected into hyperbolic orbits was about 0.25-0.27 at $i_{ej}{=}45°$ and $v_{ej}{=}16.4$ km/s if points of ejection were not in the front or back of the Earth's motion. More detailed studies of the probabilities of bodies ejected from the Earth with the Moon were discussed by Ipatov (2024), and such probabilities with planets will be published in Icarus.

6. Probabilities of collisions of bodies ejected from the Moon with the Earth

Migration of bodies ejected from the Moon was studied similar to the ejection of bodies from the Earth from point F, but greater initial distances from the center of the Earth were considered. In computer simulations, bodies started from the point F located at a distance r_{ME} from the center of the Earth on the line from the Sun to the Earth. In different calculation variants, the distance r_{ME} varied from $3r_E$ to $60r_E$, where r_E is the radius of the Earth. The semi-major axis of the orbit of the Moon is equal to $60r_E$. It is usually considered that the initial Moon formed close to the Earth, and then it increased its distance from the Earth. The escape velocity for the Moon is equal to 2.38 km/s. In considered variants of calculations, initial velocities v_{ej} varied from 2.5 to 16.4 km/s, and values of i_{ej} varied from 15° to 89°. The gravitational influence of the Moon and the velocity v_M of the motion of the Moon around the Earth were not taken into account in calculations. For initial velocities of ejected bodies relative to the Sun, only the motion of the Earth around the Sun was added to ejection velocities. The vector of v_M can have different directions as the Moon moves around the Earth. The value of v_M is about 1 km/s for the present orbit of the Moon.The velocity of the lunar surface due to its rotation about the center of mass of the Moon is small and do not exceed 4 m/s.

At $r_{ME}{=}5r_E$, the parabolic and circular velocities for the Earth are equal to 4.9 and 3.5 km/s, respectively. Note that the parabolic velocity is for eccentricity e of the orbit equal to 1. Some ejected bodies can reach the surface of the Hill sphere of the Earth at $e{<}1$ and can leave this sphere even if their initial velocities are less than the parabolic velocity. For ejection velocities less than the circular velocity for the Earth at a considered distance, all ejected bodies could fall onto the Earth or the Moon. As it is discussed in Section 2 and in (Ipatov (2024)), the considered model does not calculate the probability of a collision of a body with the Moon before this body left the Hill sphere of the Earth for the first time. The results of calculations obtained for some values of v_{ej} and i_{ej} correspond to the evolution of bodies for other values of real ejection velocities and inclinations. Such difference is smaller for greater v_{ej}. In our studies we were interested in analyzing different types of evolution of orbits of bodies ejected from the Moon, but we were not interested in understanding the exact dependence of the evolution on initial data.

At $v_{ej}{=}2.5$ km/s and $r_{ME}{=}3r_E$, the dynamical lifetime of all ejected bodies was less than 5 days. For $r_{ME}{=}5r_E$ and $v_{ej}{=}2.5$ km/s, some of the bodies left the Hill sphere of the Earth at $15°{\leq}i_{ej}{\leq}30°$, though at greater i_{ej} most of the ejected bodies quickly fell onto the Earth. At v_{ej}=5 km/s, T=T_{end}, and r_{ME}=5r_E, the value of p_E was about 0.3 for the considered ejection angles i_{ej} from 15° to 89°. Note that the parabolic velocity for the Earth at r_{ME}=5r_E is a little less than 5 km/s. In some variants (e.g., at r_{ME}=3r_E, v_{ej}=5 km/s, i_{ej}=30°; r_{ME}=5r_E, v_{ej}=2.5 km/s, 15°$\leq i_{ej} \leq$30°; r_{ME}=7r_E, v_{ej}=2.5 km/s, 15°$\leq i_{ej} \leq$45°) some ejected bodies could still move around the Earth after 100 Myr. For the model with the Moon included in integration, most of such bodies should collide with the Moon.

For the ejection of bodies from the present orbit of the Moon, T=T_{end} and 15°$\leq i_{ej} \leq$89°, the value of p_E was about 0.27-0.35, 0.2-0.25, and 0.1-0.14 at v_{ej}=2.5 km/s, v_{ej}=5 km/s, and at 12$\leq v_{ej} \leq$16.4 km/s, respectively. The probabilities of collisions of bodies ejected from the Moon with the terrestrial planets were similar to those ejected from the Earth, but for different ejection velocities.

7. Conclusions

This study analyzes the motion of bodies ejected from the Earth or the Moon. We studied the ejection of bodies from several points on the Earth's surface, as well as from the most far point of the Moon from the Sun. Different velocities and angles of ejection of bodies were considered. The dynamical lifetimes of bodies reached a few hundred million years. Over the entire considered time interval, the values of the probability of a collision of a body ejected from the Earth with the Earth were approximately 0.3, 0.2, and 0.15-0.2 at v_{ej} equaled to 11.5, 12, and 14 km/s, respectively. At $v_{ej} \leq$11.3 km/s, most of the ejected bodies fell back onto the Earth. The total number of bodies delivered to the Earth and Venus probably did not differ much. The probabilities of collisions of bodies with Mercury and Mars usually did not exceed 0.1 and 0.02, respectively. At $v_{ej} \geq$11.5 km/s, the probability of a collision of a body ejected from the Earth with the Moon was about 15-35 times less than that with the Earth, and it was about 0.01. The probability of a collision with the Earth for a body ejected from the Moon moving in its present orbit was about 0.3–0.32, 0.2–0.22, and 0.1–0.14 at v_{ej}=2.5 km/s, v_{ej}=5 km/s, and at 12$\leq v_{ej} \leq$16.4 km/s, respectively.

Acknowledgements

This work was supported by the Ministry of Science and Higher Education of the Russian Federation, within the budget theme of the Vernadsky Institute of Geochemistry and Analytical Chemistry of Russian Academy of Sciences. I thank the reviewer for helpful comments and suggestions.

References

Gladman, B., Dones, L., Levison, H.F., Burns, J.A. 2005, *Astrobiology* 5, 483

Ipatov, S.I. 2018, *Solar System Research*, 52, 401, DOI: 10.1134/S0038094618050052, http://arxiv.org/abs/2003.09925

Ipatov, S.I. 2019, *Solar System Research*, 53, 332, DOI: 10.1134/S0038094619050046, https://arxiv.org/abs/2003.11301

Ipatov, S.I. 2024, *Solar System Research*, 58, 94, DOI: 10.1134/S0038094624010040, http://arxiv.org/abs/2405.19797

Levison, H.F., Duncan, M.J. 1994, *Icarus*, 108, 18

Mileikowsky, C., Cucinotta, F.A., Wilson, J.W., Gladman, B., Horneck, J., Lindegren, L., Melosh, J., Rickman, H., Valtonen, M., Zheng, J.Q. 2000, *Icarus*, 145, 391

Reyes-Ruiz, M., Chavez, C.E., Aceves, H., Hernandez, M.S., Vazquez, R., Nunez, P.G., 2012, *Icarus*, 220, 777

Shuvalov, V.V., Trubetskaya, I.A. 2011, *Solar System Research*, 45, 392

Planetary Science and Exoplanets in the Era of JWST
Proceedings IAU Symposium No. 393, 2026
Z. Benkhaldoun, Y. Moulane & A. Szentgyorgyi, eds.
doi:10.1017/S1743921324002321

Unveiling New Parent Bodies in Meteor Showers: A Robust Methodological Approach

M. Guennoun[1], J. Vaubaillon[2], D. Čapek[3], P. Koten[3] and Z. Benkhaldoun[4]

[1]Oukaimeden Observatory, High Energy Physics and Astrophysics Laboratory, Faculty of sciences Semlalia, Cadi Ayyad University, Marrakech, Morocco
emails: meryem.guennoun@uca.ac.ma, guennoun.meryem27@gmail.com

[2]Institut de Mécanique celeste et calcul des Éphémérides, Observatoire de Paris, PSL, France

[3]Astronomical Institute of Academy of Sciences, Fricova 298, 251 65 Ondcejov, Czech Republic

[4]Sharjah University, United Arab Emirates and University Cadi Ayyad, Morocco

Abstract. The increasing addition of meteor showers to the International Astronomical Union (IAU) list underscores the need for more accurate identification of new parent bodies, especially with the growing availability of video meteor data. Current methods using a high-threshold single-linking Dsh criterion often lack accuracy, prompting our study to explore advanced techniques for linking meteors to their parent bodies. Using the Moroccan meteor observation network as a foundation, we applied and compared several methods on the SonotaCo and EDMOND databases to detect new parent bodies. We further applied rigorous statistical tests to avoid random associations. This presentation highlights the network's innovative methodology, its success in accurately tracing meteoroid origins, and its potential to predict impact zones on Earth (Guennoun et al. (2019)). Through collaborative efforts, this network provides a significant contribution in the study of meteor dynamics.

Keywords. meteors, meteor showers, parent bodies, asteroids, comets

1. Introduction

Accurate identification of meteor showers and their associated parent bodies is essential for understanding the dynamics of meteoroids and their potential impacts on Earth. Traditional methods, such as the high-threshold single-linking Dsh-criteria developed by (Drummond (1981)), have been a standard in the field. However, these methods often face significant challenges due to the high variability and uncertainty in meteor orbit data, leading to a potential overestimation or misidentification of parent bodies. Such inaccuracies are of concern in rapidly expanding meteor shower databases like those managed by the IAU (Vida et al. (2018), Rudawska & Jenniskens (2014)).

Recent advancements in video meteor measurements, such as those employed in the Moroccan meteor observation network, offer an unprecedented opportunity to refine our understanding of meteor shower associations. The availability of high-resolution meteor observation data, including datasets from SonotaCo and EDMOND (SonotaCo (2016), Egal et al. (2017)), necessitates more sophisticated methodologies to ensure accuracy in identifying the parent bodies of meteors. Addressing these challenges, our study applies an advanced statistical framework designed to improve the reliability of such associations.

Table 1. Number of asteroids associated with at least 50 meteors, for each method.

Methods	#	Criteria	Number of associations SonotaCo Database	EDMOND Database
Method 1	1.1	$Dsh < 0.05$, $\lambda_o < 8°$	2	3
	1.2	$Dsh < 0.1$, $\lambda_o < 8°$	17	28
	1.3	$Dsh < 0.2$, $\lambda_o < 8°$	850	1424
Method 2	2.1	$\Delta V < 0.5$ km/s, $Dis < 6°$, $\lambda_o < 8°$, $\alpha < 8°$, $\delta < 8°$	16	25
	2.2	$\Delta V < 1$ km/s, $Dis < 6°$, $\lambda_o < 8°$, $\alpha < 8°$, $\delta < 8°$	35	64
	2.3	$\Delta V < 3$ km/s, $Dis < 6°$, $\lambda_o < 8°$, $\alpha < 8°$, $\delta < 8°$	207	389
	2.4	$\Delta V < 0.5$ km/s, $Dis < 2°$, $\lambda_o < 8°$, $\alpha < 8°$, $\delta < 8°$	13	20
	2.5	$\Delta V < 1$ km/s, $Dis < 2°$, $\lambda_o < 8°$, $\alpha < 8°$, $\delta < 8°$	23	42
	2.6	$\Delta V < 3$ km/s, $Dis < 2°$, $\lambda_o < 8°$, $\alpha < 8°$, $\delta < 8°$	128	240
Method 3	3.1	$dv = 5\%$, $Dis < 2°$, $\lambda_o < 8°$, $\alpha < 8°$, $\delta < 8°$	36	55
	3.2	$dv = 10\%$, $Dis < 2°$, $\lambda_o < 8°$, $\alpha < 8°$, $\delta < 8°$	84	131
	3.3	$dv = 20\%$, $Dis < 2°$, $\lambda_o < 8°$, $\alpha < 8°$, $\delta < 8°$	251	367

By utilizing comprehensive datasets from the Moroccan meteor observation network, SonotaCo, and EDMOND, we aim to enhance the precision of meteor shower associations, ensuring a more accurate and refined understanding of meteoroid dynamics. This refined methodology also contributes to the broader goal of updating and improving the IAU's list of meteor showers (Vida et al. (2018)).

2. Methodology

Our methodology was built upon the comprehensive data collected from the Moroccan meteor observation network and supplemented by international datasets like those from SonotaCo and EDMOND (SonotaCo (2016)). The initial step involved aggregating meteor and asteroid data from these databases. Various methods for linking meteors to potential parent bodies were explored (table 1), with particular focus on refining association techniques developed by previous studies (Drummond (1981), Rudawska & Jenniskens (2014)).

In our analysis, we initially set various threshold values: we used 8° for the difference in solar longitude (λ_o), and 0.05, 0.1, and 0.2 for D_{sh}. This approach is designated as Method 1, as detailed in Tables 1.

We then calculated the angular distance between radiants, along with the differences in solar longitude and velocity. For the difference in solar longitude, we maintained a threshold of 8°. To optimize our findings, we explored several thresholds for both the radiant position and velocity differences. Initially, the distance to the radiant was set at 6°, which was later refined to 2°. For the velocity difference, we considered thresholds of 0.5, 3, and 5 km s^{-1}. This procedure is referred to as Method 2, yielding six unique combinations of radiant position and velocity differences.

To further investigate the impact of velocity, we introduced a variable velocity threshold based on a percentage of the measured meteoroid velocity. Notably, the UFOOrbit software provides a $dv12\%$ parameter to assess orbit quality. In our study, we identified a Near-Earth Object (NEO) as a potential parent body if the following condition was met: $V_g - V_g \cdot v\% < V_{NEO}^g < V_g + V_g \cdot V\%$, where $V\%$ is set at $5\%, 10\%$, or 20% of the meteoroid velocity. Thus, a NEO's geocentric velocity must lie within the uncertainty range of the meteoroid's geocentric velocity for the association to be deemed plausible. This methodology is referred to as Method 3.

Following data aggregation, a series of statistical tests were applied to evaluate the significance of potential associations. Our framework employs rigorous statistical analysis to eliminate random associations by calculating what we called "PCO," the probability of chance occurrence, which assesses the likelihood that an association between a meteor shower and a potential parent body is due to random chance. This is crucial given the

Table 2. Probability of chance occurrence (PCO) computed for each potential parent body associated to meteor orbits. The first three potential parent bodies have low values. PCO_S: PCO for meteors from the SonotaCo database; PCO_E: PCO for meteors from the EDMOND database. See text for further explanation.

Parent body	PCO_S	PCO_E
2015 QT_3	97%	13%
2009 HE_{21}	57 %	62%
(139359) 2001 ME1	–	77%
(251722) 1997 US_2	96%	82%
2016 BV_{14}	87 %	83%
1999 VK_{12}	90 %	84%
2003 UV_{11}	91 %	84%
2014 YQ_{34}	96 %	84%
2015 FP_{33}	–	84%
2007 RU_{17}	79 %	86%
2008 XM_1	92 %	88%
2003 WP_{21}	93 %	91%
2016 TA	–	92%
2006 SO_{198}	96 %	92%
2013 RL_{43}	–	93%
2011 UE_{256}	96 %	94%
2010 RV_3	–	95%
2016 UZ_{25}	97 %	95%
2014 HK_{197}	–	96%
(297274) 1996 SK	–	96%
2016 QB_{11}	–	97%
2001 XX_{103}	97 %	97%
2007 UL_{12}	98 %	97%
2005 UR	97 %	98%
2014 NK_{52}	98 %	98%
2016 VK	98 %	98%
(312942) 1995 EK_1	98%	98%
2015 TD_{144}	–	98%
2015 QO_3	–	98%
2007 PR_{25}	–	98%
2014 UQ_{114}	97 %	99%
2002 XM_{35}	99 %	99%
1997 VM_4	99 %	99%
2016 PZ_{39}	99 %	99%
(4197) Morpheus 1982 TA	99 %	99%
2013 BP_{45}	99 %	99%
2014 NF_{64}	–	99%
2014 BD_{33}	–	99%
2016 DF_1	–	99%
(144861) 2004 LA_{12}	–	99%
2011 GS_{60}	–	99%
1999 VR_6	–	99%
2014 OX_{299}	–	99%
2012 LL_9	–	99%
2005 CM_7	–	99%
2014 WD_7	–	99%
2015 PM_{307}	–	99%
(434677) 2006 BZ_7	–	99%
2015 XL_1	98 %	–
2010 XZ	99 %	–
(452302) 1995 YR_1	98%	–

very large number of meteor orbits calculated worldwide and the inherent errors and inaccuracies in them. A low PCO value indicates a stronger, more reliable link between a meteor shower and its parent body, which is the desired outcome (Table 2). Through this process, we were able to identify new parent bodies with a greater degree of confidence compared to traditional methods (Table 3) (Egal et al. (2017), Abedin et al. (2017)).

Table 3. Statistically not excluded new parent bodies and number of associated meteors in each database. α and δ are the location of the radiant (right ascension and declination, in *deg*. N_S: number of meteors in the SonotaCo database. N_E: number of meteors in the EDMOND Database.

Parent body	α	δ	N_S	N_E	PCO_S	PCO_E
2016 BN_{14}	306.6169	-5.3973	247	462	31 %	37 %
2015 TX_{24}	49.3551	14.0942	179	278	24 %	10 %

Table 4. Number of associations with known parent bodies.

	Parent body	Number in Sonotaco Datanase	Number in EDMOND Database
Comets	55P/Tempel-Tuttle	2257	1242
	1P/Halley	234	128
	P/2005 T4 SWAN	3008	2305
	109P/Swift-Tuttle	9758	35258
asteroids	(2011 XA3)	75	64
	(2009 WN25)	199	–
	(2004 TG10)	70	98
	(2010 TU149)	195	199
	(2015 DU180)	–	55
	196256 (2003 EH1)	2480	1976
	3200 Phaethon (1983 TB)	14682	11811

3. Results

To validate the effectiveness of our approach, we compared the results with several well-established parent bodies; asteroids and comets, (see Table 4). Notable associations were found, including 3200 Phaethon (1983 TB), the parent body of the Geminids meteor shower, which was linked to 14,682 meteors from the Sonotaco Database and 11,811 from the EDMOND Database. Similarly, the 109P/Swift-Tuttle comet, responsible for the Perseids meteor shower, was associated with 9,758 meteors from Sonotaco and 35,258 meteors from EDMOND.

Further associations include comets such as 55P/Tempel-Tuttle, the parent body of the Leonids, which was linked to 2,257 meteors from Sonotaco and 1,242 from EDMOND. Several asteroids were also confirmed as parent bodies, providing strong validation for our method, such as 196256 (2003 EH1), associated with 2,480 meteors in Sonotaco and 1,976 in EDMOND.

This comparison demonstrates the reliability of the proposed framework by aligning our results with known parent bodies, thereby reinforcing the efficacy of the methodology in accurately identifying meteor-parent body relationships.

And to summarise our statistical results, we present a histogram (figure 1) that visually summarizes the differences between the two first asteroids and the rest of the sample. The x-axis represents, PCO_S the probability of chance occurrence of orbit associations determined from the SonotaCo database, and PCO_E, the equivalent probability calculated from the EDMOND database, while the y-axis indicates asteroid names.

The first two asteroids are distinctly marked, showcasing their unique characteristics relative to the other asteroids. This visual representation facilitates an easier comparison and highlights the significant divergence in the PCO values of the first two asteroids compared to the others, effectively summarizing the findings from Tables 3 and 2.

The advanced linking methodology successfully identified several new potential parent bodies that were previously unrecognized in the IAU database. By improving the accuracy of meteor-parent body associations, our approach offers a more reliable framework for expanding the IAU's meteor shower list (Vaubaillon (2017), Pauls & Gladman

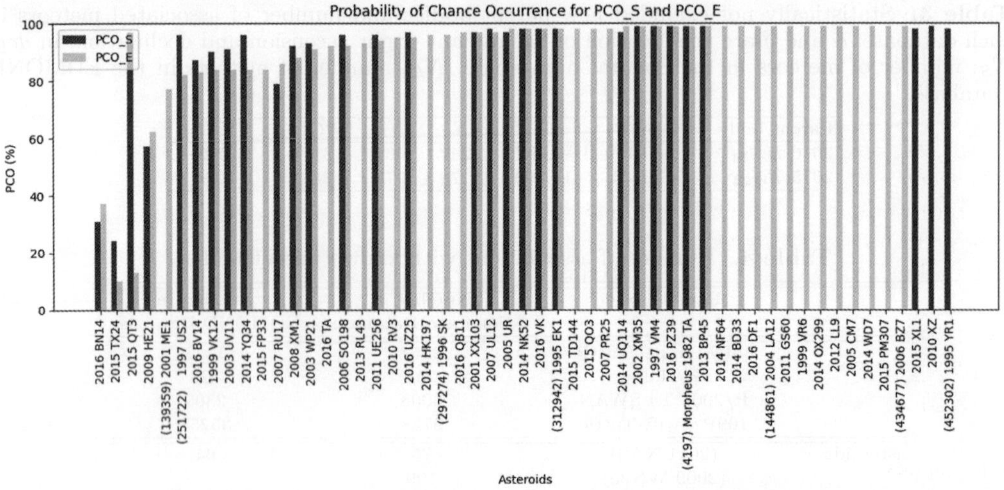

Figure 1. Probability of chance occurence for Sonotaco and EDMOND database associations.

(2005)).For example, our method effectively associated meteors with their parent bodies by accounting for statistical variations, demonstrating a marked improvement over traditional high-threshold linking techniques (Drummond (1981), Jopek (1993)).

Notably, the analysis revealed multiple new potential parent bodies not listed in the current IAU database (Vida et al. (2018)). These findings emphasize the value of our statistical framework and the effectiveness of the Moroccan meteor observation network in contributing to this area of research (Vida et al. (2018), SonotaCo (2016)).

4. Discussion

Our findings underscore the limitations of relying solely on high-threshold single-linking criteria for meteor shower associations. Traditional methods often struggle with uncertainties in meteor orbit data, which can lead to the misidentification of parent bodies (Drummond (1981)). By employing a more nuanced statistical approach, we are able to overcome these limitations and provide a clearer understanding of meteor shower origins. The Moroccan meteor observation network, with its extensive dataset and refined methodology, has proven to be a crucial resource in identifying new parent bodies and improving the accuracy of meteor shower associations (Egal et al. (2017), Vaubaillon (2017)).

The identification of new parent bodies underscores the need for continuous updates to the IAU's meteor shower list. The existing IAU lists, including the working and removed lists, reflect ongoing efforts to refine meteor shower associations, but our findings suggest that incorporating more advanced statistical methods could further enhance the reliability of these associations (Rudawska & Jenniskens (2014), Vida et al. (2018)). By improving the precision of parent body identifications, we can contribute to a better understanding of meteoroid dynamics and potential Earth impacts (Egal et al. (2017), SonotaCo (2016)).

Despite the advancements achieved in this study, future research could benefit from incorporating additional meteor and asteroid data to further validate and expand these findings. Exploring more advanced statistical techniques could also enhance the robustness of parent body associations, leading to even more accurate identifications (Hasegawa (1990), Micheli (2013)). Collaboration with international research networks would provide access to a broader dataset and enable global validation of these results (Vaubaillon (2017), SonotaCo (2016)).

5. Future Work

To build upon our results, future work should focus on enhancing data analysis by incorporating additional meteor and asteroid data from global sources. This will help validate and expand our findings. Further refinement of statistical methods, especially by exploring more advanced techniques, will also be crucial in improving the robustness of parent body associations (Egal et al. (2017), Micheli (2013)). Additionally, leveraging international collaborations, including partnerships with the IAU and other meteor observation networks, will provide the broader dataset necessary for global validation of our findings (Vida et al. (2018), Rudawska & Jenniskens (2014)).

Acknowledgements

We extend our gratitude to L. Neslusan for providing the Neslusan et al.(1998) program, which was instrumental in our data analysis.

References

Abedin, A., Wiegert, P., Pokorný, P., & Brown, P. 2017, *Icarus*, 281, 417
Andreić, Ž., Šegon, D., Korlević, K., et al. 2013, *WGN J. Int. Meteor Organ.*, 41, 103
Drummond, J. D. 1981, *Icarus*, 45, 545
Egal, A., Gural, P. S., Vaubaillon, J., Colas, F., & Thuillot, W. 2017, *Icarus*, 294, 43
Guennoun, M., Vaubaillon, J., Čapek, D., Koten, P. & Benkhaldoun, Z. 2019, *Astronomy & Astrophysics*, Volume 622, id.A84, 9 pp.
Hasegawa, I. 1990, *PASJ*, 42, 601
Jenniskens, P., & Nénon, Q. 2016, *Icarus*, 266, 371
Jopek, T. J. 1993, *Icarus*, 106, 603
Koten, P., Vaubaillon, J., Čapek, D., et al. 2014, *Icarus*, 239, 244
Lindblad, B. A. 1971a, *Smithson. Contrib. Astrophys.*, 12, 14
Micheli, M. 2013, Ph.D. Thesis, University of Hawai'i at Manoa, USA
Pauls, A., & Gladman, B. 2005, *Meteorit. Planet. Sci.*, 40, 1241
Rudawska, R., & Jenniskens, P. 2014, *Meteoroids 2013*, 217
Šegon, D., Andreić, Ž., Gural, P., et al. 2014, *WGN J. Int. Meteor Organ.*, 42, 227
SonotaCo 2016, *WGN J. Int. Meteor Organ.*, 44, 42
Vaubaillon, J. 2017, *Planet. Space Sci.*, 143, 78
Vida, D., Brown, P. G., & Campbell-Brown, M. 2018, *MNRAS*, 479, 4307

Planetary Science and Exoplanets in the Era of JWST
Proceedings IAU Symposium No. 393, 2026
Z. Benkhaldoun, Y. Moulane & A. Szentgyorgyi, eds.
doi:10.1017/S1743921325000444

Host stars and their effects on exoplanet atmospheres: Reconstructing the missing extreme-ultraviolet and Lyman-alpha radiation that exoplanets see

Jeffrey L. Linsky[1] and **Seth Redfield**[2]

[1]JILA, University of Colorado and NIST,
Campus Box 440, Boulder, Colorado, 80309-0440 USA
email: jlinsky@jila.colorado.edu

[2]Astronomy Department, Wesleyan University,
Middletown CT, 06459-1234 USA
email: sredfield@wesleyan.edu

Abstract. Whether an exoplanet can retain its atmosphere is mainly controlled by the extreme-ultraviolet (EUV) radiation received from its host star, and the photo-chemistry in its outer atmosphere is driven by the far-ultraviolet radiation, primarily the hydrogen Lyman-α line, from its host star. Since interstellar hydrogen absorbs most of this EUV and Lyman-α radiation, there is a critical need for accurate reconstruction techniques to identify the intrinsic EUV and Lyman-α radiation that impacts the outer atmosphere of an exoplanet. This paper describes and critiques the available reconstruction techniques.

Keywords. exoplanets, EUV radiation, stellar activity

1. Introduction

With the launch of the James Webb Space Telescope, the search for evidence of life on exoplanets is proceeding at high speed. Optical and especially near-infrared spectra of exoplanets during and outside of transits are being searched for the presence of atoms and molecules that could only be produced by life forms. Of particular interest are rocky Earth-mass and super-Earth-mass exoplanets that are located in the liquid water habitable zones of their host stars, because liquid water on or near the surfaces of these exoplanets is thought to be necessary for the existence of life. The presence of surface water on an exoplanet is not just a function of the total incident flux from the host star that controls the exoplanet's surface temperature and defines the liquid water habitable zone. To support life forms it is also essential that the exoplanet retains its atmosphere against the present and past photo-ionizing EUV radiation of its host star that ionizes hydrogen from dissociated water thereby heating the outer atmosphere and driving mass loss.

This paper describes and critiques the various techniques used to reconstruct the host star's EUV radiation that is absorbed by interstellar hydrogen. Interstellar hydrogen also absorbs most of the stellar Lyman-α radiation that is the dominant source of photo-chemistry in the outer atmospheres of exoplanets. This paper also describes and critiques the techniques used to reconstruct the intrinsic Lyman-α radiation from host stars. These topic and the many ways that stellar activity affects exoplanet atmospheres

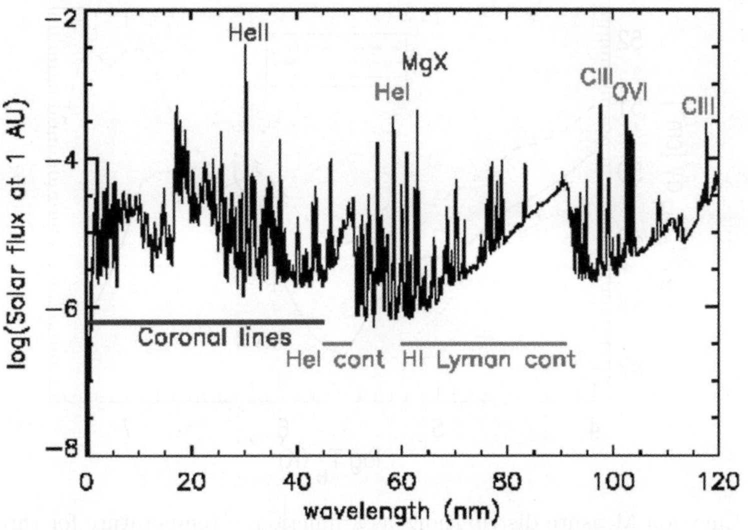

Figure 1. The solar spectrum from X-rays to 120 nm. Different spectral features and important emission lines are indicated. Red indicates spectral features formed in the chromosphere or transition region. Blue indicates spectral features formed in the corona. Figure from Linsky et al. (2014).

are described in the book "Host Stars and their Effects on Exoplanet Atmospheres" (Linsky 2024).

2. Reconstructing the intrinsic EUV spectral energy distribution of host stars

Water in the atmospheres of exoplanets is photo-dissociated into oxygen and hydrogen by the far-ultraviolet (91.2–180 nm) radiation from the host star, in particular the hydrogen Lyman-α line at 121.7 nm. Host star radiation short-ward of 91.2 nm then photo-ionizes the hydrogen, heating and expanding the atmosphere, and driving the loss of hydrogen by hydrodynamic expansion and the pickup of ionized hydrogen by the magnetic stellar wind. The efficiency of hydrodynamic mass loss depends on the spectral energy distribution of the host star's EUV radiation, the exoplanet's gravitational potential, and the fraction of the photo-ionization energy that goes into heat (leading to mass loss) rather than recombination photons (leading to no mass loss). While the efficiency factor is uncertain, observations of escaping hydrogen as seen for example in Lyman-α absorption post-transit of the mini-Neptune GJ 436b (Ehrenreich et al. 2015) support the reality of exoplanet mass loss of hydrogen and therefore water.

The only star for which we will ever observe the complete EUV spectrum with no interstellar absorption is the Sun. As shown in Figure 1, the solar EUV spectrum consists of chromospheric emission in the Lyman continuum (< 91.2 nm), He I recombination continuum (< 50.4 nm), and some emission lines, and the coronal component consists of highly ionized coronal emission lines mostly in the 10–40 nm region. Since the Sun is a relatively quiet middle-aged star, the chromospheric component dominates and the EUV emission is much stronger than the X-ray emission, but for more active stars like AU Mic and AD Leo the coronal component of the EUV emission dominates over the chromospheric component and the X-ray emission is stronger than the EUV emission (Chadney et al. 2015).

The bound-free edge of hydrogen is at 91.2 nm, and the optical depth in the hydrogen Lyman continuum, which is proportional to λ^3, rapidly decreases below 91.2 nm. As a

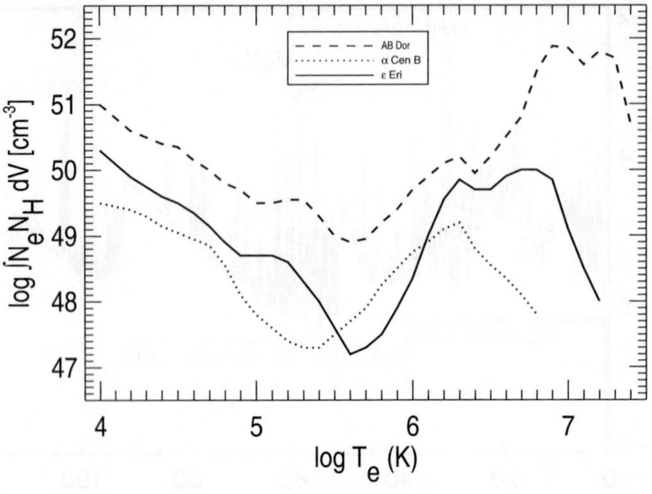

Figure 2. Emission Measure distributions as a function of temperature for three stars with increasing X-ray emission: α Cen B, ϵ Eri, and AB Dor. Figure from Sanz-Forcada et al. (2011).

result optical depth unity occurs just below 40 nm for nearby stars like ϵ Eri (d=3.22 pc) with interstellar hydrogen column density of $10^{18.0}$ cm^2. This is confirmed by the absence of observed emission lines at wavelengths above 40 nm even for the closest star (Proxima Centauri) (Craig et al. 1997). Thus reconstruction of the EUV spectral energy distribution for host stars is essential if we are to accurately determine whether an exoplanet can retain its atmosphere and thus retain surface water. We describe and critique three methods.

A rigorous but complicated method is to compute a stellar model atmosphere from the photosphere to the corona, and then synthesize the EUV spectrum from the temperature-density structure of the model using an extensive atomic physics data archive such as CHIANTI (Dere et al. 2023). These are very complex models that include non-LTE partial redistribution radiative transfer for optically thick lines and the include very many atomic and molecular spectral lines. Fontenla et al. 2016 and Tilipman et al. 2021 computed such models for two M dwarfs with coronae consistent with x-ray observations and provided both the coronal and chromospheric components of the EUV emission. Models computed by Peacock et al. 2019b, Peacock et al. 2020, and Hintz et al. 2023 fit many UV and optical lines but do not include coronal temperatures and the coronal component of the EUV emission. The M dwarf model atmospheres computed by Peacock et al. 2020, however, infer the coronal contribution to the the EUV spectrum by fitting GALAX FUV spectra. They found that the intrinsic EUV flux from young M dwarfs is 100 times larger than for older M dwarfs and that the EUV flux decreases as 1/age after a few hundred million years. Comparison of observed and computed EUV spectra indicate that this method for computing EUV spectra should provide sensible predictions for solar-like stars when coronal temperatures are included in the models. An accurate representation of the coronal component is needed, especially for active young stars, because coronal emission lines should dominate the EUV spectrum and the CHIANTI line list is likely incomplete.

A second method for computing EUV spectra is based on the many high temperature emission lines in the EUV spectrum that are emitted by the same ions responsible for emission lines at x-ray wavelengths. Thus the abundances of highly ionized atoms computed from coronal models or observed in x-ray spectra are used to compute the coronal component of EUV spectra. An example of this approach is the EUV spectrum

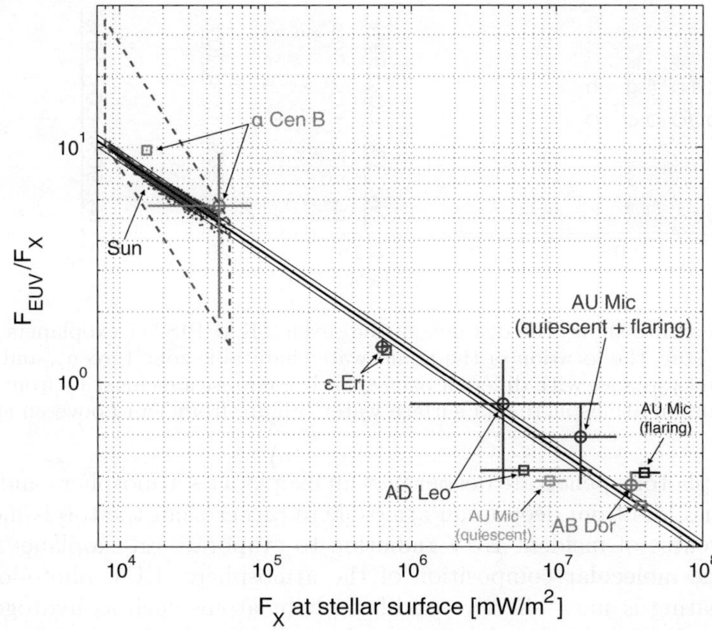

Figure 3. Emission Measure distributions as a function of temperature for three stars with increasing X-ray emission: α Cen B, ϵ Eri, and AB Dor. Figure from Chadney et al. (2015).

computed by Sanz-Forcada et al. 2011 from the emission measure distributions (EMD) (see Figure 2) as a function of temperature for three stars with increasing X-ray emission: α Cen B, ϵ Eri, and AB Dor. The EMD $= \int n_e n_H dV$ measures the emission from optically thin lines produced by electron collisional excitation of ions in the temperature range and volume where each ion is present. Duvvuri et al. 2023 computed the EUV spectrum of the young K1 host star V1298 Tau, showing that the coronal component of the EUV emission completely dominates the EUV spectrum of young active stars. The EMD technique is based on observed x-ray emission, but it is limited by the assumptions of thermal equilibrium, which may not be valid for coronae heated by flares at all energies, and by incomplete atomic data. Also, stellar EMDs often lack emission lines from ions formed below 10^6 K, which are often faint and thus hard to observe.

A third method is to correlate the EUV broad-band flux with an observable spectral feature such as the x-ray flux, Lyman-α flux, or the flux of emission lines in the FUV or optical. Correlations with the x-ray flux should be most reliable when the coronal component of the EUV emission dominates, however, this is not always the case. As shown in Figure 3, the EUV flux is much larger than the x-ray flux for less active stars like the Sun and α Cen B, but the reverse is true for more active stars like AU Mic, AD Leo, and AB Dor. The power-law relation, $\log L(\mathrm{EUV}) = 9.93 + 0.67 \log \mathrm{L(X)}$ (Shoda & Takasao 2021) is consistent with the decrease in L(EUV)/L(X) with increasing x-ray luminosity. Also, x-ray emission from stellar coronae is highly variable and x-ray observations are typically made at different times than the time frame needed for estimating the EUV emission. Correlations of EUV emission with Lyman-α (e.g., Linsky et al. 2014) or Ca II emission (e.g., Sreejith et al. 2020) better fits inactive stars as chromospheric emission dominates the EUV spectrum for inactive stars. A problem with these correlation methods is that they estimate integrated EUV fluxes rather than the EUV spectral energy distribution. The shape of the EUV spectral energy distribution is important, because

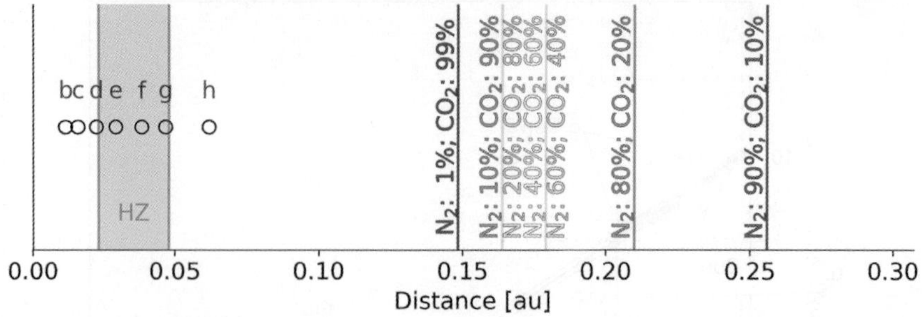

Figure 4. The figure shows the distance of the seven TRAPPIST-1 exoplanets (b through h) from their host star, the location of the liquid water habitable zone (green), and the minimum distance that an exoplanet with different atmospheric compositions must be from TRAPPIST-1 in order to retain an atmosphere and surface water. Figure from Van Looveren et al. (2024).

higher energy photons penetrate deeper into an exoplanet's atmosphere and deliver more energy but mass loss from deep in an atmosphere rather than the top is more difficult.

The effectiveness of incident EUV radiation to evaporate an exoplanet's atmosphere depends on the molecular composition of the atmosphere. EUV photo-ionization and subsequent heating is most effective on lifting light atoms such as hydrogen out of the exoplanet's gravitational potential, but far less effective for heavier atoms such as C, N, and O produced by photo-dissociation of H_2O, CO_2, and CH_4. A recent study of mass-loss from the TRAPPIST-1 exoplanets by Van Looveren et al. 2024 computes the mass-loss rates for different atmospheric mixtures of N_2 and CO_2 by simply scaling the solar EUV flux received at Earth to the distances of the TRAPPIST-1 exoplanets. Since they all lie close to their host star, the EUV flux is 1928 times that received by the Earth for the closest exoplanet (b) and 63 times that received by the Earth for the furthest exoplanet (h). Figure 4 shows the distances of the TRAPPIST-1 exoplanets from their host star, the liquid water habitable zone, and the distances from the host star where the exoplanets would have to be in order to retain their atmospheres and water for different chemical compositions of their atmospheres. As shown in the figure, for any mixture of N_2 and CO_2 all of these exoplanets should have lost their entire atmospheres and surface water in less than 1 Gyr and probably in a much shorter time since the EUV flux from young TRAPPIST-1 would have been far larger than present. All of the TRAPPIST-1 exoplanets, even those further away than the liquid water habitable zone should be bare rocks. This result implies that all Earth-mass exoplanets of M dwarf stars inside their respective liquid water habitable zones should be bare rocks and sterile. We shall see whether JWST's large observing programs of M dwarf exoplanets confirms this prognosis.

3. Reconstructing the intrinsic Lyman-α flux illuminating exoplanet atmospheres

The hydrogen Lyman-α line dominates the FUV spectra of solar-like and cooler stars and is the primary source for photo-dissociating the important molecules H_2O, CH_4 in the upper atmospheres of exoplanets. This is illustrated in Figure 5 which shows the cumulative fraction of these and other molecules that are photo-dissociated with increasing wavelengths due to the overlap of the molecular cross-sections with the spectra of a high FUV flux star (ϵ Eri), the Sun, and a low FUV flux star (GJ 581 M3V). Note the steep rise in the fractional dissociation at the Lyman-α (121.5 nm) line.

Unfortunately, the line center optical depth of the interstellar Lyman-α line has an optical depth of 10^5 to 10^6 even for the closest stars, which completely removes any

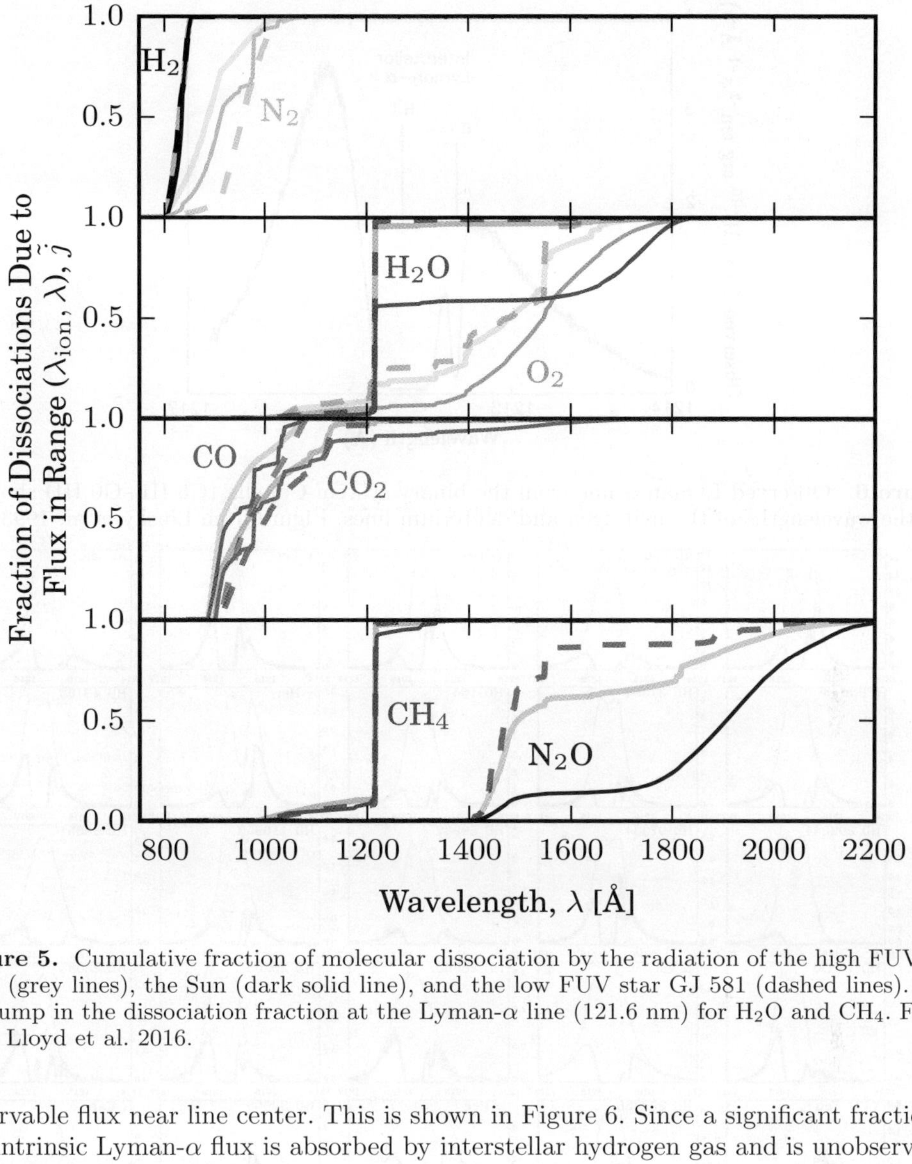

Figure 5. Cumulative fraction of molecular dissociation by the radiation of the high FUV star ϵ Eri (grey lines), the Sun (dark solid line), and the low FUV star GJ 581 (dashed lines). Note the jump in the dissociation fraction at the Lyman-α line (121.6 nm) for H_2O and CH_4. Figure from Lloyd et al. 2016.

observable flux near line center. This is shown in Figure 6. Since a significant fraction of the intrinsic Lyman-α flux is absorbed by interstellar hydrogen gas and is unobservable, the intrinsic line profile and flux must be reconstructed. There are several techniques for doing so.

The first technique takes advantage of the deuterium Lyman-α line located at –0.33 Å (–0.033 nm) from the center of the hydrogen line. Since the D/H ratio within about 100 pc of the Sun is $(1.56 \pm 0.4) \times 10^{-5}$ (Wood et al. 2004), the deuterium line is optically thin or nearly so for sight lines to nearby stars. The measured flux in the deuterium line divided by the D/H ratio results in the flux of the intrinsic hydrogen line. Since the D/H mass ratio is 2, the thermal and turbulent broadening parameters of the hydrogen and deuterium lines must be known from analysis of heavier mass elements such as Mg II and Fe II. Using this technique, (Wood et al. 2005) reconstructed the hydrogen Lyman-α lines for many stars as shown in Figure 7. In all cases, the flux in the reconstructed hydrogen lines is at least a factor of 2 and often much larger than the observed lines. This deuterium line technique only works when the deuterium Lyman-α line profile can be measured against the interstellar hydrogen absorption which broadens with increasing

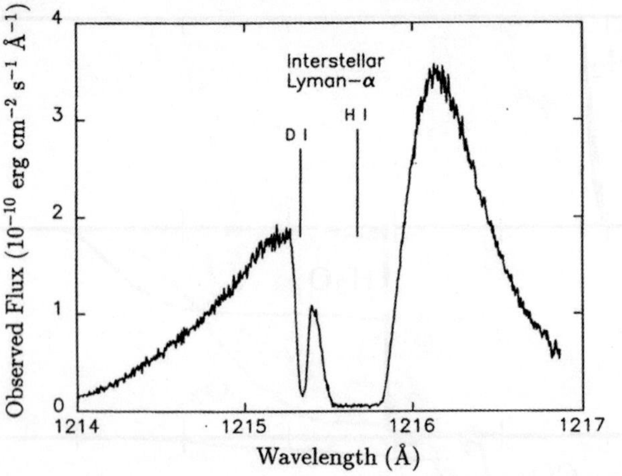

Figure 6. Observed Lyman-α line from the binary system Capella (G5 III+G0 III) showing the wavelengths of the hydrogen and deuterium lines. Figure from Linsky et al. 1993.

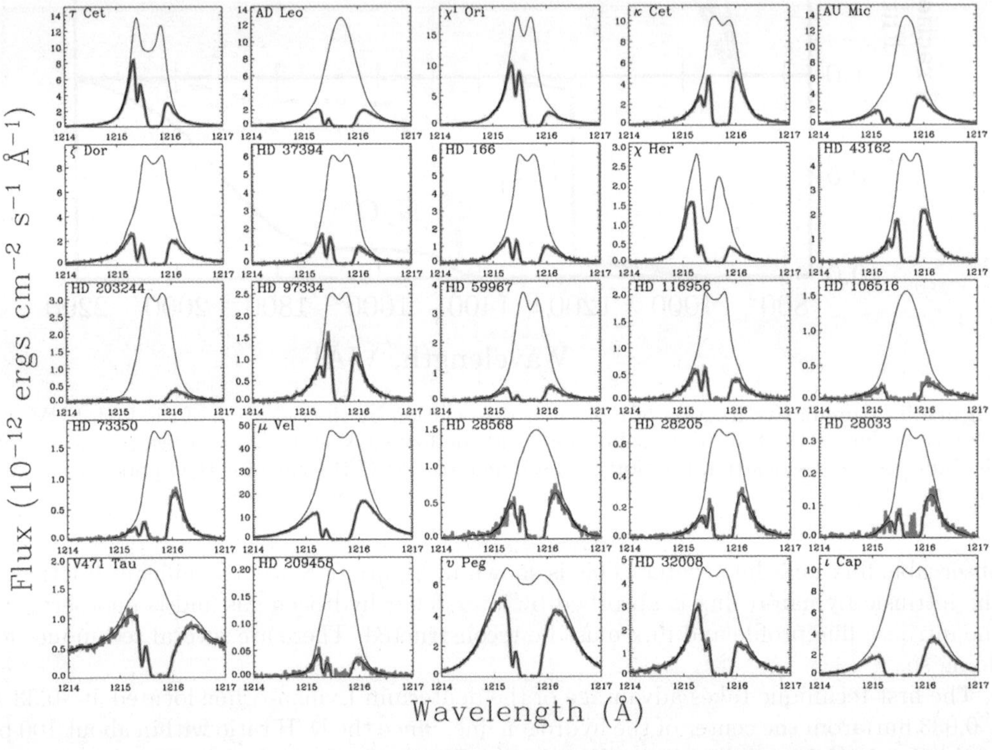

Figure 7. Comparison of reconstructed Lyman-α line profiles (thin lines) with observed line profiles (thick lines) for 20 stars using the deuterium line technique. Figure from (Wood et al. 2005).

H(HI). Reliable hydrogen Lyman-α line fluxes and line profiles are only possible with this technique when N(HI) $\leq 5 \times 10^{18}$ cm^2. The sight lines to most stars within 50 pc meet this constraint.

A second technique involves simultaneously solving for the hydrogen Lyman-α line shape, line flux, and the interstellar hydrogen column density in the sight line to the star at the same time. The parameters for the intrinsic hydrogen line (width and peak flux)

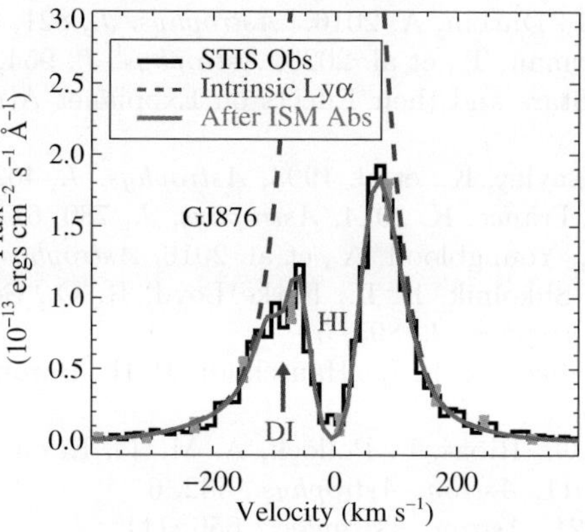

Figure 8. An example of the multi-parameter fit technique is shown for the star GJ 876. The black line is the observed line profile, the dashed line is the reconstructed intrinsic Lyman-α line profile, and the red line is the reconstructed line profile after interstellar absorption. Figure from (France et al. 2012).

and interstellar N(HI) are iterated until an intrinsic line profile matches the observed profile after inclusion of interstellar absorption. (France et al. 2012) showed that this technique provided sensible results for the star GJ 876 (M3.5 V, d=4.67). As shown in Figure 8 the reconstructed Lyman-α line profile after absorption by interstellar hydrogen with the inferred value of N(HI) matches the observed line profile very well. This multi-parameter fit technique has also been applied to 8 K and M dwarf stars by (Youngblood et al. 2016) and to Kapteyn's star (M1 VI, d=3.93 pc) by (Guinan et al. 2016). The accuracy of the technique is limited by possible parameter degeneracies, insufficient spectral resolution, and by the assumption of a single interstellar velocity component in the sight line to the star. As described by (Youngblood et al. 2016), a comparison of the interstellar parameters and inferred Lyman-α flux for the well observed star AU Mic (M1 V,d=9.71 pc) show excellent agreement between the multi-parameter fit and the deuterium line techniques.

Acknowledgements

We acknowledge support from the NASA Outer Heliosphere Guest Investigators Program to Wesleyan University and the University of Colorado for grant 80NSSC20K0785. The spectra used in this paper were obtained by *HST* primarily from the SNAP programs 10236, 11568, 13332, 13658, 14084, and 16487 but also from programs 13650 and 15071. The data were extracted from MAST.

References

Chadney, J. M., Galand, M., Unruh, Y. C., Koskinen, T. T., Sanz-Forcada, J. 2015, *Icarus*, 250, 357

Craig, N., Abbott, M., Finley, D., Jessop, H., Howell, S. B., Mathioudakis, M., Sommers, J., Vallerga, J. V., Malina, R. F. 1997, *Astrophys J. Supp. Ser.*, 113, 131

Dere, K. P., Del Zanna, G., Young. P. R., Landi, E., Mason, H. E. 2023, *Astrophys. J. Supp. Ser.*, 268, 52

Duvvuri, G. M., Cauley, P. W., Aguirre, F. C., et al. 2023, *Astron. J.*, 166, 196

Ehrenreich, D., Bourrier, V., Wheatley, P. J., et al. 2015, *Nature*, 522, 459

Fontenla, J. M., Linsky, J. L., Garrison, J., et al. 2016, *Astrophys. J.*, 830, 154

France, K., Linsky, J. L., Tian, F., Froning, C. S., Roberge, A. 2012, *Astrophys. J. Letters*, 750, L32

Guinan, E. F., Engle, S. G., Durbin, A. 2016, *Astrophys. J.*, 821, 81
Hintz, D., Peacock, S., Barman, T., et al. 2023, *Astrophys. J.*, 954, 1
Linsky, J. L. 2024 "Host Stars and their Effects on Exoplanet Atmospheres", to be published by Springer
Linsky, J. L., Brown, A., Gayley, K., et al. 1993, *Astrophys. J.*, 402, 694
Linsky, J. L., Fontenla, J., France, K. 2014, *Astrophys. J.*, 780, 61
Loyd, R. O. P., France, K,, Youngblood, A., et al. 2016, *Astrophys. J.*, 82, 102
Peacock, S., Barman, T., Shkolnik, E. L., Parke Loyd, R. O., Schneider, A. C., Pagano, I., Meadows, S. 2020, *Astrophys. J.*, 895, 5
Peacock, S., Barman, T., Shkolnik, E. L., Hauschildt, P. H., Baron, E., Fuhrmeister, B. (2019) *Astrophys. J.*, 886, 77
Sanz-Forcada, J., Micela, G., Ribas, I., Pollock, A. M. T., Eiroa, C., Velasco, A., Solano, E., García-Álvarez, D. 2011, *Astron. Astrophys.*, 532, 6
Shoda, M., Takasao, S. 2021, *Astron. Astrophys.*, 656, 111
Sreejith, A. G., Fossati, L., Youngblood, A., France, K., Ambily, S. 2020, *Astron. Astrophys.*, 644, A67
Tilipman, D., Vieytes, M., Linsky, J. L., Buccino, A. P., France, K. 2021, *Astrophys. J.* 909, 61
Van Looveren, G., Güdel, M., Boro Saikia, S., Kislyakova, K. 2024, *Astron. Astrophys.*, 683, A153
Wood, B. E., Linsky, J. L., Hébrard, G., Williger, G. M., Moos, H. W., Blair, W. P. 2004, *Astrophys. J.*, 609, 838
Wood, B. E., Redfield, S., Linsky, J. L., Müller, H.-R., Zank, G. P. 2005, *Astrophys. J. Supp. Ser.*, 159, 118
Youngblood, A., France, K., Loyd, R. O. P., et al. 2016, *Astrophys. J.*, 824, 101

Planetary Science and Exoplanets in the Era of JWST
Proceedings IAU Symposium No. 393, 2026
Z. Benkhaldoun, Y. Moulane & A. Szentgyorgyi, eds.
doi:10.1017/S1743921324001546

Migration of bodies in the Proxima Centauri and Trappist 1 planetary systems

Sergei I. Ipatov

Vernadsky Institute of Geochemistry and Analytical Chemistry of RAS,
119991, 19 Kosygin st., Moscow, Russia
email: siipatov@hotmail.com

Abstract. The motion of planetesimals was studied in the Proxima Centauri and TRAPPIST 1 exoplanetary systems. The size of the feeding zone of planet Proxima Centauri c is discussed. It was noted that after hundreds of Myrs, some planetesimals could still move in elliptical resonant orbits inside the feeding zone of this planet that had been mainly cleared from planetesimals. The probability of a collision of a planetesimal initially located in the feeding zone of planet c with inner planet b was obtained to be about 0.0002 and 0.001 at initial eccentricity of orbits of planetesimals equal to 0.02 or 0.15, respectively. A lot of icy material and volatiles could be delivered from the icy zone near the orbit of planet c to inner planets b and d. The inclinations of orbits of 80% of the planetesimals that moved between 500 or 1200 AU from the star did not exceed 10°. It was obtained that several planets in the TRAPPIST-1 system accumulated planetesimals initially located at the same distance. Outer layers of neighbouring TRAPPIST-1 planets can include similar material.

Keywords. methods: n-body simulations, planets and satellites: formation

1. Initial data for calculations of the motion of planetesimals in the Proxima Centauri planetary system

The motion of planetesimals and other bodies in the Solar System was studied in many papers (see e.g. the review in (Marov, Ipatov 2023)). Ipatov (2023a,b,c) studied the motion of planetesimals in the Proxima Centauri planetary system. In these calculations, the gravitational influence of the star (with a mass equal to 0.122 of the solar mass) and two planets: b (a_b=0.04857 AU, e_b=0.11, m_b=1.17m_E, m_E is the mass of the Earth) and c (a_c=1.489 AU, e_c=0.04, m_c=7m_E) was taken into account. Calculations with other values of m_c (0.7m_E, 3.5m_E, and 12m_E) were also made. The planetesimals were excluded from integration when they collided with the star or the planets or reached 1200 AU (the Hill radius of the star) from the star. The symplectic code from (Levison, Duncan, 1994) was used for integrations. In most calculations, the integration time step t_s equaled to 1 day. The obtained results were about the same for different considered t_s equaled to 0.2 day, or 0.5 day, or 2 days. The considered time interval usually exceeded 100 Myr. For some variants it reached 1000 Myr. The late gas-less stage of formation of planets was considered. Migration of planetesimals from the feeding zone of planet c was calculated. In each calculation variant, initial semi-major axes a_o of orbits of N_o=250 planetesimals were in the range from a_{min} to a_{min}+0.1 AU. In different variants, the values of a_{min} were in the range from 0.9 to 2.2 AU. Initial eccentricities e_o of orbits of planetesimals equaled to 0.02 or 0.15. Initial inclinations of orbits of the planetesimals were equal to $e_o/2$ rad. Not small considered eccentricities of planetesimals could be a result of the previous evolution of the disk of planetesimals and could be reached due to mutual gravitational influence of planetesimals.

2. The feeding zone of Proxima Centauri c

After hundreds of million years, some planetesimals could still move in elliptical orbits inside the feeding zone of planet c that had been mainly cleared from planetesimals. Initial semi-major axes of orbits of the planetesimals that still had elliptical orbits at the end of considered evolution are presented in (Ipatov 2023a). For the range (a_{mine}, a_{maxe}) of initial semi-major axes a_o of orbits for which planetesimals were mainly ejected into hyperbolic orbits or collided with planets, Ipatov (2023a) obtained a_{mine}=a_c(1-e_c)-$e_o a_{mine}$-$k_{min} a_c \mu^{1/3}$ and a_{maxe}=a_c(1+e_c)+$e_o a_{maxe}$+$k_{max} a_c \mu^{1/3}$, where a_c and e_c=0.04 are the semi-major axis and eccentricity of the orbit of planet c, and μ is the ratio of the mass of planet c to the star mass. In these formulas, k_{min}=2.54 and k_{max}=2.40 at e_o=0.02, and k_{min}=2.23 and k_{max}=4.3 at e_o=0.15. Some planetesimals could still move inside the feeding zone, often in some resonances with the planet. For example, they moved in the resonances 1:1 (as Jupiter Trojans), 5:4, and 3:4. The number of such left planetesimals was greater at smaller eccentricities. Ipatov (2023a) presented the examples of time evolution of a semi-major axis a, perihelion and aphelion distances q and Q of an orbit of a planetesimal moving in such resonances. Before getting for a long time in these resonances, some planetesimals could move in nonresonant orbits. Some planetesimals could be ejected into hyperbolic orbits or could collide with planets if they started from some (typically resonant) subregions of a_o located outside the main feeding zone of planet c.

The ratio of the mass of planet Proxima Centauri c to the mass of the star is smaller than that for Jupiter. The ratio of the semi-major axes of the orbits of the planets c and b is larger than that for Jupiter and Mars. There is only one large planet in the Proxima Centauri system. Therefore there can be more analogues of the asteroid and trans-Neptunian belts in the planetary system near Proxima Centauri than in the Solar System, and there can be a planet (planets) between the orbits of the planets Proxima Centauri b and c.

3. Probabilities of ejection of planetesimals into hyperbolic orbits and the total initial mass of planetesimals in the feeding zone of Proxima Centauri c

Ipatov (2023b) presented the fraction of initial planetesimals that were left in elliptical orbits after times T equal to 1, 10, and 100 Myr at different initial semi-major axes a_o of their orbits. For T=100 Myr and e_o=0.15, the fraction was not more than 0.3, 0.06, 0.2, 0.06, and 0.1 at 1.1≤a_o≤1.2 AU, at 1.2≤a_o≤1.7 AU, at 1.7≤a_o≤1.9 AU, at 2.0≤a_o≤2.1 AU, and at 2.1≤a_o≤2.2 AU, respectively. For e_o=0.02, this fraction was not more than 0.25 at 1.2≤a_o≤1.8 AU. If initial orbits of planetesimals were not far from the orbit of the planet c, then the probability of a collision of a planetesimal during its dynamical lifetime with planet c was about 0.5 for e_o=0.02 and 0.25-0.3 for e_o=0.15. For e_o=0.02 the probability was about 0.05, 0.4-0.55, 0.3, and 0.02 at 1.1≤a_o≤1.2 AU, at 1.2≤a_o≤1.7 AU, at 1.7≤a_o≤1.8 AU, and at 1.8≤a_o≤1.9 AU, respectively. For e_o=0.15, this probability was about 0.15, 0.3, and 0.04-0.05 at 1.0≤a_o≤1.1 AU, at 1.1≤a_o≤1.9 AU, and at 2.0≤a_o≤2.2 AU, respectively. Most collisions of planetesimals with planet c took place during the first 10 Myr.

If initial orbits of planetesimals were not far from the orbit of the planet c, then during a whole considered time interval the ratio p_{cej} of the number of planetesimals collided with planet c to that ejected into hyperbolic orbits was about 1 for e_o=0.02 and about

0.5 for e_o=0.15. For e_o=0.02, the ratio p_{cej} was about 2 at 1.1≤a_o≤1.2 AU, about 1.2 at 1.2≤a_o≤1.4 AU, about 0.85-1 at 1.4≤a_o≤1.6 AU, about 1.3 at 1.6≤a_o≤1.7 AU, about 0.5 at 1.7≤a_o≤1.8 AU, and about 0.2 at 1.8≤a_o≤1.9 AU. For e_o=0.15, this ratio was about 0.9 at 1.0≤a_o≤1.2 AU, 0.5-0.6 at 1.2≤a_o≤1.4 AU, 0.4 at 1.4≤a_o≤1.8 AU, 0.05 at 2.0≤a_o≤2.2 AU, and 0.02 at 2.2≤a_o≤2.3 AU. The ratio of the fraction of planetesimals ejected into hyperbolic orbits to that collided with exoplanets is greater for a greater mass of a planet moving in the orbit of planet c and for greater initial eccentricities of orbits of planetesimals.

For a whole feeding zone of planet c and the present mass of planet c ($7m_E$), Ipatov (2023b) concluded that the ratio $p_{cej}=p_c/p_{ej}$ of the probability of a collision of a planetesimal with planet c to the probability p_{ej} of its ejection into a hyperbolic orbit was about 0.8-1.3 and 0.4-0.6 at e_o=0.02 and e_o=0.15, respectively. At calculations with a mass of planet c equal to a half of its present mass, such ratio was about 1.3-1.5 and 0.5-0.6. Based on the above ratios, it is possible to conclude that the total mass of planetesimals ejected into hyperbolic orbits could be about (3.5-7)m_E. The total mass of planetesimals in the feeding zone of planet c could exceed $10m_E$ and $15m_E$ at e_o=0.2 and e_o=0.15, respectively. Based on estimates of the amount of planetesimals ejected into hyperbolic orbits and on the integral of energy, Ipatov (2023b) concluded that the semi-major axis of the orbit of planet c could decrease by at least a factor of 1.5 during the growth of the mass of planet c from $3.5m_E$ to $7m_E$.

4. Delivery of planetesimals from the vicinity of the orbit of planet Proxima Centauri c to the inner planets b and d

The probability of a collision of a planetesimal initially located in the feeding zone of planet c with planet b was about 0.0002 and 0.001 at e_o equal to 0.02 or 0.15, respectively (Ipatov 2023b). The above values of the probabilities of collisions of planetesimals with planet b were greater than the probability of a collision with the Earth of a planetesimal migrated from the zone of the giant planets in the Solar System. The latter probability (per one planetesimal) was typically less than 10^{-5} (Marov, Ipatov 2023). Planet d was not included in integrations, and the probability of collisions of planetesimals with this planet was calculated based on some formulas and on the arrays of orbital elements of migrated planetesimals during their evolution. The amount of material delivered from the feeding zone of planet c to planet d could be about twice less than that delivered to planet b. A lot of icy material and volatiles could be delivered to planets b and d.

5. Motion of planetesimals in the outer part of the Hill sphere of the star Proxima Centauri

About 90% of the planetesimals that first reached 500 AU from the star, for the first time reached 1200 AU from the star in less than 1 million years (Ipatov 2023c). The inclinations of orbits of 80% of the planetesimals that moved between 500 or 1200 AU from the star did not exceed 10°. Mainly only the bodies that came into the Hill sphere of Proxima Centauri from outside can move in strongly inclined orbits of bodies in the outer part of the Hill sphere of this star. The radius of the Hill sphere of the star Proxima Centauri is by an order of magnitude smaller than the radius of the outer boundary of the Hills cloud in the Solar System and is two orders of magnitude smaller than the radius of the Hill sphere of the Sun. Therefore, it is difficult to expect the existence of such a massive analogue of the Oort cloud near this star as near the Sun.

6. Probabilities of collisions of planetesimals with planets in the TRAPPIST 1 exoplanetary system

The exoplanetary system TRAPPIST-1 consists of a star with a mass equal to 0.0898 of the mass of the Sun and 7 planets (from b to h) with masses from 0.33 to 1.37 Earth's masses. The semi-major axes of the planets' orbits are in the range from 0.012 to 0.062 AU. The motion of planetesimals under the gravitational influence of the star and seven TRAPPIST-1 planets (from b to h) was calculated with the use of the symplectic code from (Levison, Duncan, 1994) similar to my studies of migration of planetesimals in the Proxima Centauri system. In each considered variant of calculations, the semi-major axes of orbits of N_o planetesimals were near the semi-major axis of one of the planets. Initial eccentricities of orbits of planetesimals were equal to 0.02 or 0.15. Their initial inclinations equaled to $e_o/2$ rad. Some calculations with N_o=250 were discussed in (Ipatov 2022). Below I present the results for N_o=1000 and a time integration step t_s=0.01 or t_s=0.02 days.

Most of collisions of planetesimals with planets took place in less than 10 Kyr. For disks b-g more than a half of collisions were during the first 1 Kyr. However, last collisions of planetesimals with planets could be after a few million years. There were no collisions of planetesimals with the host star. Not more than 3% of planetesimals were ejected into hyperbolic orbits. The fraction of planetesimals collided with the 'host' planet (compared to collisions with all planets) typically was smaller for a greater considered time interval. For disks c-h, the fraction f_1 of planetesimals that collided with the host planet was about 0.37-0.63 and 0.27-0.53 at e_o=0.02 and e_o=0.15, respectively. The second and third values of the fraction of planetesimals that collided with neighbouring planets are designated as f_2 and f_3. The values of f_2 and f_3 were in the ranges 0.17-0.28 and 0.11-0.17 at e_o=0.02 and were 0.2-0.32 and 0.08-0.19 at e_o=0.15. For disk b and e_o=0.02, the value of f_1 was in the range 0.78-0.8, and f_2 and f_3 were between 0.18-0.2 and 0.01-0.012, respectively (the range is for calculations with an integration step t_s equaled to 0.01 and 0.02). For disk b and e_o=0.15, the values of f_1, f_2, and f_3 were 0.74, 0.22-0.23, and 0.015-0.021, respectively. The fraction of collisions of planetesimals with the 'host' planet was usually smaller for disks located farther from the star. In each variant, including the migration from a disk located beyond the outer disk h, collisions were with all planets. Therefore, the outer layers of neighboring planets in the TRAPPIST-1 system can include similar material if there were many planetesimals near their orbits at the late stages of planetary formation. Ipatov (1993, 2019) noted that each terrestrial planet could incorporate planetesimals from the feeding zones of all terrestrial planets.

7. Conclusions

The motion of planetesimals was studied in the Proxima Centauri and TRAPPIST 1 exoplanetary systems. The size of the feeding zone of planet Proxima Centauri c is discussed. It was noted that after hundreds of Myrs, some planetesimals could still move in elliptical resonant orbits inside the feeding zone of this planet that had been mainly cleared from planetesimals. The probability of a collision of a planetesimal initially located in the feeding zone of planet c with inner planet b was obtained to be about 0.0002 and 0.001 at initial eccentricity of orbits of planetesimals equal to 0.02 or 0.15, respectively. A lot of icy material and volatiles could be delivered from the icy zone near the orbit of planet c to inner planets b and d. The inclinations of orbits of 80% of the planetesimals that moved between 500 or 1200 AU from the star did not exceed 10°. It was obtained that several planets in the TRAPPIST-1 system accumulated planetesimals initially located at the same distance. Outer layers of neighbouring TRAPPIST-1 planets can include similar material.

Acknowledgements

The studies were carried out under government-financed research project for the Vernadsky Institute of Geochemistry and Analytical Chemistry of Russian Academy of Sciences.

References

Ipatov, S.I. 1993, *Solar System Research*, 27, 65, `https://www.academia.edu/44448077/Migration_of_bodies_in_the_accretion_of_planets`.

Ipatov, S.I. 2019, *Solar System Research*, 53, 332, DOI: 10.1134/S0038094619050046, https://arxiv.org/abs/2003.11301

Ipatov, S.I. 2022, Thirteenth Moscow Solar System Symposium (13M-S3). 13MS3-EP-PS-02. P. 378-380. `https://elibrary.ru/item.asp?id=50086725`.

Ipatov, S.I. 2023a, *Solar System Research*, 57, 236, DOI: 10.1134/S0038094623030036, https://arxiv.org/abs/2309.00492.

Ipatov, S.I. 2023b, *Meteoritics and Planetary Science*, 58. 752. DOI: 10.1134/S0038094623030036, https://arxiv.org/abs/2309.00695.

Ipatov, S.I. 2023c, *Solar System Research*, 57, 612, DOI: 10.1134/S0038094623060047, http://arxiv.org/abs/2401.09086.

Levison, H.F., Duncan, M.J. 1994, *Icarus*, 108, 18

Marov, M.Ya., Ipatov, S.I. 2023, *Physics-Uspekhi*, 66. 2. DOI: 10.3367/UFNe.2021.08.039044, https://arxiv.org/abs/2309.00716.

Planetary Science and Exoplanets in the Era of JWST
Proceedings IAU Symposium No. 393, 2026
Z. Benkhaldoun, Y. Moulane & A. Szentgyorgyi, eds.
doi:10.1017/S1743921324001601

The magnetic field and space environment of known exoplanet hosts: GJ 436 and HD 63433

S. Bellotti[1,2], S. Boro Saikia[3], V. Bourrier[4], C. Danielski[5], J. F. Donati[2], D. Evensberget[1], R. Fares[6], É. M. Hébrard[2], G. A. J. Hussain[7], A. Lavail[2], T. Lüftinger[7], G. Micela[8], J. Morin[9], C. Moutou[2], P. Petit[2] and A. A. Vidotto[1]

[1]Leiden Observatory, Leiden University, PO Box 9513, 2300 RA Leiden, The Netherlands
email: bellotti@strw.leidenuniv.nl

[2]Institut de Recherche en Astrophysique et Planétologie, Université de Toulouse, CNRS, IRAP/UMR 5277, 14 avenue Edouard Belin, F-31400, Toulouse, France

[3]University of Vienna, Department of Astrophysics, Türkenschanzstrasse 17, A-1180 Vienna, Austria

[4]Observatoire Astronomique de l'Université de Genéve, Chemin Pegasi 51b, 1290 Versoix, Switzerland

[5]INAF - Osservatorio Astrofisico di Arcetri, Largo E. Fermi 5, 50125, Firenze, Italy

[6]Department of Physics, College of Science, United Arab Emirates University, P.O. Box No. 15551, Al Ain, UAE

[7]Science Division, Directorate of Science, European Space Research and Technology Centre (ESA/ESTEC), Keplerlaan 1, 2201 AZ, Noordwijk, The Netherlands

[8]INAF - Osservatorio Astronomico di Palermo, Piazza del Parlamento 1, 90134, Palermo, Italy

[9]Laboratoire Univers et Particules de Montpellier, Université de Montpellier, CNRS, F-34095, Montpellier, France

Abstract. The characterisation of stellar magnetism of planetary host stars has increased momentum, especially for transmission spectroscopy investigations of exoplanets. Indeed, the magnetic field regulates how irradiated planets are, and the presence of inhomogeneities on the stellar surface hinder the precise extraction of the planetary atmospheric absorption signal. We are conducting a spectropolarimetric campaign to unveil the magnetic field properties of known exoplanet hosting stars included in the current list of *Ariel* targets. Here, we will focus on two of them: GJ 436 and HD 63433. The former hosts a warm-Neptune experiencing substantial atmospheric loss, modelled as a comet-like trail of hydrogen atoms. The latter hosts two sub-Neptunes and an Earth-sized planet that have likely experienced different atmospheric evolution paths. We reconstructed the stellar large-scale magnetic field via Zeeman-Doppler imaging, and used it as boundary condition to simulate the stellar magnetised wind and environment at the planetary orbits. For GJ 436, the planet motion is sub-Alfvénic, meaning that star-planet magnetic connections can occur. We derived the power released by star-planet interactions to be 10^{22}–10^{23} erg/s, consistent with the upper limit of 10^{26} erg/s measured from ultraviolet lines. For HD63433, we identified 10% of the innermost planetary orbit as sub-Alfvénic, while the outer planets are outside the Alfvén surface, and a bow shock between the stellar wind and the planetary magnetosphere could form.

Keywords. Stars: magnetic field - Stars: activity - Techniques: polarimetric

---

1. Introduction

The field of exoplanetology has flourished significantly in the past thirty years. With more than 5, 500 confirmed detections (`www.exoplanet.eu`, August 2024), the main goal is progressively focusing on a comprehensive characterisation of the planetary population. When it comes to exoplanets atmospheric characterisation surveys, stellar magnetic activity is a crucial aspect to be taken into account to avoid a biased interpretation of the planet nature. In fact, stellar activity is responsible for photospheric inhomogeneities like spots and faculae, to which one is sensitive within the spectral ranges commonly examined for exoplanet detection and characterisation, namely the optical and near-infrared. In the case of transmission spectroscopy, stellar activity can bias the observed spectrum causing the transit chord to differ from the disk-integrated stellar spectrum, generating ambiguity in the extraction of the planetary atmospheric absorption signal (e.g. Rackham et al. 2018).

In addition, magnetic activity shapes the environment in which exoplanets orbit, and it affects their atmospheres, especially during the early stages of their formation (e.g. Ribas et al. 2005; Allan and Vidotto 2019). Energetic phenomena associated with intense magnetic activity such as frequent flares can alter the chemical properties of the planetary atmosphere (Segura et al. 2010; Louca et al. 2023), with hazardous consequences for habitability (e.g., Tilley et al. 2019). In particular for planets orbiting closer to the host star, energetic stellar irradiation (X-rays and extreme ultra-violet) heats and expands the upper regions of the atmosphere, resulting in hydrodynamic escape (e.g., Lammer et al. 2003; Owen and Jackson 2012). For these reasons, there is increased attention towards a comprehensive and homogeneous characterisation of exoplanet hosts (e.g. Danielski et al. 2022) and their magnetic activity (Fares et al. 2013).

Characterising the space weather for a specific system and modelling the interaction between the magnetised stellar wind and a close-in planet requires robust knowledge of the stellar magnetic field (Vidotto et al. 2014). Our assumptions on its topology and strength indeed impact the extent of the planetary magnetosphere (Villarreal D'Angelo et al. 2018) and predictions of transits duration (Llama et al. 2013). Stellar magnetic fields are most effectively studied using spectropolarimetry, with which we can analyse the Zeeman effect, i.e. the splitting of spectral lines in distinct components characterised by specific polarisation properties. From time series of polarised spectra we can map the large-scale magnetic field by means of Zeeman-Doppler Imaging (ZDI; Donati and Brown 1997) and obtain a global picture of the magnetic environment. Zeeman-Doppler imaging has been applied extensively in spectropolarimetric studies, and revealed a variety of field geometries for low-mass stars (e.g., Donati et al. 2008; Morin et al. 2010; Fares et al. 2013), as well as long-term evolution (e.g., Boro Saikia et al. 2016; Bellotti et al. 2023b).

Here, we present the analysis of the large-scale magnetic field and stellar wind environment of GJ 436 and HD 63433. This work has been conducted in a framework dedicated to the preparation of the *Ariel* mission, but finds wide applicability in the *JWST* era as well. More specifically, to obtain a complete picture of the magnetic activity of stars in the current list of potential *Ariel* targets (Edwards et al. 2019; Edwards and Tinetti 2022), we initiated a dedicated spectropolarimetric programme divided into three steps: 1) a snapshot campaign to assess the detectability of the large-scale magnetic field and optimise further observations, 2) an observing campaign to reconstruct the topology of the surface stellar magnetic field, and 3) a long-term monitoring to constrain the evolution and variability of the field. For full details of these studies, we refer the reader to Bellotti et al. (2023a), Bellotti et al. (2024), and Vidotto et al. (2023).

2. Observations

GJ 436 is a moderately inactive M2.5 dwarf, with a stellar rotation period around 40-44 d (Bourrier et al. 2018). The star hosts a hot-Neptune at 0.0285 AU, corresponding to an orbital period of 2.644 d (Butler et al. 2004). Because of intense irradiation, the planetary atmosphere is subject to hydrodynamic evaporation, which forms a comet-like cloud of hydrogen atoms (Kulow et al. 2014; Ehrenreich et al. 2015). GJ 436 b lies on a polar eccentric orbit (Bourrier et al. 2018, 2022) to which it may have migrated via interactions with an undetected outer companion (Bourrier et al. 2018).

HD 63433 is an active G5 dwarf, with an estimated age of 414 ± 23 Myr, and a rotation period of 6.45 d (Mann et al. 2020). Two sub-Neptunes were discovered to orbit the star via photometric transits (Mann et al. 2020): HD 63433 b with an orbital period of 7.11 d, and HD 63433 c, on a 20.55 d orbit. A third planet was also reported recently via photometric transit, with an orbital period of 4.2 d (Capistrant et al. 2024). As reported by (Zhang et al. 2022), planets b and c do not exhibit helium absorption, whereas only planet c shows Lyα absorption. Considering the architecture of the system, the age and activity of the star, this suggests that HD 63433 b has probably lost most of its primordial H/He atmosphere, while HD 63433 c is still experiencing atmospheric evaporation. Since planet d has the innermost orbit, we can assume that its primordial atmosphere has been lost similarly to planet b (Capistrant et al. 2024).

We used spectropolarimetric observations collected with Narval, and its recent upgrade Neo-Narval. Narval is the optical spectropolarimeter on the 2 m Télescope Bernard Lyot (TBL) at the Pic du Midi Observatory in France, covering a 360-1050 nm spectral range at a resolving power R of 65 000 (Donati 2003). The Neo-Narval upgrade, implemented in 2019, included the installation of a new detector and improved velocimetric capabilities (López Ariste et al. 2022). Issues with the extraction of blue spectral orders occurred during the period of our observations (López Ariste et al. 2022), affecting especially those with very low S/N (below approximately 400 nm). The data reduction of Narval was performed with `LIBRE-ESPRIT` (Donati et al. 1997), and for Neo-Narval a different reduction pipeline was used (López Ariste et al. 2022).

We applied least-squares deconvolution (LSD) (Donati et al. 1997; Kochukhov et al. 2010) to compute high signal-to-noise ratio (S/N) Stokes I and V profiles. Both the unpolarised and polarised spectra are deconvolved with a line mask, which is a list of photospheric absorption lines for the specific spectral type examined. Least-squares deconvolution results in high-S/N line profiles encapsulating the average information of thousands of lines (See Figure 1). We used line lists synthesised using the Vienna Atomic Line Database (VALD, Ryabchikova et al. 2015), characterised by $\log g = 5.0$ [cm s^{-2}] and $T_{\rm eff} = 3500$ K for GJ 436, and $\log g = 4.5$ [cm s^{-2}] and $T_{\rm eff} = 5500$ K for HD 63433. The lists include atomic lines between 360–1080 nm, with known sensitivity to the Zeeman effect (also known as Landé factor and indicated as $g_{\rm eff}$) and an absorption depth greater than 40 % of the continuum level (Moutou et al. 2007). In total, the list of GJ 436 contained 3240 lines and 4216 for HD 63433.

In the following, the observations are phased according to the ephemeris

$$\mathrm{HJD}_{\mathrm{GJ436}} = 2457464.4967 + \mathrm{P}_{\mathrm{rot,GJ436}} \cdot n_{\mathrm{cyc}}, \quad (1)$$
$$\mathrm{HJD}_{\mathrm{HD63433}} = 2460010.37186 + \mathrm{P}_{\mathrm{rot,HD63433}} \cdot n_{\mathrm{cyc}}, \quad (2)$$

where we used the first collected observation date as reference for each star, $\mathrm{P}_{\mathrm{rot}}$ is the stellar rotation period, and n_{cyc} is the rotation cycle.

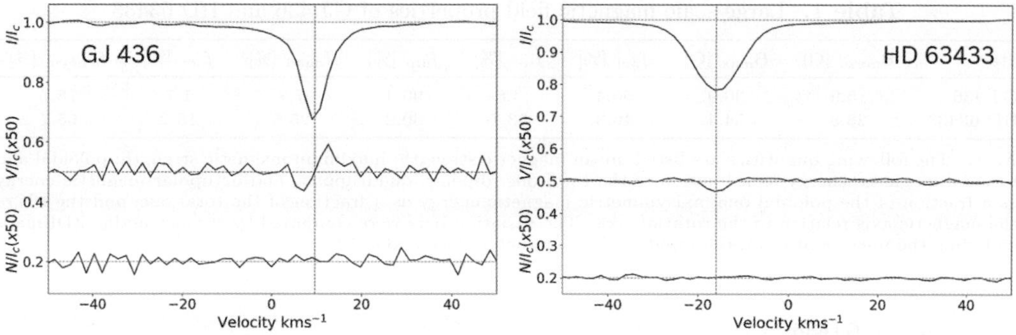

Figure 1. Example least-squares deconvolution profiles of GJ 436 and HD 63433. The panels show the Stokes I (top), V (middle) and N (bottom) profiles. The Stokes V and N profiles have been multiplied by 50 and shifted vertically for visualisation purposes. We note the presence of a clear Zeeman signature in Stokes V, indicating that the large-scale magnetic field is amenable to characterisation.

3. Zeeman-Doppler imaging

We applied ZDI to reconstruct the large-scale magnetic field at the surface of GJ 436 and HD 63433. The field is expressed as the sum of a poloidal and toroidal component, and both components are described by means of spherical harmonic decomposition (Lehmann and Donati 2022). With ZDI, we synthesise and adjust Stokes V profiles in an iterative fashion, until a maximum-entropy solution at a fixed reduced χ^2 is achieved (Skilling and Bryan 1984; Donati and Brown 1997; Folsom et al. 2018). The iterative process aims to fit the spherical harmonics' coefficients $\alpha_{\ell,m}$, $\beta_{\ell,m}$, and $\gamma_{\ell,m}$ (with ℓ and m being the degree and order of the mode, respectively).

The adopted input stellar parameters were the following. For GJ 436, the inclination is 40°, the equatorial projected velocity ($v_{\rm eq}\sin(i)$) is $0.33\,{\rm km\,s^{-1}}$ (Bourrier et al. 2022), and the rotation period is $P_{\rm rot} = 40.13 \pm 1.29\,{\rm d}$. For HD 63433, the inclination is 70°, $v_{\rm eq}\sin(i)$ is $7.3\,{\rm km\,s^{-1}}$, and the rotation period is 6.45 d. For both stars, we assumed solid body rotation and a linear limb darkening of 0.7 (Claret and Bloemen 2011). The maximum degree of harmonic expansion $\ell_{\rm max}$ was set to 5. Although the stars have distinct $v_{\rm eq}\sin(i)$, allowing larger values of $\ell_{\rm max}$ for faster rotation, we did not note a substantial fraction of magnetic energy in modes with $\ell > 5$.

The magnetic maps are illustrated in Fig. 2 and the properties are listed in Table 1. For GJ 436, the mean magnetic field strength is $B_{\rm mean} = 16\,{\rm G}$, with the poloidal component accounting for 96% of the magnetic energy. The dipolar and quadrupolar modes store 90% and 8% energy, and the field is mostly axisymmetric (79%), with an obliquity of its axis of 15.5°. For HD 63433, the mean magnetic field strength is $B_{\rm mean} = 24\,{\rm G}$, with the toroidal component accounting for 54% of the magnetic energy and the poloidal component for 46%. The dipolar mode accounts for 30% of the poloidal energy and both the quadrupolar and quadrupolar modes store a significant fraction, namely, 26% and 15%.

4. Stellar wind modelling

To characterise the stellar environment around GJ 436 and HD 63433, we performed data-driven 3D MHD simulations of the stellar wind. We created the stellar wind models using the Alfvén wave solar model (AWSoM, Sokolov et al. 2013; van der Holst et al. 2014), implemented in the numerical code BATS-R-US (Powell et al. 1999; Tóth et al. 2012). This model solves the ideal two-temperature magnetohydrodynamic equations, supplemented by a pair of equations describing the propagation of Alfvén wave energy

Table 1. Large-scale magnetic field properties of GJ 436 and HD 63433

Star	B_{mean} [G]	B_{max} [G]	f_{pol} [%]	f_{tor} [%]	f_{dip} [%]	f_{quad} [%]	f_{oct} [%]	f_{axisym} [%]
GJ 436	15.9	30.9	96.4	3.6	90.4	7.8	1.7	78.7
HD 63433	23.8	54.4	46.3	53.7	30.2	25.6	15.2	65.7

Notes: The following quantities are listed: mean magnetic strength; maximum magnetic strength; poloidal and toroidal magnetic energy as a fraction of the total one; dipolar, quadrupolar, and octupolar magnetic energy as a fraction of the poloidal one; axisymmetric magnetic energy as a fraction of the total one; and the tilt of the magnetic axis relative to the rotation axis. The variation bars were computed by reconstructing ZDI maps including the uncertainties on the input stellar parameters (see text).

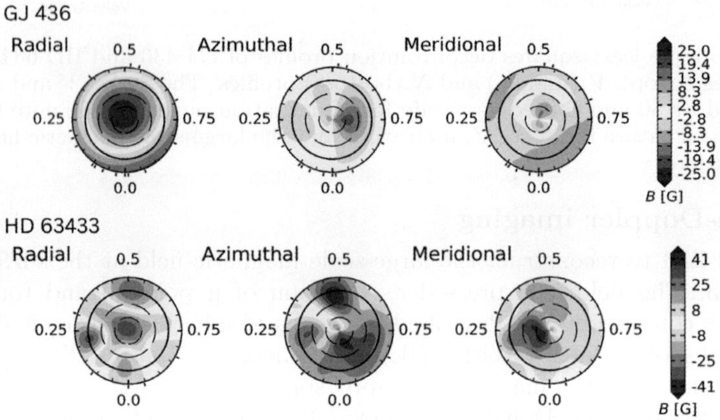

Figure 2. ZDI reconstruction in flattened polar view of the large-scale field of GJ 436 and HD 63433. From the left, the radial, azimuthal, and meridional components of the magnetic field vector are displayed. The radial ticks are located at the rotational phases when the observations were collected, while the concentric circles represent different stellar latitudes: –30 °, +30 °, and +60 ° (dashed lines), as well as the equator (solid line). The figures were adapted from (Bellotti et al. 2023a, 2024).

along magnetic field lines. Because winds of cool dwarf stars are ultimately driven by stellar magnetism, at the boundary of our model (which is located close to the stellar surface, at the transition region), we include the reconstructed surface magnetic fields shown in Fig. 2.

In addition to the stellar magnetic maps, other model inputs are stellar fundamental parameters (mass, radius and rotation period), the density and temperature at the transition region, and the Alfvén wave properties. As they propagate through magnetic field lines, Alfvén waves transfer their energy and momentum to the wind plasma, thus causing its heating and acceleration. In AWSoM, the wave dissipation is modelled through a turbulent cascade process. Its correlation length is assumed to increase with flux tube radius; thus $L_\perp = L_{\perp,0}\sqrt{B_\star/B}$. Here, we assume the proportionality factor is the same as that used in the AWSoM solar wind models, namely $L_{\perp,0}\sqrt{B_\star} = 1.5\times10^5$ m$\sqrt{\mathrm{T}}$, for both stars. The ratio between Alfvén wave flux and magnetic field ($S_{A,0}/B_\star$) was set to the typical value used in AWSoM solar wind simulations of 1.1×10^6 erg (cm^2 s G)$^{-1}$ (Sokolov et al. 2013; van der Holst et al. 2014) for HD 63433, and one order of magnitude lower for GJ 436. We also tested a solar value for GJ 436, but the results of the simulations were not compatible with observational constraints. For more details on the adopted input parameters, see Vidotto et al. (2023) and Bellotti et al. (2024). We show the 3D outputs of our simulations in Fig. 3.

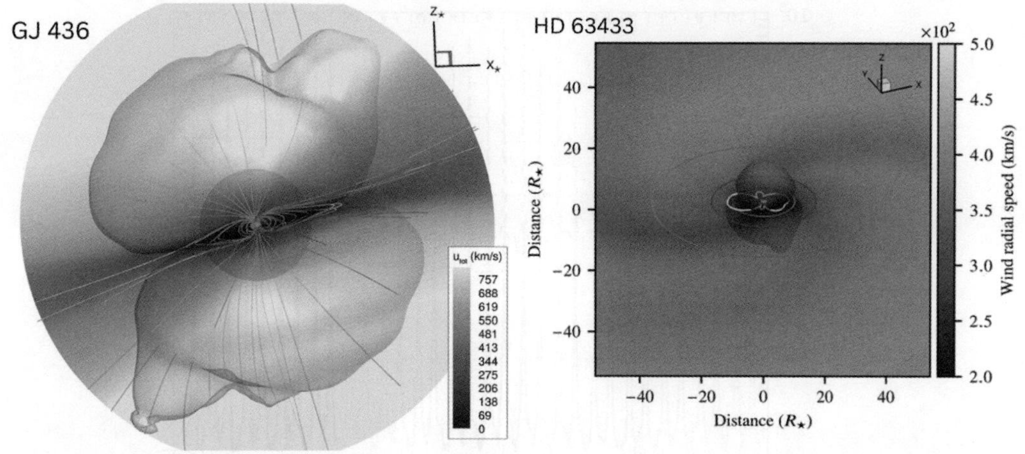

Figure 3. Magnetised stellar wind for GJ 436 (left) and HD 63433 (right). In each panel, the star is at the centre, and the colour bar encodes the wind speed. The orbit of the planets is shown either as a translucent circle (for GJ 436) or coloured lines (for HD 63433). The irregular surface is the Alfvén surface. The figures were adapted from (Vidotto et al. 2023; Bellotti et al. 2024).

4.1. *Stellar environment of GJ 436*

For GJ 436, we note that the magnetic field geometry is approximately that of an aligned dipole close to the stellar surface, with elongated closed field lines at large distances and open fields around the stellar rotation poles. Figure 3 also shows a cut in the $y_\star z_\star$ plane of GJ 436 simulation, where we see that along open field lines the wind is accelerated more quickly, with low-speed streams around closed field line regions.

An example of the simulated stellar wind properties at the position of GJ 436 b is shown in Fig. 4. The local proton density of the stellar wind is constantly changing, owing to the very misaligned orbit of GJ 436 b (spin-orbit angle is 103.2° Bourrier et al. 2022). The range of wind densities at the orbit of the planet is $[0.12, 5.0] \times 10^3$ cm^{-3}, and its time average over one stellar rotation period is 510 cm^{-3} (horizontal grey lines shown in Fig. 4). When modelling the transmission spectroscopic transit of GJ 436 b, Bourrier et al. (2016) derived densities of $2^{+2.2}_{-1.2} \times 10^3$ cm^{-3}. We computed the stellar wind mass-loss rate to be 1.1×10^{-15} $M_\odot$ yr^{-1}, which is in agreement with the observational estimates of $1.2^{+1.3}_{-0.75} \times 10^{-15}$ $M_\odot$ yr^{-1} (Vidotto and Bourrier 2017).

4.2. *Stellar environment of HD 63433*

For HD 63433, the large scale coronal magnetic field resembles a dipole with closed regions pushed towards the positive z axis. The Alfvén surface is two-lobed, which is characteristic for a dipole-dominated coronal magnetic field. We also see arch-like structures where the magnetic field lines connect regions of opposite polarity, which are the result of quadrupolar and octupolar components of the magnetic field as may be seen also in the radial field plot of Fig. 2.

The orbits of planet d, b, and c are shown in Fig. 3 as concentric curves at 9.9, 16.8, and 36.1 $R_\star$ (Mann et al. 2020; Capistrant et al. 2024). The wind conditions at the planets orbit are shown in Fig. 5. In particular, the wind speed curve shows three dips, which are correlated to the three arms of the wind spiral shape. These, in turn, are due to stellar rotation (see Fig. 3). Such spiral features were already reported, and tend to build more arms if the field complexity is larger (Evensberget et al. 2021). In our model, the wind mass loss rate is 1.1×10^{-13} $M_\odot$/yr, which is 2.9 times smaller than the value derived

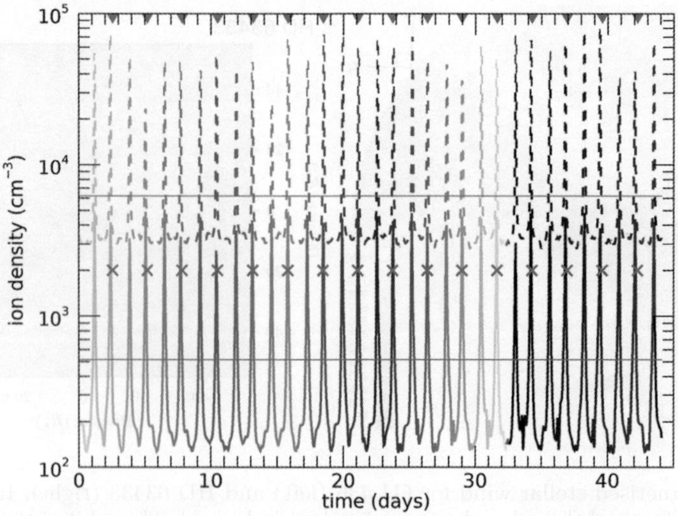

Figure 4. Stellar wind proton density at the orbit of GJ 436 b over multiple orbital cycles. The dashed and solid lines are for the model with solar Alfvén wave flux and one order of magnitude lower. The horizontal grey lines are the time average of the stellar wind properties over one stellar rotation period. The grey arrows in the upper horizontal axes indicate approximate times of mid-transits. The crosses indicate the wind values derived in the work of Bourrier et al. (2016).

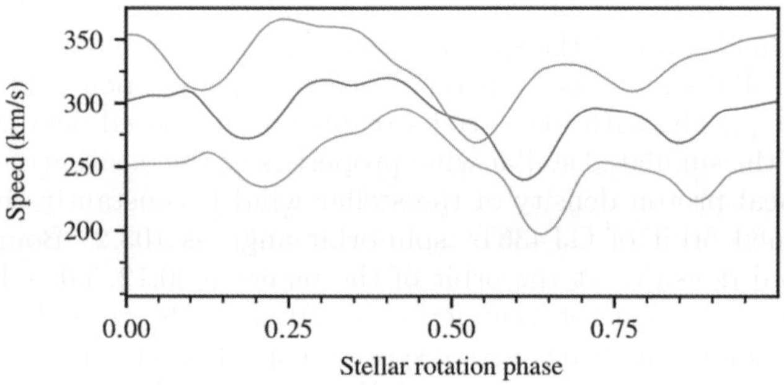

Figure 5. Stellar wind speed of HD 63433 in the planetary frame. Planets b, c, and d are colour-coded in red, pink, and orange. The wind speed exhibits three dips corresponding to spiral arms in the wind spiral shape due to the complexity of the large-scale magnetic field.

from the simulations of Zhang et al. (2022), who modelled the escaping atmospheres of HD 63433 b and c, including the effects of the interaction with the stellar wind.

5. Summary

We have presented the characterisation of the large-scale magnetic field and stellar wind environment of two known exoplanet hosting stars: GJ 436 and HD 63433. We reconstructed the large-scale magnetic field of both stars by means of ZDI and used it as boundary condition for the simulation of the system's stellar wind from the corona to the orbital distances of the respective planets. In general, the observations and analyses were conducted as part of a survey dedicated to the spectropolarimetric characterisation of bright stars in the current list of potential *Ariel* targets, in order to efficiently construct the final target list before the launch and, consequently, optimise the science outcome of the mission.

The reconstructed magnetic field topology of GJ 436 is predominantly dipolar and axisymmetric, with an average, unsigned field strength of 16 G. Our 3D numerical simulations of the wind of GJ 436 show that its planet experience fast-changing conditions at the location of the orbit. The motion of GJ 436 b through the stellar wind is sub-Alfvénic, implying that the planet has a direct magnetic connectivity with its host star. Two different star-planet interaction scenarios may apply: the Alfvén wing scenario, in which perturbations caused in the stellar magnetic field line by the planet travel towards the star as MHD waves (Saur et al. 2013), and the magnetic reconnection scenario, in which magnetic reconnection between stellar and planetary magnetic fields releases energy (Lanza 2009). For both scenarios, we assumed that the planet is magnetised with an equatorial dipolar magnetic field of 1 G, leading to a magnetospheric size of about 5.2 R_p. Using the formalism outlined in (Lanza 2009; Saur et al. 2013), our estimated maximum powers released through star-planet interactions are of the order of 10^{22} to 10^{23} erg s^{-1}, with the largest values associated with the reconnection scenario. These values are compatible with the upper limit power of 10^{26} erg s^{-1} derived by Loyd et al. (2023).

The reconstructed magnetic field topology of HD 63433 is complex, with a significant amount of energy stored in the toroidal and poloidal components, and with an average, unsigned field strength of 24 G. In this case, we found that the outermost two planets orbit in the super-Alfvénic region. In this regime, a bow shock between the stellar wind and a planetary magnetosphere (if present) can be generated and the stellar wind is deflected around the planet (Vidotto et al. 2010). Also, the formation of a tail-like structure of evaporated planetary material can occur (e.g. Villarreal D'Angelo et al. 2018). The innermost Earth-sized planet is located in the trans-Alfvénic region, as our modelling indicates that the stellar wind conditions along its orbit may switch between sub- and super-Alfvénic. During the sub-Alfvénic regime, similar star-planet interactions as those described for GJ 436 may occur.

References

Allan, A. & Vidotto, A. A. 2019, Evolution of atmospheric escape in close-in giant planets and their associated Ly α and H α transit predictions. *MNRAS*, 490(3), 3760–3771

Bellotti, S., Evensberget, D., Vidotto, A. A., Lavail, A., Lüftinger, T., Hussain, G. A. J., Morin, J., Petit, P., Boro Saikia, S., Danielski, C., & Micela, G. 2024, Spectropolarimetric characterisation of exoplanet host stars in preparation of the Ariel mission. Magnetic environment of HD 63433. *A&A*, 688, A63

Bellotti, S., Fares, R., Vidotto, A. A., Morin, J., Petit, P., Hussain, G. A. J., Bourrier, V., Donati, J. F., Moutou, C., & Hébrard, É. M. 2023,a The space weather around the exoplanet GJ 436b. I. The large-scale stellar magnetic field. *A&A*, 676a, A139

Bellotti, S., Morin, J., Lehmann, L. T., Folsom, C. P., Hussain, G. A. J., Petit, P., Donati, J. F., Lavail, A., Carmona, A., Martioli, E., Romano Zaire, B., Alecian, E., Moutou, C., Fouqué, P., Alencar, S., Artigau, E., Boisse, I., Bouchy, F., Cadieux, C., Cloutier, R., Cook, N. J., Delfosse, X., Doyon, R., Hébrard, G., Kochukhov, O., & Wade, G. A. 2023,b Monitoring the large-scale magnetic field of AD Leo with SPIRou, ESPaDOnS, and Narval. Towards a magnetic polarity reversal? *A&A*, 676b, A56

Boro Saikia, S., Jeffers, S. V., Morin, J., Petit, P., Folsom, C. P., Marsden, S. C., Donati, J. F., Cameron, R., Hall, J. C., Perdelwitz, V., Reiners, A., & Vidotto, A. A. 2016, A solar-like magnetic cycle on the mature K-dwarf 61 Cygni A (HD 201091). *A&A*, 594, A29

Bourrier, V., Lecavelier des Etangs, A., Ehrenreich, D., Tanaka, Y. A., & Vidotto, A. A. 2016, An evaporating planet in the wind: stellar wind interactions with the radiatively braked exosphere of GJ 436 b. *A&A*, 591, A121

Bourrier, V., Lovis, C., Beust, H., Ehrenreich, D., Henry, G. W., Astudillo-Defru, N., Allart, R., Bonfils, X., Ségransan, D., Delfosse, X., Cegla, H. M., Wyttenbach, A., Heng, K., Lavie,

B., & Pepe, F. 2018, Orbital misalignment of the Neptune-mass exoplanet GJ 436b with the spin of its cool star. *Nature*, 553(7689), 477–480

Bourrier, V., Zapatero Osorio, M. R., Allart, R., Attia, O., Cretignier, M., Dumusque, X., Lovis, C., Adibekyan, V., Borsa, F., Figueira, P., González Hernández, J. I., Mehner, A., Santos, N. C., Schmidt, T., Seidel, J. V., Sozzetti, A., Alibert, Y., Casasayas-Barris, N., Ehrenreich, D., Lo Curto, G., Martins, C. J. A. P., Di Marcantonio, P., Mégevand, D., Nunes, N. J., Palle, E., Poretti, E., & Sousa, S. G. 2022, The polar orbit of the warm Neptune GJ 436b seen with VLT/ESPRESSO. *A&A*, 663, A160

Butler, R. P., Vogt, S. S., Marcy, G. W., Fischer, D. A., Wright, J. T., Henry, G. W., Laughlin, G., & Lissauer, J. J. 2004, A Neptune-Mass Planet Orbiting the Nearby M Dwarf GJ 436. *ApJ*, 617(1), 580–588

Capistrant, B. K., Soares-Furtado, M., Vanderburg, A., Jankowski, A., Mann, A. W., Ross, G., Srdoc, G., Hinkel, N. R., Becker, J., Magliano, C., Limbach, M. A., Stephan, A. P., Nine, A. C., Tofflemire, B. M., Kraus, A. L., Giacalone, S., Winn, J. N., Bieryla, A., Bouma, L. G., Ciardi, D. R., Collins, K. A., Covone, G., de Beurs, Z. L., Huang, C. X., Jenkins, J. M., Kreidberg, L., Latham, D. W., Quinn, S. N., Seager, S., Shporer, A., Twicken, J. D., Wohler, B., Vanderspek, R. K., Yarza, R., & Ziegler, C. 2024, TESS Hunt for Young and Maturing Exoplanets (THYME). XI. An Earth-sized Planet Orbiting a Nearby, Solar-like Host in the 400 Myr Ursa Major Moving Group. *AJ*, 167(2), 54

Claret, A. & Bloemen, S. 2011, Gravity and limb-darkening coefficients for the Kepler, CoRoT, Spitzer, uvby, UBVRIJHK, and Sloan photometric systems. *A&A*, 529, A75

Danielski, C., Brucalassi, A., Benatti, S., Campante, T., Delgado-Mena, E., Rainer, M., Sacco, G., Adibekyan, V., Biazzo, K., Bossini, D., Bruno, G., Casali, G., Kabath, P., Magrini, L., Micela, G., Morello, G., Palladino, P., Sanna, N., Sarkar, S., Sousa, S., Tsantaki, M., Turrini, D., & Van der Swaelmen, M. 2022, The homogeneous characterisation of Ariel host stars. *Experimental Astronomy*, 53(2), 473–510

Donati, J. F. ESPaDOnS: An Echelle SpectroPolarimetric Device for the Observation of Stars at CFHT. In Trujillo-Bueno, J. & Sanchez Almeida, J., editors, *Solar Polarization* 2003,, volume 307 of *Astronomical Society of the Pacific Conference Series*, 41

Donati, J. F. & Brown, S. F. 1997, Zeeman-Doppler imaging of active stars. V. Sensitivity of maximum entropy magnetic maps to field orientation. *A&A*, 326, 1135–1142

Donati, J. F., Morin, J., Petit, P., Delfosse, X., Forveille, T., Aurière, M., Cabanac, R., Dintrans, B., Fares, R., Gastine, T., Jardine, M. M., Lignières, F., Paletou, F., Ramirez Velez, J. C., & Théado, S. 2008, Large-scale magnetic topologies of early M dwarfs. *MNRAS*, 390(2), 545–560

Donati, J. F., Semel, M., Carter, B. D., Rees, D. E., & Collier Cameron, A. 1997, Spectropolarimetric observations of active stars. *MNRAS*, 291(4), 658–682

Edwards, B., Mugnai, L., Tinetti, G., Pascale, E., & Sarkar, S. 2019, An Updated Study of Potential Targets for Ariel. *AJ*, 157(6), 242

Edwards, B. & Tinetti, G. 2022, The Ariel Target List: The Impact of TESS and the Potential for Characterizing Multiple Planets within a System. *AJ*, 164(1), 15

Ehrenreich, D., Bourrier, V., Wheatley, P. J., Lecavelier des Etangs, A., Hébrard, G., Udry, S., Bonfils, X., Delfosse, X., Désert, J.-M., Sing, D. K., & Vidal-Madjar, A. 2015, A giant comet-like cloud of hydrogen escaping the warm Neptune-mass exoplanet GJ 436b. *Nature*, 522(7557), 459–461

Evensberget, D., Carter, B. D., Marsden, S. C., Brookshaw, L., & Folsom, C. P. 2021, The winds of young Solar-type stars in the Hyades. *MNRAS*, 506(2), 2309–2335

Fares, R., Moutou, C., Donati, J. F., Catala, C., Shkolnik, E. L., Jardine, M. M., Cameron, A. C., & Deleuil, M. 2013, A small survey of the magnetic fields of planet-host stars. *MNRAS*, 435(2), 1451–1462

Folsom, C. P., Bouvier, J., Petit, P., Lèbre, A., Amard, L., Palacios, A., Morin, J., Donati, J. F., & Vidotto, A. A. 2018, The evolution of surface magnetic fields in young solar-type stars II: the early main sequence (250-650 Myr). *MNRAS*, 474(4), 4956–4987

Kochukhov, O., Makaganiuk, V., & Piskunov, N. 2010, Least-squares deconvolution of the stellar intensity and polarization spectra. *A&A*, 524, A5

Kulow, J. R., France, K., Linsky, J., & Loyd, R. O. P. 2014, Lyα Transit Spectroscopy and the Neutral Hydrogen Tail of the Hot Neptune GJ 436b. *ApJ*, 786(2), 132

Lammer, H., Selsis, F., Ribas, I., Guinan, E. F., Bauer, S. J., & Weiss, W. W. 2003, Atmospheric Loss of Exoplanets Resulting from Stellar X-Ray and Extreme-Ultraviolet Heating. *ApJl*, 598(2), L121–L124

Lanza, A. F. 2009, Stellar coronal magnetic fields and star-planet interaction. *A&A*, 505(1), 339–350

Lehmann, L. T. & Donati, J. F. 2022, Diagnosing large-scale stellar magnetic fields using PCA on spectropolarimetric data. *MNRAS*, 514(2), 2333–2345

Llama, J., Vidotto, A. A., Jardine, M., Wood, K., Fares, R., & Gombosi, T. I. 2013, Exoplanet transit variability: bow shocks and winds around HD 189733b. *MNRAS*, 436(3), 2179–2187

López Ariste, A., Georgiev, S., Mathias, P., Lèbre, A., Wavasseur, M., Josselin, E., Konstantinova-Antova, R., & Roudier, T. 2022, Three-dimensional imaging of convective cells in the photosphere of Betelgeuse. *A&A*, 661, A91

Louca, A. J., Miguel, Y., Tsai, S.-M., Froning, C. S., Loyd, R. O. P., & France, K. 2023, The impact of time-dependent stellar activity on exoplanet atmospheres. *MNRAS*, 521(3), 3333–3347

Loyd, R. O. P., Schneider, P. C., Jackman, J. A. G., France, K., Shkolnik, E. L., Arulanantham, N., Cauley, P. W., Llama, J., & Schneider, A. C. 2023, Flares, Rotation, Activity Cycles, and a Magnetic Star-Planet Interaction Hypothesis for the Far-ultraviolet Emission of GJ 436. *AJ*, 165(4), 146

Mann, A. W., Johnson, M. C., Vanderburg, A., Kraus, A. L., Rizzuto, A. C., Wood, M. L., Bush, J. L., Rockcliffe, K., Newton, E. R., Latham, D. W., Mamajek, E. E., Zhou, G., Quinn, S. N., Thao, P. C., Benatti, S., Cosentino, R., Desidera, S., Harutyunyan, A., Lovis, C., Mortier, A., Pepe, F. A., Poretti, E., Wilson, T. G., Kristiansen, M. H., Gagliano, R., Jacobs, T., LaCourse, D. M., Omohundro, M., Schwengeler, H. M., Terentev, I. A., Kane, S. R., Hill, M. L., Rabus, M., Esquerdo, G. A., Berlind, P., Collins, K. A., Murawski, G., Sallam, N. H., Aitken, M. M., Massey, B., Ricker, G. R., Vanderspek, R., Seager, S., Winn, J. N., Jenkins, J. M., Barclay, T., Caldwell, D. A., Dragomir, D., Doty, J. P., Glidden, A., Tenenbaum, P., Torres, G., Twicken, J. D., & Villanueva, Steven, J. 2020, TESS Hunt for Young and Maturing Exoplanets (THYME). III. A Two-planet System in the 400 Myr Ursa Major Group. *AJ*, 160(4), 179

Morin, J., Donati, J. F., Petit, P., Delfosse, X., Forveille, T., & Jardine, M. M. 2010, Large-scale magnetic topologies of late M dwarfs*. *MNRAS*, 407(4), 2269–2286

Moutou, C., Donati, J. F., Savalle, R., Hussain, G., Alecian, E., Bouchy, F., Catala, C., Collier Cameron, A., Udry, S., & Vidal-Madjar, A. 2007, Spectropolarimetric observations of the transiting planetary system of the K dwarf HD 189733. *A&A*, 473(2), 651–660

Owen, J. E. & Jackson, A. P. 2012, Planetary evaporation by UV & X-ray radiation: basic hydrodynamics. *MNRAS*, 425(4), 2931–2947

Powell, K. G., Roe, P. L., Linde, T. J., Gombosi, T. I., & De Zeeuw, D. L. 1999, A Solution-Adaptive Upwind Scheme for Ideal Magnetohydrodynamics.*Journal of Computational Physics*, 154(2), 284–309

Rackham, B. V., Apai, D., & Giampapa, M. S. 2018, The Transit Light Source Effect: False Spectral Features and Incorrect Densities for M-dwarf Transiting Planets. *ApJ*, 853(2), 122

Ribas, I., Guinan, E. F., Güdel, M., & Audard, M. 2005, Evolution of the Solar Activity over Time and Effects on Planetary Atmospheres. I. High-Energy Irradiances (1-1700 Å). *ApJ*, 622(1), 680–694

Ryabchikova, T., Piskunov, N., Kurucz, R. L., Stempels, H. C., Heiter, U., Pakhomov, Y., & Barklem, P. S. 2015, A major upgrade of the VALD database. *Phys. Scr.*, 90(5), 054005

Saur, J., Grambusch, T., Duling, S., Neubauer, F. M., & Simon, S. 2013, Magnetic energy fluxes in sub-Alfvénic planet star and moon planet interactions. *A&A*, 552, A119

Segura, A., Walkowicz, L. M., Meadows, V., Kasting, J., & Hawley, S. 2010, The Effect of a Strong Stellar Flare on the Atmospheric Chemistry of an Earth-like Planet Orbiting an M Dwarf. *Astrobiology*, 10(7), 751–771

Skilling, J. & Bryan, R. K. 1984, Maximum Entropy Image Reconstruction - General Algorithm. *MNRAS*, 211, 111

Sokolov, I. V., van der Holst, B., Oran, R., Downs, C., Roussev, I. I., Jin, M., Manchester, IV, W. B., Evans, R. M., & Gombosi, T. I. 2013, Magnetohydrodynamic Waves and Coronal Heating: Unifying Empirical and MHD Turbulence Models. *ApJ*, 764, 23

Tilley, M. A., Segura, A., Meadows, V., Hawley, S., & Davenport, J. 2019, Modeling Repeated M Dwarf Flaring at an Earth-like Planet in the Habitable Zone: Atmospheric Effects for an Unmagnetized Planet. *Astrobiology*, 19(1), 64–86

Tóth, G., van der Holst, B., Sokolov, I. V., De Zeeuw, D. L., Gombosi, T. I., Fang, F., Manchester, W. B., Meng, X., Najib, D., Powell, K. G., Stout, Q. F., Glocer, A., Ma, Y.-J., & Opher, M. 2012, Adaptive numerical algorithms in space weather modeling. *Journal of Computational Physics*, 231, 870–903

van der Holst, B., Sokolov, I. V., Meng, X., Jin, M., Manchester, W. B., I., Tóth, G., & Gombosi, T. I. 2014, Alfvén Wave Solar Model (AWSoM): Coronal Heating. *The Astrophysical Journal*, 782(2), 81

Vidotto, A. A. & Bourrier, V. 2017, Exoplanets as probes of the winds of host stars: the case of the M dwarf GJ 436. *MNRAS*, 470(4), 4026–4033

Vidotto, A. A., Bourrier, V., Fares, R., Bellotti, S., Donati, J. F., Petit, P., Hussain, G. A. J., & Morin, J. 2023, The space weather around the exoplanet GJ 436b. II. Stellar wind-exoplanet interactions. *A&A*, 678, A152

Vidotto, A. A., Jardine, M., & Helling, C. 2010, Early UV Ingress in WASP-12b: Measuring Planetary Magnetic Fields. *ApJl*, 722(2), L168–L172

Vidotto, A. A., Jardine, M., Morin, J., Donati, J. F., Opher, M., & Gombosi, T. I. 2014, M-dwarf stellar winds: the effects of realistic magnetic geometry on rotational evolution and planets. *MNRAS*, 438(2), 1162–1175

Villarreal D'Angelo, C., Esquivel, A., Schneiter, M., & Sgró, M. A. 2018, Magnetized winds and their influence in the escaping upper atmosphere of HD 209458b. *MNRAS*, 479(3), 3115–3125

Zhang, M., Knutson, H. A., Wang, L., Dai, F., dos Santos, L. A., Fossati, L., Henry, G. W., Ehrenreich, D., Alibert, Y., Hoyer, S., Wilson, T. G., & Bonfanti, A. 2022, Detection of Ongoing Mass Loss from HD 63433c, a Young Mini-Neptune. *AJ*, 163(2), 68

Planetary Science and Exoplanets in the Era of JWST
Proceedings IAU Symposium No. 393, 2026
Z. Benkhaldoun, Y. Moulane & A. Szentgyorgyi, eds.
doi:10.1017/S1743921324002230

Observing Exoplanet Biosignatures outside the Conservative Habitable Zone

Amri Wandel

Racah Inst. of Physics, The Hebrew University of Jerusalem
email: amri@huji.ac.il

Abstract. Liquid water can exist on exoplanets in a circumstellar region significantly wider than the Conservative Habitable Zone (CHZ). In particular, tidally locked planets orbiting closer inwards than the inner CHZ border could have liquid water on their night side provided the heat transport between the day and night side is low. Planets further away than the outer CHZ boundary can maintain subglacial liquid water if they have geothermal heat sources or in the form of subglacial lakes similar to the Martian polar lakes. Pressurized water can reach the planetary surface through faults and cracks in the ice cover, creating plumes and enriching the atmosphere with molecules that may be present in subglacial oceans. We discuss the model and the conditions needed to detect water and other molecules in the transmission spectrum by JWST, as have recently been detected in small hot exoplanets such as GJ287 b.

Keywords. Habitable Zone, exoplanets, M-stars, tidally locked planets

1. The Habitable Zone

The HZ is the circumstellar region where a planet can maintain liquid water on the surface given the stellar irradiation, or instellation, for appropriate atmospheric conditions, Albedo, etc. The inner boundary of the CHZ is determined by the point at which water would evaporate due to excessive heat even before the boiling point is reached, due to processes such as the moist greenhouse effect (Kasting *et al.* 1993), while the outer boundary is determined by water freezing. These boundaries are consistent with the 1D climate models (Kopparapu *et al.* 2013; Yang *et al.* 2014). Alternatively, wider empirical HZ boundaries may be drawn by assuming that early Mars and recent Venus had liquid water Fig. 1.

Like the Habitable Zone (HZ) of the Sun, which extends roughly between the orbits of Venus and Mars, also for other stars the HZ can be defined as the region where a planet with a suitable atmosphere could maintain surface liquid water. However, the distance of the HZ from its host star depends on the star's luminosity, as the planet's temperature depends on the stellar irradiation, or instellation (that is, irradiated energy per unit area at the planet's surface). For Main Sequence stars, the luminosity is related to the stellar type or surface temperature. Hence, the HZ's location and extent strongly correlate with the host's stellar type, as shown in Fig. 1.

This Conservative Habitable Zone (CHZ) may be significantly extended, both inwards and outwards, by considering subglacial liquid water. The obvious outward extension could be driven by internal ("geothermal") heat, either due to radiogenic elements (as in the Earth) or due to tidal heating (as in the case of Europa and Enceladus in our Solar System). However, this may not be considered a "proper" HZ, as it is unrelated to the host star. A proper extension would be a lower limit on the radiative flux from the host star, which can be estimated empirically by the solar radiation at the Martian polar

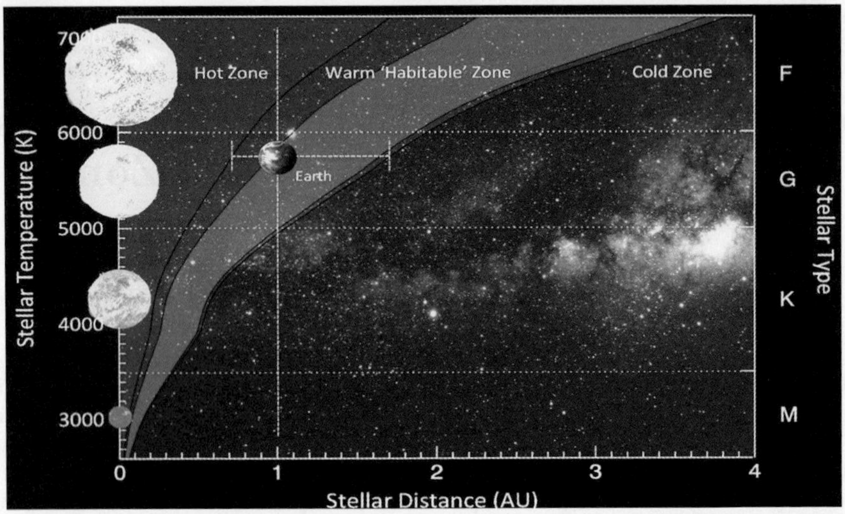

Figure 1. The Habitable Zone (green) in terms of distance from the host star (horizontal axis), for various host star types (left vertical axis) denoted by their surface temperature (left vertical axis).

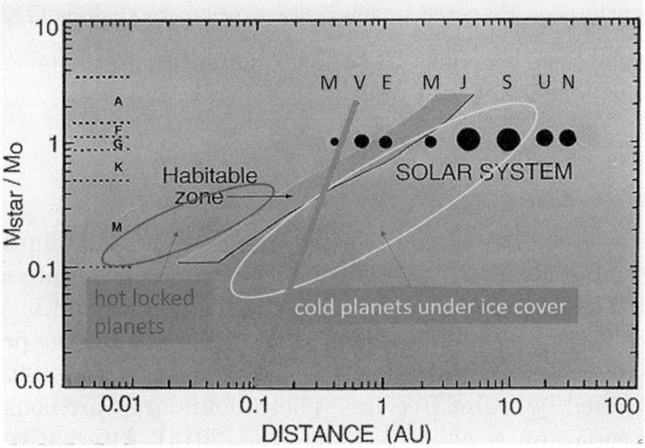

Figure 2. Extended Boundaries of the Conservative Habitable Zone (CHZ, in orange) as a function of distance from the host star and the mass of the Main Sequence host star (vertical axis). The purple diagonal line is the tidal locking radius. The light-blue regime represents the extension outwards of the CHZ, while the red one - the extension inwards of the CHZ due to night-side ice on tidally locked planets.

regions, sufficient to maintain intraglacial lakes (Arnold *et al.* 2022). The HZ-extension inwards is more complex. While increased albedo due to a permanent cloud layer such as in Venus could considerably lower the equilibrium surface temperature, a more extended habitability condition may be obtained by considering the large temperature gradient on tidally locked planets (Wandel 2018) and subglacial liquid water on the night side tidally locked planets (Wandel 2023a,b). Tidally locked planets in the inwards extended HZ exist mostly around M-dwarf stars (the red ellipse in Fig. 2). Their night side may have a low enough surface temperature for liquid water. It is also sheltered from the intense XUV radiation from the eruptions in the early evolutionary stages of those stars. If the planet is not too small it could maintain a magnetosphere, shielding the planetary atmosphere against erosion by stellar winds and giant mass ejections at the early phase

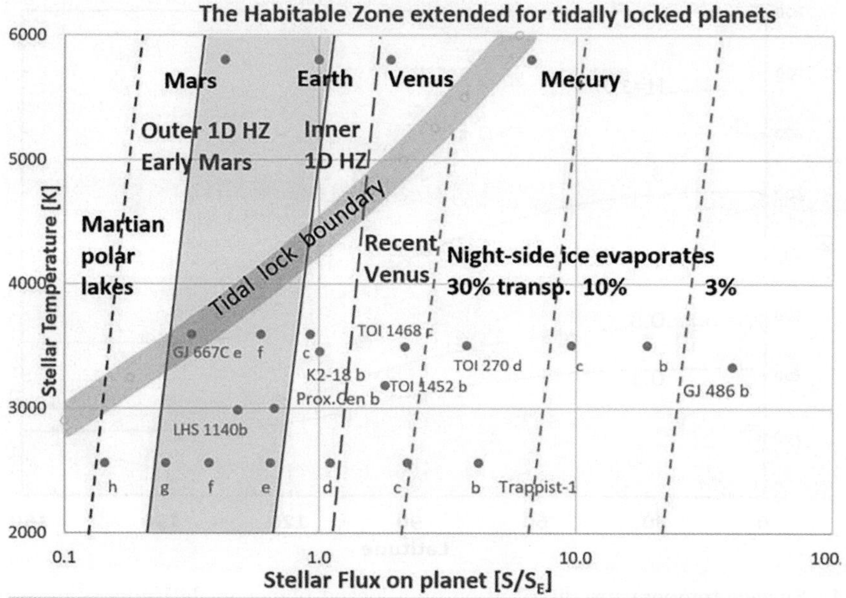

Figure 3. Boundaries of the Habitable Zone as a function of radiative flux received from the host star, relative to Earth (Wandel 2023b). Solid green lines denote the conservative HZ boundaries. Dashed blue lines represent the extended inner HZ boundaries for tidally locked planets. The dashed violet line is the outer HZ extended boundary (Martian polar lake analog). The orange curve marks the tidal locking radius. Blue dots denote the solar system and some well-known exoplanets orbiting M-dwarfs.

by, (Wandel and Gale 2020). On the other hand, if the surviving atmosphere is too dense, it can induce efficient heat transport from the day side to the night side, causing an evaporation of the night side ice. This determines the inner boundary of subglacial habitability as shown by the blue dashed lines in Fig. 3.

The inner extension of the HZ is more clearly described in Fig. 3, where the instellation (horizontal axis) is shown logarithmically. The blue dashed lines correspond to subglacial liquid water on the night side ice on locked planets. The percentage marks the amount of heat transported from the day side, which when high enough would evaporate potential ice. This percentage depends on the instellation and is calculated using the 1D inner boundary of the CHZ (Yang *et al.* 2014). Fig. 3 demonstrates that subglacial liquid water (which may reach the atmosphere as water vapor through cracks forming plumes like those recently observed in Enceladus) may exist on many exoplanets of M-dwarfs which were considered way too close to their parent star to maintain atmospheric water, such as Trappist-1 b and c, TOI 270 b, and GJ 486 b.

2. Habitability of locked planets

Liquid water is essential for the evolution and presence of life as we know it on Earth, but additional elements are required–such as nutrient sources, chemical diversity, and stable energy sources. In this work, however, we only concentrate on liquid water. The energy required to keep the water liquid can come from the host star, or, in the case of planets further away from their host star than the outer boundary of the HZ, from geothermal heat. Another factor is protection from energetic radiation that may harm complex biology and life, or erode the atmosphere and the water content on the planetary surface. It is often stated that planets orbiting M stars may be affected by the host's energetic outbursts in the early evolutionary stages, producing intense UV

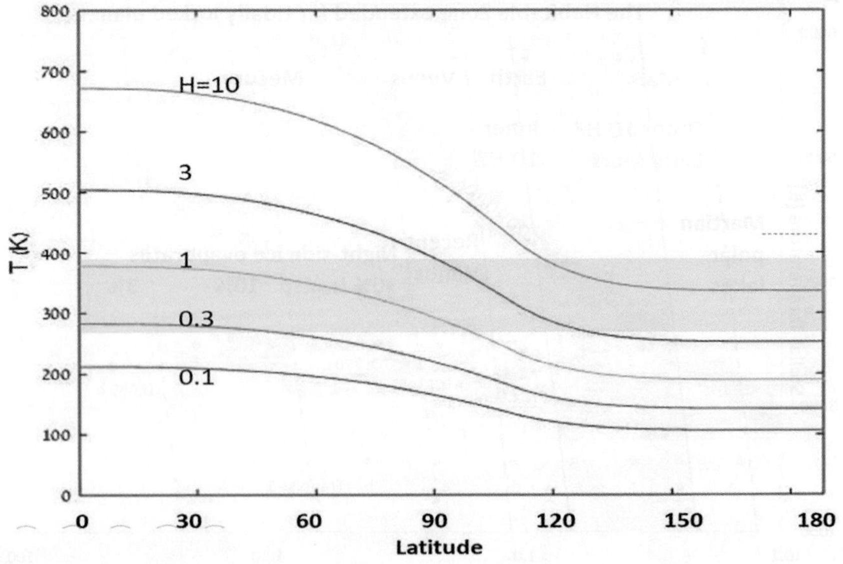

Figure 4. Surface temperature distribution on a locked planet vs. latitude, as measured from the sub-stellar point (Wandel 2018). Curves are marked by the heating parameter H. Global heat transport f=0.2 is assumed. The light blue stripe marks the liquid water regime (for a 1 bar atmosphere).

and X-ray radiation and stellar eruptions that may disintegrate water molecules (e.g. Luger and Barnes 2015). While super-Earth and larger planets are likely to have a thick H/He atmosphere that may survive their M-dwarf host's early active evolutionary phase, the survival of water and atmosphere on smaller, Earth-sized planets orbiting M-dwarf stars was considered less certain. However, other works argue that atmosphere and water may survive ((Wandel and Gale 2020) and refs. therein). An additional path to the habitability of such planets is proposed by (Wandel 2023a): sub-glacial liquid water could survive and replenish the water in the atmosphere once the M-star host has calmed down. At that stage, liquid water could exist on the surface, at least at some latitude (Wandel 2018). Planets in the CHZ of M-dwarf stars are likely to be tidally locked (Fig. 2) and may have surface temperatures adequate for liquid water, at least on part of their surface, under a broad range of atmospheric conditions (Fig. 4 and Fig. 6). The atmospheric impact on the surface temperature distribution of a locked planet can be described by mainly two parameters: the heating factor H and the heat transport parameter f (Wandel 2018). H is the effective surface heating by the radiation from the host star combined with the planet's surface properties (albedo and atmosphere) given by

$$H = (1 - A)H_g S/S_E. \quad (1)$$

Here A is the planet's average Albedo, H_g is the greenhouse factor, S is the radiative flux from the host star (instellation) and SE is the solar flux on Earth. The heat transport parameter f is the fraction of the instellation transported by advection and distributed over the entire planet. For example, setting $f = 0.1$ we assume that at any point on the planet's surface, 10% of the stellar energy radiated onto the planet is advected and finally distributed equally all over the planet, while $f = 1$ means that at any point on the planet, the entire incoming stellar radiation energy is distributed over the whole planet, leading to an isothermal surface. Fig. 4 shows the temperature distribution as a function of "latitude" (defined as the angle from the sub-stellar point of a locked planet) for various

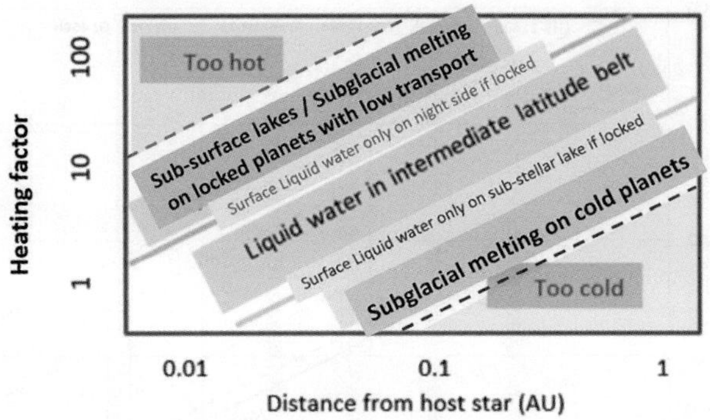

Figure 5. A schematic presentation of the extended HZ (in terms of the allowed heating factor, on the vertical axis) for surface liquid water on locked planets (green, between solid lines) and for subglacial liquid water (pink and blue, between dashed lines). The distance of the planet from the host star is marked on the horizontal axis and assumes a host luminosity of $10^{-4}L_{\odot}$.

values of the heating factor, assuming a heat transport parameter $f = 0.2$. Note that for a wide range of atmospheric heating, there can be liquid water somewhere on the planet.

The heating parameter is closely related to the instellation and the atmospheric density and composition, the latter two being represented by the greenhouse factor H_g. For example, let us assume an albedo of rock ($A \sim 0.3$). Then we can use eq. (1) to demonstrate the relation between the heating parameter H and the instellation S. For an atmosphere with a low greenhouse effect ($H_g \approx 1$) we get $H = 0.7S/S_E$, while for a large greenhouse effect (e.g. $H_g \sim 10$, we have $H = 7S/S_E$. For comparison, the greenhouse factor on Venus is $H_g \approx 40$. The usual definition of the HZ is expressed in terms of distance and surface temperature (namely, the range of distances from the host star between the boiling or evaporation temperature at the inner edge and the freezing at the outer one). This definition does not consider planetary features such as albedo and atmospheric heating (the greenhouse effect). One may define the HZ boundaries in terms of the heating parameter. At the lower boundary, the temperature at a given distance would be the freezing point, and at the upper boundary - the evaporation point. This is schematically shown in Fig. 5. The green region corresponds to surface liquid water at some intermediate latitude on a locked planet with low global heat transport. The blue region denotes subglacial liquid water on planets beyond the outer boundary of the tidal locking HZ, where significant heating (e.g. by a high greenhouse effect) is required even to maintain subglacial liquid water. On the other hand, the pink region denotes subglacial liquid water on planets beyond the inner boundary of the tidal locking HZ, where lower heating (e.g. by a high albedo) is required to maintain subglacial liquid water on the night side of the locked planet. The distance of the planet from the host star is marked on the horizontal axis assuming a host luminosity of $10^{-4}L_{\odot}$. For stars with a different luminosity the distance scales as $d \sim L^{1/2}$. For example, in the case of a host luminosity of $10^{-2}L_{\odot}$, the marks would be 0.1,1 and 10 AU.

While Fig. 5 shows the HZ (in terms of heating factor) for specific host star luminosity, we can construct a "universal" diagram of the "habitability range" of the heating factor by switching from distance to flux on the horizontal axis (Fig. 6). The light green vertical bars indicate a few locked exoplanets with various values of instellation, marked in Fig. 3. For any given value of the global heat transport parameter (f) one can estimate the heating range allowed by habitability for a planet with a given instellation. For example, the green vertical rectangle shows the range of the heating factor allowing liquid water (for $f = 1$) on part of the surface of K2-18 b, assuming a 1 bar atmospheric pressure.

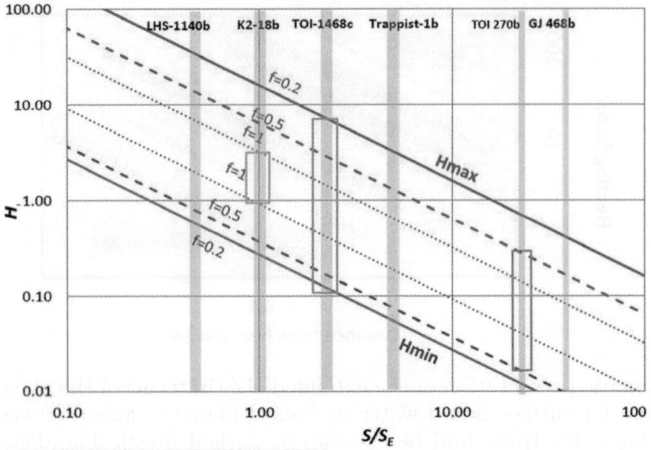

Figure 6. A "universal heating-habitability range" diagram. Liquid water range of the stellar + atmospheric heating factor H vs. stellar irradiation of the planet (instellation) relative to Earth. Curves are marked by their value of the global heat transport f. The lines indicate maximal (red) and minimal (blue) values of H allowing surface liquid water for a given instellation. (Adapted from (Wandel, A. and Tal-Or 2019)).

Assuming less heat transport, e.g. $f = 0.2$, the habitability range of the heating factor is significantly extended, as shown by the orange vertical rectangle for TOI-1468 c. Planets with high instellation could be too hot for liquid water even on their night side (unless the heat transport is very low), as shown by the blue rectangle for TOI 270 b. With the assumed $f = 0.5$ the required H-range is much less than 1. This range is apparently too low: because of the high installation ($S \sim 20S_E$) and as the greenhouse factor cannot be less than 1, eq. 1 gives $H \geq 20(1-A)$. However, if there is no atmosphere the effective transport parameter may be very low (e.g. on Mercury $f \sim 0.05$), and the allowed heating correspondingly higher. The presence of an atmosphere can be inferred by comparing the calculated and observed temperatures. E.g. in recent work (Greene *et al.* 2023) suggests that Trappist-1 b may not have an atmosphere or a very thin one.

We can relate the extremal values of the surface temperature on locked planets (Fig. 4) to the instellation, and, for a given host star, to the distance of the planet from its star. A useful relation can be derived between the host star properties (luminosity, or equivalently, surface temperature T_s and radius R_s), the planet's distance, d and surface equilibrium temperature T_p (e.g. (Wandel and Gale 2024) Ch. 14):

$$T_p = T_s[(1-A)^{1/2} R_s/2d]^{1/2}. \tag{2}$$

Fig. 7 shows the range of temperatures on the surface of locked planets, as a function of distance from the host, for several values of the transport parameter f. The host star is assumed to be an M-type star with a surface temperature of 3,400 K. For example, the green vertical rectangle shows the temperature range on the surface of a planet orbiting at 0.2 AU. Assuming the transport parameter is 0.5 (50% of the heat due to the stellar radiation is advected and equally distributed over the entire planet), the temperature at the substellar point is approximately 310 K, while the temperature at the coldest point on the night side is 60K. Increasing the transport parameter would reduce the temperature difference: e.g. for $f = 1$, the planet becomes isothermal (black curve in Fig. 4) at ≈ 250K. In the context of night-side surface or subglacial water, as we have shown earlier, a high heat of transport would lower the flux limit for evaporating the water or ice on the night side.

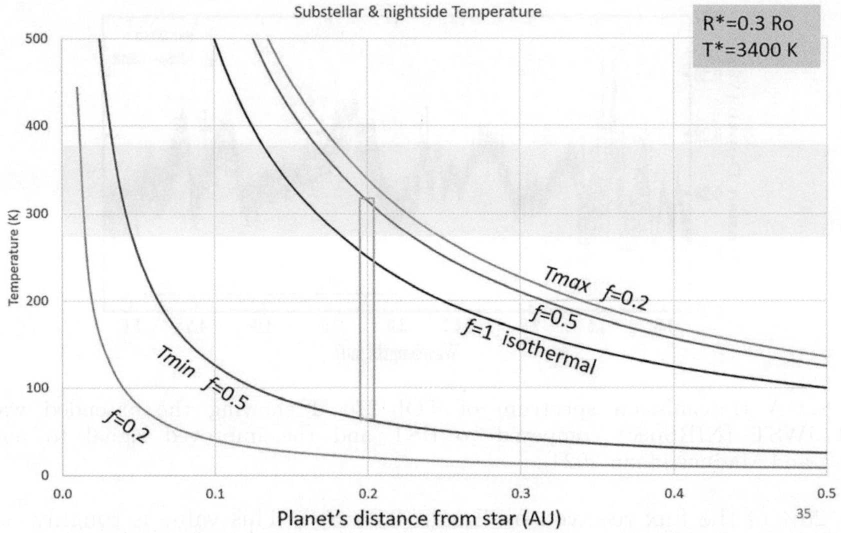

Figure 7. Maximum (sub-stellar point) and minimum (night-side off-stellar point) temperatures for locked planets orbiting an M-dwarf of T=3400K. Curves are marked by their value of the global heat transport f. The light blue stripe marks the liquid water regime (for a 1 bar atmosphere).

The habitability-analyses can be combined with the observable parameters of exoplanets (mass and radius) and with the inferred internal structure to estimate the atmosphere and the habitability (Daspute *et al.* 2022).

3. Implications for the search of biosignatures

The discussion in the previous section presents a challenge: could water and biosignatures be detected on hot, tidally locked small exoplanets, beyond the inner boundary of the CHZ? Actually, signs for water vapor on such planets may have already been detected by JWST, e.g. GJ 486 b (Moran *et al.* 2023) and TOI 270 d, perhaps also TOI 270 b (Holmberg and Madhusudhan 2024). A typical transmission spectrum (TOI 270 d) is shown in Fig. 8. Water detections could challenge the survival of atmosphere and water under extreme conditions, as even the quiescent radiative flux from the host M-dwarf on these planets is much higher than on Earth and the planetary calculated average surface temperature is hundreds of degrees C. Being much closer to their host than the inner boundary of the CHZ, these and similar planets with transmission spectra consistent with the presence of water challenge the CHZ model. If confirmed, the empirical evidence of water vapor supports the extended HZ for locked planets of M-stars (Wandel 2018); (Wandel and Gale 2020) and the further extended HZ due to subglacial water on the night side (Wandel 2023a).

Water vapor and eventually other molecules in subglacial oceans would be transported to the planet's atmosphere by plumes fed through cracks in the ice layer, similarly to molecules like HCN and Hydrocarbons observed in the plume of Enceladus (Peter *et al.* 2024). A similar mechanism of plumes and biosignature detection could exist also outwards of the CHZ (that is further out of the outer HZ boundary), where subglacial liquid water may exist on geothermally heated rocky exoplanets even with low surface temperatures. This is true also for K- and G-stars, but their contribution to the abundance of habitable planets is less significant than M-dwarfs, as they are less frequent. As a conservative outer limit for the HZ extended by subglacial basal melting, one may use the flux on Trappist-1 g, the coldest one of the exoplanets simulated by (Ojha *et al.* 2022),

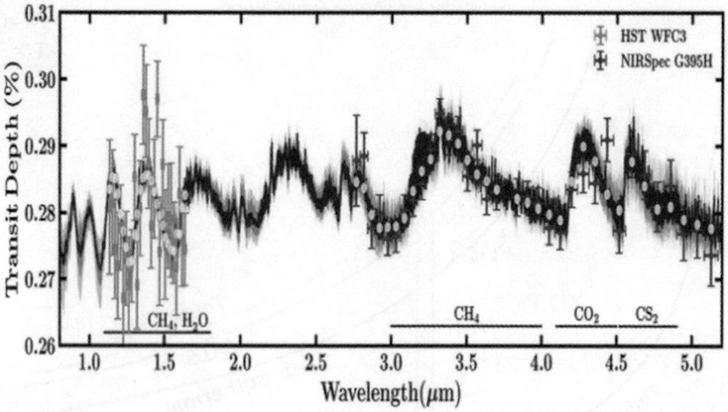

Figure 8. A transmission spectrum of TOI 270 d showing the extended wavelength range of JWST (NIRSpec) compared to HST and the improved signal to noise ratio (Holmberg and Madhusudhan 2024).

which is 25% of the flux received by Earth (0.25 S_E). This value is roughly consistent with the outer boundary set by the 1D model, as well as with the Early Mars hypothesis. A lower flux limit inferring a further out boundary of the HZ may be derived from the evidence for an intra-glacial lake in Mars' south pole (Arnold *et al.* 2022). Using for the radiative flux received by Martian polar region $cos(i) = 0.2$ gives a flux of $\sim 0.1 S_E$, which is denoted in Fig. 3 by the violet line marked "Martian polar lakes".

Acknowledgements

This research has been supported by the Minerva Foundation, at the Center for Studying the Planetary Emergence of Life.

References

Arnold, N. S., Butcher F.E.G., and Conway, S.J., *et al.* 2022, *Nature Astronomy*, 6, 1256
Daspute, M., Wandel, A., Tal-Or, L., Kopparapu R. K., 2024, submitted to *Astrophys. J.*
Kasting J. F., Whitmire D. P. and Reynolds R. T. 1993 *Icarus* 101, 108
Kopparapu R. K., Ramirez R., Kasting J. F. et al. 2013 *Astrophys. J.* 770, 82
Greene, T.P., Bell, T.J., Ducrot, E., et al. 2023, *Nature* 618, 39
Luger, R. and Barnes, R. 2015, *Astrobiology*, 15, 2
Holmberg, M. and Madhusudhan, N. 2024 *A& A* 683, L2
Menou, K. 2013 *Astrophys. J.* 774, 51
Moran, S. E., Stevenson, K. B., Sing, D. K., et al. 2023, *Astrophys. J.* 948, L11
Ojha L., Troncone B., Buffo J. et al. 2022 *Nature Comm.* 13, 7521.
Peter, J. S., Nordheim, T. A. and Hand, K. P. 2024, *Nature Astronomy* 8, 164
Shields A. L., Ballard S. and Johnson J. A. 2016 *Phys. Rev.* 663, 1
Wandel A. 2018 *Astrophys. J.* 858, 1
Wandel A. 2023a *Nature Comm.* 14, 2125
Wandel A. 2023b *Astron. J.* 166, 222
Wandel A. 2024 *Porc. IAUS* 387
Wandel A. and Gale J. 2020 *Int. J. of Astrobiology* 19, 126W
Wandel, A. and Tal-Or, L. 2019, *Nature Astronomy Lett.* 880, L21
Wandel and Gale 2024, Life in Space, Springer, ISBN 978-3-031-64638-6
Yang, J., Boué, G., Fabrycky, D. C. and Abbot, D. S. 2014 *Nature Astronomy Lett.* 787, L2

Planetary Science and Exoplanets in the Era of JWST
Proceedings IAU Symposium No. 393, 2026
Z. Benkhaldoun, Y. Moulane & A. Szentgyorgyi, eds.
doi:10.1017/S1743921324002679

Impact of Surface Pressure on Nitrogen-Dominated USP Super-Earths Atmospheres

J. Chouqar[1], J. Lustig-Yaeger[2,3], Z. Benkhaldoun[5], A. Szentgyorgyi[4], A. Jabiri[1] and A. Soubkiou[1]

[1]Oukaimeden Observatory, PHEA Laboratory, Cadi Ayyad University, Box 2390, Marrakech Morocco
email: jamila.chouqar@ced.uca.ma

[2]Johns Hopkins University Applied Physics Laboratory, Laurel, MD 20723, USA

[3]NASA NExSS Virtual Planetary Laboratory, Box 351580, University of Washington, Seattle, Washington 98195, USA

[4]Center for Astrophysics — Harvard & Smithsonian, 60 Garden Street, Cambridge, MA 02138, USA

[5]Sharjah University, United Arab Emirates and University Cadi Ayyad, Morocco

Abstract. In this work, we compared the chemistry and emission spectra of nitrogen-dominated Ultra-Short-Period (USP) super-Earth atmospheres at various surface pressures (0.1 to 10 bar) in and out of chemical equilibrium. We used the VULCAN chemical kinetic code, which includes thermochemical kinetics, vertical transport, and photochemistry, linked with the petitRADTRANS radiative transfer model to predict emission spectra. Radiative-convective temperature-pressure profiles were computed with the HELIOS code. Using PandExo noise simulations, we explored the observability of disequilibrium effects with JWST. Our results showed that surface pressure significantly impacted temperature profiles, atmospheric abundances, and emission spectra. Disequilibrium signatures in cooler planets were detectable by targeting HCN, C_2H_4, and CO, while warmer planets showed CH_4 and HCN features using JWST's NIRSpec and MIRI instruments. These species also provided insights into the presence of surfaces in nitrogen-dominated atmospheres, revealing details about atmospheric thickness.

Keywords. planets and satellites: atmospheres – planets and satellites, composition – planets and satellites, individual (HD-213885b; GJ-1252b & LP791-18b)

1. Introduction

Over the past two decades, more than 5,000 exoplanets have been discovered, revealing a diverse population. Among these are Ultra-Short-Period (USP) exoplanets, which orbit at extremely short periods, from about one day down to less than 10 hours (e.g., Howard (2013); Dai *et al.* (2017)), and even as short as $\approx$ 4 hours, especially around M-dwarfs (Ofir & Dreizler (2013); Rappaport *et al.* (2013)). Planets in this group tend to be smaller than 2 $R_\oplus$, and appear to have bulk compositions similar to that of the Earth (Winn *et al.* (2018)). There is also evidence that some of them might have iron-rich compositions (Santerne *et al.* (2018)). These planets experience extreme surface temperatures due to their proximity to their stars, which allows for the detection of thermal emission (Demory *et al.* (2012), Sanchis-Ojeda *et al.* (2014)). Although many USP planets were identified by the Kepler mission, only a few have both precise mass and radius measurements, as the stars in the Kepler field are typically much too faint for precise radial velocity (RV) follow-up. However, transit surveys like TESS, provide the perfect strategy to find these planets as they are designed to find short-period transiting exoplanets around bright stellar hosts, allowing us to explore the yet poorly understood property of mass and, thus, the bulk composition of these interesting extrasolar worlds.

Since its science operations began in July 2018, NASA's Transiting Exoplanet Survey Satellite (TESS; Ricker (2015)) has significantly expanded the size of this sample. An additional 24 planets with $R_p < 2\ R_\oplus$, P < 1 day and host stars amenable to James Webb Space Telescope (JWST) observations have been discovered, with thirteen of them having an accurate determination of mass and radius (`https://exoplanetarchive.ipac.caltech.edu/`). Our ability to characterize short-period terrestrial planets will improve drastically with the JWST, which will allow for the characterization of exoplanet atmospheres and surface properties via transmission and emission spectroscopy (Greene *et al.* (2016)). These planets are among the Cycle 1 observations and will continue to be observed in Cycle 2 and beyond. However, we currently lack a reliable set of models to guide the planning of such observations and help their interpretation.

Nitrogen-dominated atmospheres of USP super-Earth planets have been widely studied with the benchmark 55 Cnc e. Effects of orbital distances and compositions with different abundances of hydrogen, C/O, and N/O ratios have been explored (Zilinskas *et al.* (2020)). A broader exploration as a function of surface pressure is also required. Surface pressure variations affect the thermal profile of the planet, hence the chemistry and the spectra. In this work, we present a grid of thermal profiles, photochemical mixing ratios, and emission spectra for different surface pressure scenarios for cool (LP791-18b), warm (GJ-1252b), and hot (HD-213885b) USP super-Earths in order to explore the effects of surfaces and what to expect in current and upcoming observations. For each planet, we consider three scenarios that cover three different surface pressures ranging from 10^{-1} bar to 10 bar. These scenarios are chosen in light of previous studies on USP super-Earths. So far, only four USP super-Earths, 55 Cnc e (Angelo *et al.* (2017)), GJ-1252b (Crossfield *et al.* (2022)), LHS-3844b (Kreidberg *et al.* (2019)) and K2-141b (Zieba *et al.* (2022)), have had their thermal infrared radiation measured. Spitzer 4.5 μm observations of these planets' eclipse and phase curves revealed no phase offset and suggested an upper limit to the atmospheric surface pressure. For example, the data set for LHS-3844b indicates $P_{sur} \leq 10$ bar (Kreidberg *et al.* (2019); Whittaker *et al.* (2022)). Crossfield *et al.* (2022) compared the measurement of GJ-1252b to a large suite of atmospheric models and concluded that any atmosphere on GJ 1252b likely has a surface pressure of ≤ 10 bar. Furthermore, Angelo *et al.* (2017) have shown that the 1.4 bar of photospheric pressure is required to explain the planet's heat redistribution efficiency value constrained by the phase curve of 55 Cnc e.

2. Planet Properties

For this work, we selected three confirmed TESS USP planetary candidates, with a radius smaller than $2R_\oplus$, (see Table 1). GJ 1252b and HD 213885b have measured masses while the mass of LP791-18b is estimated using Forecaster (`https://github.com/chenjj2/forecaster`). In Fig. 1, we plotted the masses and radii of all the known USP planets; with radius $< 2R_\oplus$; having these parameters measured with a precision better than 25%. We compared the masses and radii of our targets with the theoretical and empirical mass-radius relationship of Zeng *et al.* (2016), and we found that their compositions are consistent with Earth-like/rocky planets.

Table 1. Relevant stellar and planetary properties.

	R (R_*)	M (M_*)	T_{eff}(K)	$logg_s(ms^{-2})$	$[M/H]$
LP791-18[1]	0.171	0.139	2960	5.11	-0.09
GJ1252[2]	0.391	0.381	3458	4.83	0.1
HD213885[3]	1.1011	1.068	5978	4.38	-0.04

	R ($R_\oplus$)	M ($M_\oplus$)	g (ms^{-2})	T_{eq} (K)	Period (d)	f^b	f^c	f^d
LP791-18b[1]	1.12	1.5^a	14	599	0.95	0.60	0.39	0.27
GJ-1252b[2]	1.19	2.09	14.4	1089	0.52	0.63	0.48	0.29
HD-213885b[3]	1.74	8.83	28.5	2130	1.00	0.66	0.61	0.39

Notes. The equilibrium Temperatures are calculated assuming zero Bond albedo and complete heat circulation between the day and night hemispheres.
[a] Masses are estimated based on Chen *et al.* (2016).
[b] Heat redistribution factor, for 0.1 bar surface pressure.
[c] Heat redistribution factor, for 1 bar surface pressure.
[d] Heat redistribution factor, for 10 bar surface pressure.
[1] Crossfield *et al.* (2019); [2] Shporer *et al.* (2020); [3] Espinoza *et al.* (2019)

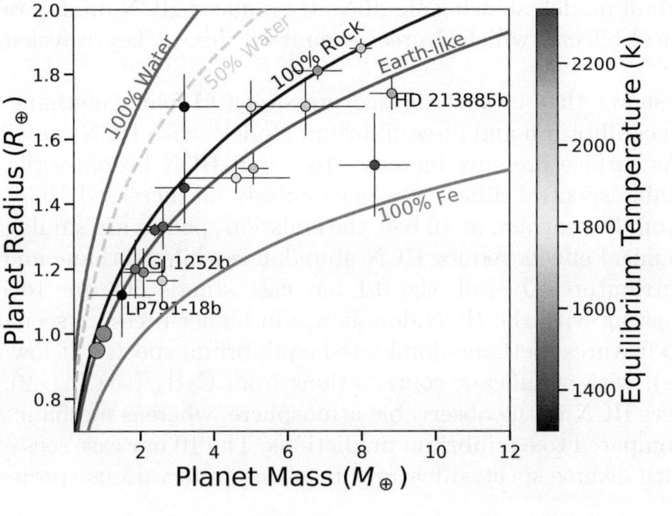

Figure 1. Mass-radius for USP planets, with mass-radius measurements better than 25% from `http://www.astro.keele.ac.uk/jkt/tepcat/` and for our USP TESS candidates (see Tab. 1), color-coded by their equilibrium temperature. Two-layer models from Zeng *et al.* (2016) are displayed with different lines and colors. “Earth-like” here means a composition of 30% Fe and 70% $MgSiO_3$, whereas “100% Rock” means a composition of 100% $MgSiO_3$. Earth and Venus are identified in this plot as pale blue and orange circles resp.

3. Atmospheric Modeling Methodology

In this study, we used the open-source photochemical kinetic code VULCAN (Tsai *et al.* (2017)) to simulate the chemistry in exoplanetary atmospheres assuming a Titan-like composition (Miguel (2018)). The radiative-convective temperature-pressure profiles were computed using the HELIOS code (Malik *et al.* (2017)). The one-dimensional chemical kinetic code VULCAN, in which thermochemical kinetic, vertical transport and photochemistry were taken into account, was then linked to the one-dimensional radiative transfer model petitRADTRANS (Mollière *et al.* (2019)) to predict the emission spectra of our planets for various surface pressure scenarios ranging from 10^{-1} bar to 10 bar. I refer the reader to Chouqar *et al.* (2023) for more details about the modeling Methodology.

4. Modeling Results

Here, we explored how disequilibrium processes can affect the predicted spectra at different surface pressures. We ran two model scenarios: equilibrium (i.e. no-diffusion, no-photochemistry) and disequilibrium (i.e. diffusion and photochemistry) with surfaces located at three different pressure levels (0.1 bar, 1 bar, and 10 bar). The resulting thermal spectra of our hot super-Earth N_2 dominated atmospheres with varying surface pressure are shown in Fig. 2. The predicted spectra of our hot-super-Earth exoplanets depend strongly on the adopted thermal structure. Different surface levels with different surface pressures and temperatures are shown to have significant impacts on the spectra of the atmospheres (e.g. see Fig. 3).

HD-213885b exhibits similar equilibrium and disequilibrium emission spectra across various surface pressures, dominated by HCN and CN spectral features (see Fig. 2). At lower pressures (0.1 bar and 1 bar), both spectra are nearly identical due to comparable HCN and CN abundances. However, at a higher pressure of 10 bar, deviations occur in the disequilibrium model, showing significantly reduced HCN mixing ratios (VMRs) due to photochemical effects, which decreases emitted flux at key wavelengths (3.2, 8, and 13.9 μm).

GJ-1252b shows that at low surface pressure (0.1 bar), methane is the dominant species in both equilibrium and disequilibrium models, with HCN contributing minimally (see Fig. 2). As surface pressure increases to 1 bar, HCN becomes the primary species, leading to notable spectral differences, particularly an increased HCN absorption band depth at 13.9 μm. In contrast, at 10 bar, the emission spectra are similar to the 1 bar case, but thermochemical effects reduce HCN abundance while methane and ethylene remain significant contributors. Overall, the 0.1 bar case stands out due to its high methane content, contrasting with the HCN dominance in higher-pressure scenarios.

LP791-18b features methane-dominated equilibrium spectra at low surface pressures (0.1 and 1 bar), with significant contributions from C_2H_4 (see Fig. 2). Photochemistry greatly enhances HCN in the observable atmosphere, whereas methane and other species are depleted compared to equilibrium predictions. The 10 bar case sees a shift to methane dominance, with diverse spectral features resulting from various species including HCN, and NH_3.

Furthermore, Fig. 3 is included to show the sensitivity of disequilibrium models to different surface pressures. For the GJ-1252b planet, the 0.1 bar surface pressure case displays a notable departure from the remaining cases (1 bar and 10 bar). This is due to the predominance of CH_4 in the 0.1 bar surface pressure atmosphere, in contrast to the prevalence of HCN in the higher-surface pressure atmospheres. While for the LP791-18b planet, the models of thinner atmospheres (0.1 bar and 1 bar surface pressure cases) are primarily affected by HCN, resulting in large spectral features between 2 to 15 μm. Conversely, the thicker atmospheric case (10 bar surface pressure) is dominated by CH_4 and C_2H_4, allowing for its differentiation from thinner atmospheres.

5. Constraining Disequilibrium effect with JWST

In this section, we investigated whether the difference between equilibrium and disequilibrium chemistry can be observed in the atmospheres of USP super-Earth planets with the JWST. To do, we used the JWST instrument noise simulator PandExo (`https://natashabatalha.github.io/PandExo`; Batalha *et al.* (2017)) to simulate secondary eclipse observations of the planets GJ-1252b and LP791-18b. For hotter planets like HD-213885b, distinguishing disequilibrium chemistry is challenging because their high temperatures allow thermochemical equilibrium to dominate over disequilibrium. These planets (GJ-1252b and LP791-18b) are bright enough to offer high S/N while faint enough

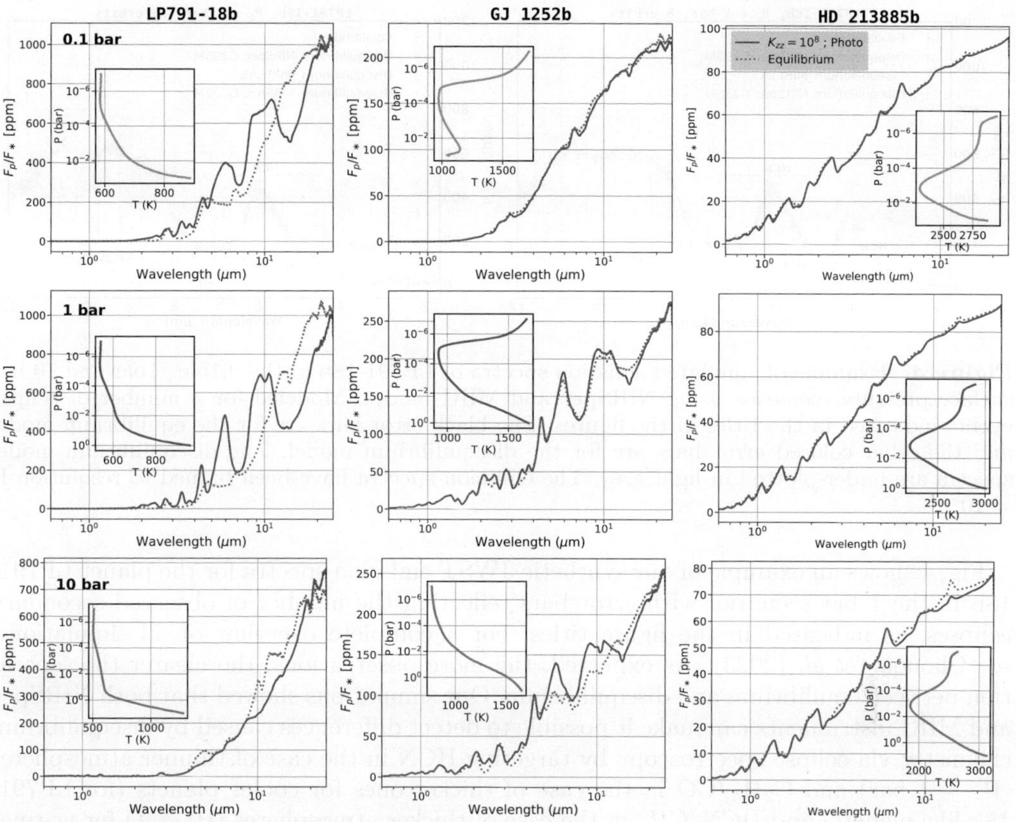

Figure 2. Effect of photochemical kinetic and vertical transport on the synthetic emission spectra of HD-213885b, GJ-1252b, and LP791-18b. Each row represents a different surface pressure. Each column corresponds to a different planet. **Inlaid:** temperature-pressure profiles. The dotted profiles are the thermochemical equilibrium abundances. The solid profiles are the disequilibrium chemical abundances calculated with vertical mixing ($K_{zz} = 10^8$ cm^2 s^{-1}) and photochemistry. The models shown are smoothed for clarity.

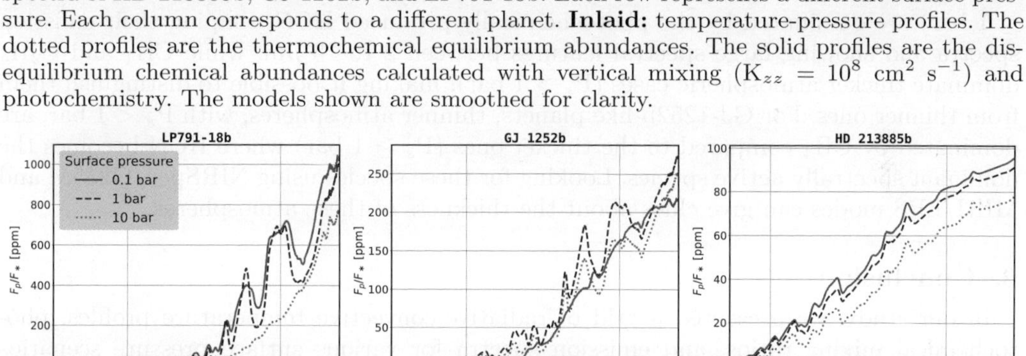

Figure 3. Sensitivity of disequilibrium spectra to different surface pressures for each planet.

to be observable with all of JWST's instruments. We, therefore, choose to observe each of our planets with NIRSpec/G235M from 1.66 to 3 μm (R=1000), NIRSpec/G395M mode from 2.9 to 5 μm (R=1000), and MIRI/LRS mode from 5 to 11 μm (R=100). NIRSpec Prism would saturate on GJ-1252b and was therefore not considered in this analysis. These observing modes were chosen because they offer nearly full wavelength coverage from 2 to 11 μm, where the disequilibrium features can be distinguished from equilibrium ones.

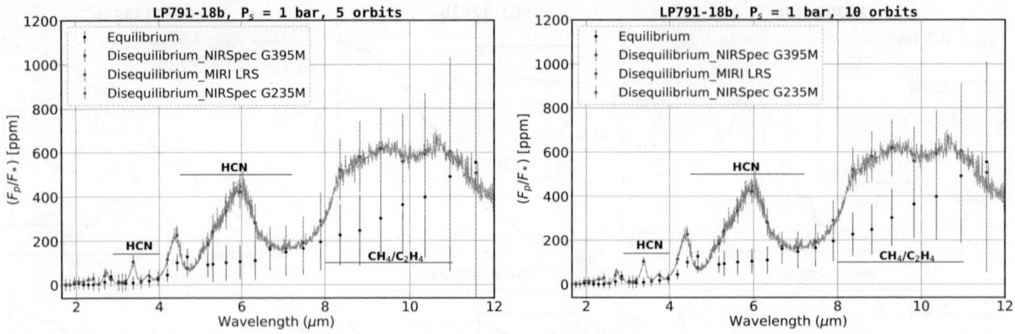

Figure 4. Example of simulated emission spectra of LP791-18b in the 0.1bar, 1bar and 10 bar surface pressure scenarios using NIRSpec and MIRI modes. Modeled for a number of eclipse events indicated in the titles of the figures. The black error bars are for the equilibrium model and the other colored error bars are for the disequilibrium model. The disequilibrium model spectra are under-plotted in light grey. The emission spectra have been binned to resolution R = 10.

Fig. 4 shows an example of our synthetic JWST emission spectra for the planet LP791-18b in the 1 bar scenario, with error bars reflecting the number of observed secondary eclipses, as indicated in the figure titles. For a complete overview of all simulations, see Chouqar *et al.* (2023). As expected, the more observations, the clearer the separation between equilibrium and disequilibrium. Our simulations showed that both NIRSpec and MIRI instruments can make it possible to detect differences caused by disequilibrium chemistry via eclipse spectroscopy by targeting HCN in the case of thinner atmospheres ($P_s \leq 1$ bar) and C_2H_4/CO in the case of thicker ones for cooler planets (for LP791-18b-like planets) and HCN/CH_4 in the case of thicker atmospheres ($P_s \geq 1$) for warmer planets (for GJ-1252b-like planets). Additionally, our grids show that also surface pressure scenarios can be distinguished. LP791-18b-like planets with thinner atmospheres ($P_s \leq 1$ bar) are dominated by the photochemically produced HCN, shaping their emission spectra and showing large spectral features between 2 to 15 μm, while CH_4 and C_2H_4 dominate thicker atmospheric cases ($P_s > 1$ bar), making it possible to distinguish them from thinner ones. For GJ-1252b-like planets, thinner atmospheres, with $P_s < 1$ bar, are dominated by CH_4 compared to the thicker ones ($P_s \geq 1$ bar) where HCN becomes the dominant spectrally active species. Looking for these species using NIRSpec/G235M and MIRI/LRS modes can give clues about the thickness of these atmospheres.

6. Conclusion

In our study, we presented a grid of radiative-convective temperature profiles, photochemical mixing ratios, and emission spectra for various surface pressure scenarios ranging from 10^{-1} bar to 10 bar; on cool (LP791-18b), warm (GJ-1252b), and hot (HD-213885b) USP super-Earths (see Chouqar *et al.* (2023) for a comprehensive review of all models). We explored how various surface pressures affect thermal profiles. We also performed a case study comparing the atmospheric abundances and the day-side emission spectra of these planets, in and out of chemical equilibrium at the different surface pressure scenarios in order to explore the observability of differences due to disequilibrium processes with JWST. Our grids demonstrate that various surface levels can significantly affect the thermal profiles, and alter the atmospheric abundances, and thus, the emission spectra of a nitrogen-dominated USP super-Earth atmosphere. We found that along with hotter USP rocky planets, warmer planets with $T_{eq} \geq 990$ K orbiting M-dwarfs may also show inversions in a nitrogen-dominated atmosphere. Modeling atmospheres of

USP super-Earths orbiting M-dwarfs requires additional attention in future work. We found that in addition to disequilibrium processes, HCN, CH_4, and C_2H_4 abundances are also sensitive to different surface pressures, making it possible to distinguish equilibrium from disequilibrium models, particularly for lower surface levels in cooler planets and higher surface pressures in warmer planets. These species can help distinguish the existence and role of photochemical reactions. Furthermore, these species can be used to differentiate between the different surface levels, giving clues about the thickness of a nitrogen-dominated USP super-Earth atmosphere. We showed that both NIRSpec and MIRI instruments are viable for the purpose of distinguishing disequilibrium chemistry from equilibrium chemistry via eclipse spectroscopy on cooler planets (LP791-18b-like planets), and on warmer planets with thicker atmospheres ($P_s \geq 1$ bar for GJ-1252b-like planets). With these instruments, we cannot readily observe the contribution from disequilibrium chemistry in HD 213885b-like planets given our inputs because it is too hot. At such high temperatures, thermochemical equilibrium dominates over disequilibrium processes making it difficult to distinguish them.

References

Angelo, I., Hu, R., et al. 2017, *The Astronomical Journal*, 154, 232

Batalha, N. E., Mandell, A., Pontoppidan, K., et al. 2017, *Publications of the Astronomical Society of the Pacific*, 129(976), 064501

Chen, J., Kipping, D., et al. 2016, *The Astrophysical Journal*, 834(1), 17

Chouqar, J., Lustig-Yaeger, J., Benkhaldoun, Z., Szentgyorgyi, A., Jabiri, A., & Soubkiou, A. 2023, *Monthly Notices of the Royal Astronomical Society*, 522(1), 648–659

Crossfield, I. J. M., Waalkes, W., Newton, E. R., et al. 2019, *The Astrophysical Journal*, 883(1), L16

Crossfield, I. J. M., Malik, M., Hill, M. L., et al. 2022, *arXiv preprint arXiv:2208.09479*

Dai, F., Winn, J. N., Gandolfi, D., et al. 2017, *The Astronomical Journal*, 154, 226

Demory, B.-O., Gillon, M., Seager, S., et al. 2012, *The Astrophysical Journal*, 751, L28

Espinoza, N., Brahm, R., Henning, T., et al. 2019, *Monthly Notices of the Royal Astronomical Society*, 491(2), 2982–2999

Greene, T. P., Line, M. R., Montero, C., et al. 2016, *The Astrophysical Journal*, 817, 17

Howard, A. W., Sanchis-Ojeda, R., Marcy, G. W., et al. 2013, *Nature*, 503, 381–384

Kreidberg, L., Koll, D. D. B., Morley, C., et al. 2019, *Nature*, 573, 87–90

Malik, M., Grosheintz, L., Mendonca, J. M., et al. 2017, *The Astronomical Journal*, 153(2), 56

Miguel, Y. 2018, *Monthly Notices of the Royal Astronomical Society*, 482(3), 2893–2901

Mollière, P., Wardenier, J. P., van Boekel, R., et al. 2019, *Astronomy & Astrophysics*, 627, A67

Ofir, A., Dreizler, S., et al. 2013, *Astronomy & Astrophysics*, 555, A58

Rappaport, S., Sanchis-Ojeda, R., Rogers, L., et al. 2013, *The Astrophysical Journal Letters*, 773, L15

Ricker, G. R. 2015, *Journal of Astronomical Telescopes, Instruments, and Systems*

Sanchis-Ojeda, R., Rappaport, S., Winn, J. N., et al. 2014, *The Astrophysical Journal*, 787, 47

Santerne, A., Brugger, B., Armstrong, D. J., et al. 2018, *Nature Astronomy*, 2, 393–400

Shporer, A., Collins, K. A., Astudillo-Defru, N., et al. 2020, *The Astrophysical Journal*, 890(1), L7

Tsai, S. M., Lyons, J. R., Grosheintz, L., et al. 2017, *The Astrophysical Journal Supplement Series*, 228(2), 20

Winn, J. N., Sanchis-Ojeda, R., Rappaport, S., et al. 2018, *New Astronomy Reviews*, 83, 37–48

Whittaker, E. A., Malik, M., Ih, J., et al. 2022, *arXiv preprint arXiv:2207.08889*

Zeng, L., Sasselov, D. D., Jacobsen, S. B., et al. 2016, *The Astrophysical Journal*, 819(2), 127

Zieba, S., Zilinskas, M., Kreidberg, L., et al. 2022, *arXiv preprint arXiv:2203.00370*

Zilinskas, M., Miguel, Y., Mollière, P., et al. 2020, *Monthly Notices of the Royal Astronomical Society*, 494, 1490–1506

Zilinskas, M., Miguel, Y., Lyu, Y., et al. 2020, *Monthly Notices of the Royal Astronomical Society*, 500, 2197–2208

Planetary Science and Exoplanets in the Era of JWST
Proceedings IAU Symposium No. 393, 2026
Z. Benkhaldoun, Y. Moulane & A. Szentgyorgyi, eds.
doi:10.1017/S174392132400190X

Water at the earliest stages of star and planet formation

Itziar de Gregorio-Monsalvo[1], Santiago Gil[2], Alejandro Santamaría-Miranda[3], José Francisco Gómez[4], Andrés Pérez-Sánchez[5], Adele Plunkett[6], Elizabeth Artur de la Villarmois[1], Viviana Guzmán[2] and Matthias R. Schreiber[7]

[1]European Southern Observatory, Alonso de Cordova 3107, Casilla 19, Vitacura, Santiago, Chile
email: idegrego@eso.org

[2]Instituto de Astrofísica, Pontificia Universidad Católica de Chile, Av. Vicuña Mackenna 4860, 7820436 Macul, Santiago, Chile

[3]Departamento de Astronomía; Universidad de Chile; Camino El Observatorio 1515, Las Condes, Santiago, Chile

[4]Instituto de Astrofísica de Andalucía, (IAA-CSIC), Glorieta de la Astronomía s/n, 18008 Granada, Spain

[5]Joint ALMA Observatory, Alonso de Cordova 3107, Casilla 19, Vitacura, Santiago, Chile

[6]National Radio Astronomy Observatory, 520 Edgemont Rd, Charlottesville, VA, 22903, USA

[7]Departamento de Física, Universidad Técnica Federico Santamaría, Av. España 1680, Valparaíso, Chile

Abstract. Water, an essential molecule for sustaining life on Earth, plays a crucial role in the physical and chemical processes governing the formation of stars and planets, serving as a primary coolant in the environments surrounding emerging stars. The journey of water from the earliest stages of star and planet formation to its eventual incorporation to exoplanetary atmospheres and its potential oceans is still a matter of debate. Very recent works using ALMA and JWST data reveals that water can be found in the inner astronomical units in protoplanetary disks, in the terrestrial planet-forming zone. Detecting water molecules from ground-based telescopes is challenging due to the water vapor content within Earth's atmosphere, severely limiting atmospheric transmissivity. Nevertheless, water is one of the few compounds capable of generating bright maser emission within star-forming regions. Their exceptional brightness, narrow spectral profiles, and their origin in highly compact regions offer a unique set of characteristics that render them invaluable tools for investigating circumstellar structure and dynamics at sub-arcsecond spatial scales. In this work we present our efforts to search and characterize water maser emission at submillimeter frequencies in a sample of young stellar objects.

Keywords. star formation, molecular outflows, protoplanetary disks, masers

1. Introduction

Water is a very common compound in the Universe and it is a key molecule for sustaining life as we know it. It plays a crucial role in both the physical and chemical processes involved in star and planet formation, where it acts as a primary coolant in the environments around young stars.

The water snowline in protoplanetary disks, which marks the midplane radius where water transitions from ice to vapor, is critical for both planet formation and their chemical composition. At the snowline, dust particles can coagulate more efficiently, increasing the surface density of the disk and enhancing planet formation. Moreover it plays an important role in setting the chemical composition of planets since different locations of the water snowline can alter the C/O ratio both in the gas and in the solid phases (Oberg et al. 2011).

In addition to that, water is relevant to understand the origins of our oceans and, ultimately, the origing of life itself.

2. Observations of water from Earth and from Space

Good progress has been made in the last two decades to characterize water surrounding young protostars thanks to science space missions. Twenty years ago, the Infrared Space Observatory (ISO) identified low-mass young Class 0 protostars as prominent water emitters pointing to an origining from shocks at the base and along the molecular outflow. However, water emission was not detected in Class I sources, suggesting changes in the interaction between the protostellar jet and the surrounding environment (Nisini et al. 2002). Later, the Herschel Space Observatory through its key programme WISH (Water In Star-forming regions with Herschel), made advances in understanding the star formation processes by using water to study the physics and chemistry of young stellar objects, from the earliest stages to more evolved protoplanetry disks (Van dishoeck et al. 2011).

With the advent of the James Webb Space Telescope, we can now advance studies of water in the inner regions of protoplanetary disks, where sub-Neptune and terrestrial planets form. A good example is the detection of water in the inner cavity of the disk around PDS 70 (Perotti et al. 2023), a system where protoplanets are currently forming (Muller et al. 2018; Haffert et al. 2019). This discovery highlights the potential for terrestrial planets forming in such regions to access water reservoirs. Despite the progress made by space missions in characterizing water emission in young star-forming regions, their limited spatial resolution compared to ground-based interferometers makes the latter an attractive alternative for conducting such studies.

In the millimeter/submillimeter regime, the ALMA (Atacama Large Millimeter Array) interferometer is yielding promising results regarding the origin of water on planets. By studying deuterium-enriched water in the disk surrounding V883 Ori, Tobin et al. (2023) concluded that disks inherit water from their parent molecular cloud, which is then incorporated into large icy bodies, such as comets, with minimal chemical alteration. In another study, Facchini et al. (2024) spatially resolved water vapor in the inner region of the disk around HL Tau using the main water isotopologue.

3. Maser emission as a tool to study water in star and planet forming regions

Despite ALMA's advancements in characterizing water in planet-forming regions, detecting thermal water molecule emission from ground-based telescopes remains challenging.

The water vapor content in Earth's atmosphere strongly affects millimiter and submillimeter frequencies, limiting the atmospheric transmissivity at these wavelegths. The troposphere, located at an altitude of $\sim$ 11 km, contains 70–80% of the total atmospheric mass and nearly all of the water vapor. Therefore, placing millimeter/submillimeter telescopes at high altitudes in extremely dry environments is essential for conducting high-frequency observations. This is the case for the Atacama Pathfinder EXperiment

(APEX) antenna and ALMA, both located on the Llano de Chajnantor, at approximately 5000 meters elevation, in the Atacama Desert, the driest place on Earth. However, given the faint emission of thermal water and the atmospheric conditions, detecting this molecule within a reasonable integration time remains highly challenging even for a high sensitive instrument like ALMA.

Fortunately, water is one of the few molecules able to producing maser emission in star-forming regions, which enhances its detectability (Reid & Moran 1981).

Water masers (Microwave Amplification by Stimulated Emission of Radiation) occur naturally in space for some molecular transitions. These masers can be pumped either by collisions or radiation and are tipically found in regions of high density and high temperature.

Given the technology available in previous decades, most studies focused on the 22 GHz water maser transition. This transition has proven to be a powerful tool to study circumstellar structure and dynamics at sub-arcsecond spatial scales, as it is bright, spectrally narrow, and originates from very compact regions. Observations with single-dish telescopes and interferometers operating at centimeter wavelengths have shown that the 22 GHz water maser transition is an effective tracer of mass-loss activity in young stars across all mass ranges (De Buizer et al. 2005; Furuya et al. 2003) and in a few intermediate and high mass sources they trace disk-like structures (Torrelles et al. 1998; Patel et al. 2007).

At submillimeter wavelengths, water exhibits dozens of maser transitions. Some of these can be detected by telescopes such as APEX and ALMA, making these facilities ideal for extending the studies conducted at centimeter wavelengths in recent decades into the submillimeter regime.

4. A survey of water masers at 321 and 325 GHz in star Forming Regions

The observation of multiple maser lines is essential to constrain the physical parameters of the masing region, since different water maser transitions can be pumped under a broad range of physical conditions.

Theoretical models predict that water masers at 22 GHz, 321 GHz, and 325 GHz can be pumped by both radiative and collisional processes under rather different conditions. While the 22 GHz water masers are excited in regions of high density ($n > 10^8$ cm^{-3}) and high temperature ($T > 500$ K), the 321 GHz transition requires even warmer gas ($T > 1000$ K) for its excitation, and the 325 GHz maser transition is excited in lower density environments (Gray et al. 2016).

So far only a few high-angular resolution studies of submillimeter water masers in high-mass star-forming regions have been published (Patel et al. 2007; Hirota et al. 2021; Niederhofer et al. 2012). These studies reveal that the 321 and 325 GHz masers trace different regions compared to those probed by the 22 GHz transition.

Using the APEX antenna, we conducted a survey of water masers at 321 and 325 GHz in a sample of twenty young stellar objects that are strong emitters of the 22 GHz water maser transition. The sample covered a wide range of young stellar objects luminosities, from 3 $L_\odot$ to 10^6 $L_\odot$. We detected emission at 321 GHz in approximately 30% of the sources and at 325 GHz in about 25% of the sources, all of which are high-mass young stellar objects. Another significant finding is that the maser luminosity in both transitions correlates with the bolometric luminosity of the central young stellar object, with no emission detected from sources below 750 $L_\odot$ (see Fig. 1).

Future ALMA observations will be essential for resolving the morphology and kinematics of water masers at these frequencies, as well as for characterizing mass-loss phenomena and disk-like structures at scales where the outflow collimation process occurs.

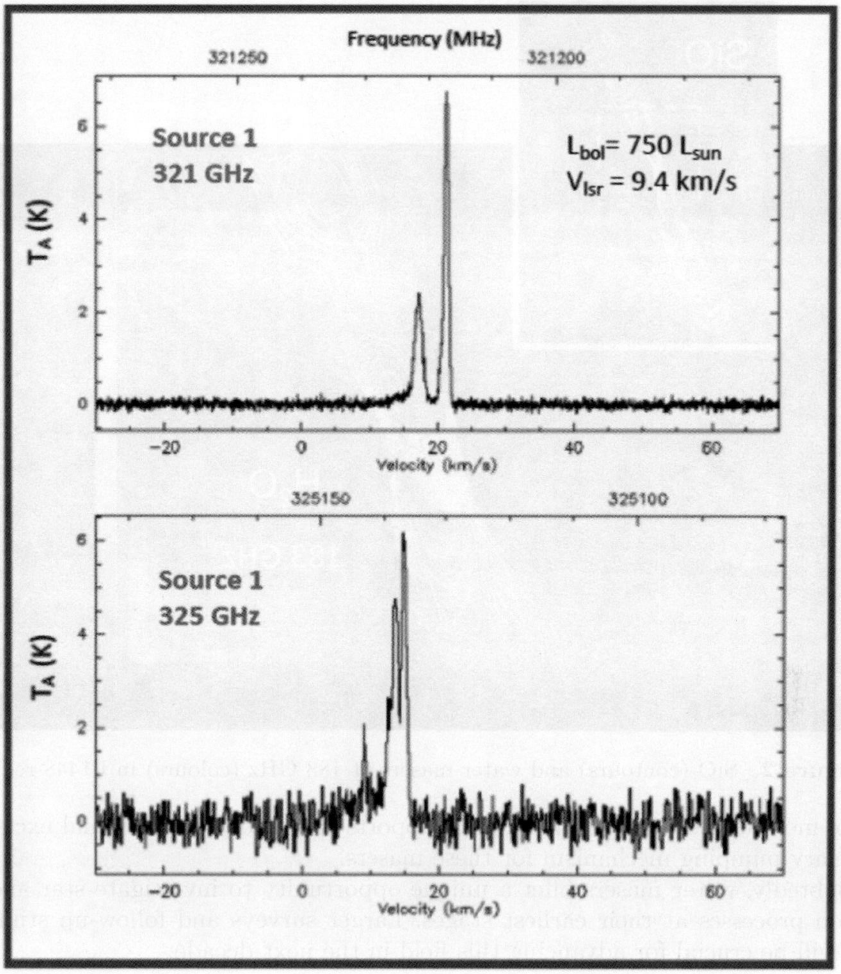

Figure 1. Water maser emission at 321 and 325 GHz detected in the lowest luminosity source of our sample.

5. Water masers at 183 GHz as a probe of shocks in low-mass protostars

The physical conditions necessary to excite the 22 GHz water maser transition (high density ($n > 10^8$ cm^{-3}) and high temperature ($T > 500$ K)) are relatively uncommon in low-mass star-forming regions, making its detection more challenging compared to the high-mass regime.

Nevertheless, the 183 GHz water maser transition presents an excellent alternative for probing water in low-mass protostars. It can be excited under a wide range of physical conditions common in the interstellar medium, requiring lower temperatures ($T > 150$ K) and densities ($n > 10^5$ cm^{-3}) to be pumped (Gray et al. 2016). Theoretical models predict that the 183 GHz water maser transition is primarily pumped through collisional mechanisms.

Using ALMA observations at subarcsecond spatial resolution, we have detected 183 GHz water masers in a prototypical Class 0 protostar in the L1448 region (Fig. 2). One of the most intriguing outcomes of this study is the location of these masers, which coincides with the region where thermal SiO emission traces shock-compressed material

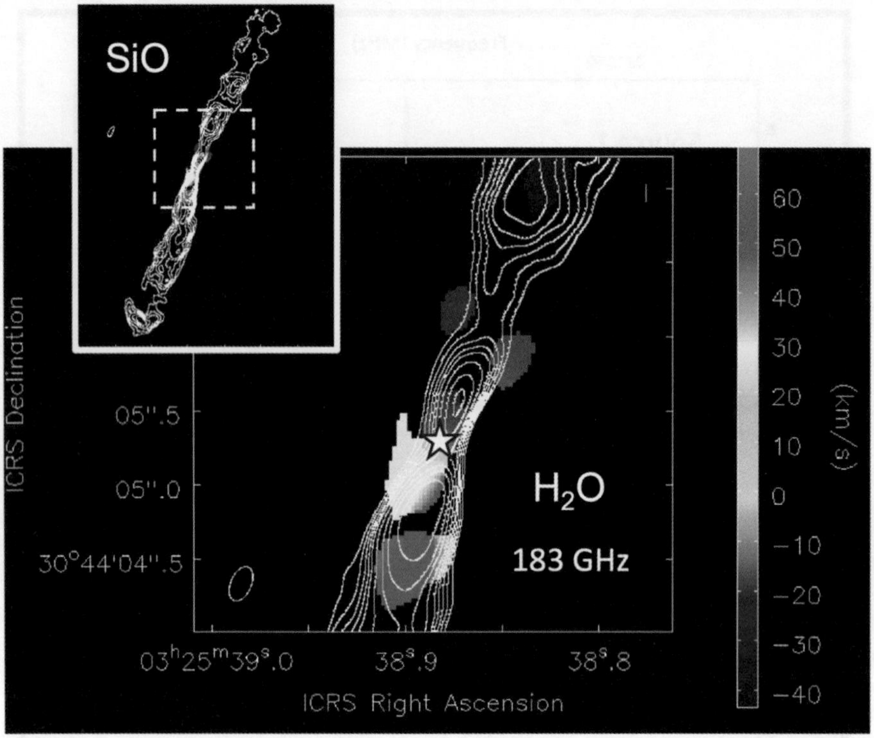

Figure 2. SiO (contours) and water masers at 183 GHz (colours) in L1448 region

from the molecular outflow. This finding supports the theory of collisional excitation as the primary pumping mechanism for these masers.

Undoubtedly, water masers offer a unique opportunity to investigate star and planet formation processes at their earliest stages. Larger surveys and follow-up studies with ALMA will be crucial for advancing this field in the next decade.

References

De Buizer, J. M., Radomski, J. T., Telesco, C. M., et al. 2005, *ApJS*, 156, 179
Facchini et al. 2024, *Nature Astronomy*, Volume 8, pp. 587–595
Furuya, R. S., Kitamura, Y., Wootten, A., et al. 2003, *ApJS*, 144, 71
Gray, M. D., Baudry, A., Richards, A. M. S., et al. 2016, *MNRAS*, 456, 374
Haffert, S. Y., Bohn, A. J., de Boer, J., et al. 2019, *Nature Astronomy*, 3, 74
Hirota, T., Cesaroni, R., Moscadelli, L., et al. 2021, *A&A*, 647, A23
Muller, A., Keppler, M., Henning, T., et al. 2018, *A&A*, 617, L2
Niederhofer, F., Humphreys, E., Goddi, C., et al. 2012, *Cosmic Masers - from OH to H0*, IAU Symp. 287, 184
Nisini et al. 2002, *The Astrophysical Journal*, Volume 574, Issue 1, pp. 246–257
Oberg, K. I., Murray-Clay, R., & Bergin, E. A. 2011, *ApJL*, 743, L16
Patel, N. A., Curiel, S., Zhang, Q., et al. 2007, *ApJL*, 658, L55
Perotti et al. 2023, *Nature*, Volume 620, Issue 7974, pp. 516–520
Reid, M. J. & Moran, J. M. 1981, *ARAA*, 19, 231. doi:10.1146/annurev.aa.19.090181.001311
Shirley, Y. L., Claussen, M. J., Bourke, T. L., et al. 2007, *ApJ*, 667, 329
Tobin et al. 2023, *Nature*, Volume 615, Issue 7951, pp.227–230
Torrelles, J. M., Gómez, J. F., Rodriguez, L. F., et al. 1998, *ApJ*, 505, 756
Van dishoeck et al. 2011, *Publications of the Astronomical Society of the Pacific*, Volume 123, Issue 900, p. 138 (2011)

Planetary Science and Exoplanets in the Era of JWST
Proceedings IAU Symposium No. 393, 2026
Z. Benkhaldoun, Y. Moulane & A. Szentgyorgyi, eds.
doi:10.1017/S1743921324003405

Probing the gas that builds planets: Results from the JWST MINDS program

Ewine F. van Dishoeck and the MINDS team

Leiden Observatory, Leiden University, the Netherlands
email: ewine@strw.leidenuniv.nl

Abstract. Infrared observations with JWST open up a new window into the chemical composition of the gas in the inner disk (<few au) where planets are built. Results from the MIRI GTO program MINDS are presented for several disks around T Tauri and lower-mass stars. A large diversity in spectra is found. Some disks are very rich in H_2O lines whereas other disks show prominent CO_2. The spectra of disks around very low-mass stars (<0.3 $M_\odot$, late-M type stars like Trappist-1) are dominated by C_2H_2 and other hydrocarbon features including those of benzene, suggesting volatile C/O>1. Together these data point to a rich chemistry in the inner regions that is linked to the physical structure of these disks (e.g., dust traps) and that may be affected by processes such as radial drift of icy pebbles from the outer to the inner disk.

Keywords. protoplanetary disks, planet formation, molecules, infrared spectroscopy

1. Introduction

Disks around young stars are the birthplaces of planets, but the kind of planets that are formed and their composition varies significantly from star to star and is largely determined by the structure and evolution of their parent disks (Öberg & Bergin 2021). Over the past decade, new observational facilities at millimeter and infrared wavelengths have allowed to study the chemical composition of disks with unprecedented detail. At millimeter wavelengths, the pure rotational transitions of molecules take place, and the Atacama Large Millimeter/submillimeter Array (ALMA) has been key to image a growing number of species (Öberg et al. 2023). Both simple and more complex molecules (Booth et al. 2024) have been probed down to abundances as low as 10^{-12} with respected to hydrogen and with fully resolved line profiles ($R = \lambda/\Delta\lambda = 10^7$). However, ALMA is limited to the cold outer regions of disks, typically > 10 au.

Infrared observations probe the vibrational transitions of molecules, and, in a few cases (H_2O and OH), also very high-lying pure rotational transitions. The *James Webb Space Telescope* (JWST) provides much higher sensitivity and spectral resolution ($R \approx 3000$) over the 1–28 μm range than previous space missions and uniquely probes the warm inner part of disks (<few au) where terrestrial planets are thought to form. At infrared wavelengths, molecules without a permanent dipole moment can be observed; this includes H_2 as well as carbon-bearing species (e.g., CH_4, C_2H_2, CO_2, ...) down to abundances of $\sim 10^{-8}$. Also Polycyclic Aromatic Hydrocarbons (PAHs), ices, silicates and other solid-state species are uniquely probed with infrared data (van Dishoeck 2004). Thus, JWST and ALMA strongly complement each other in studies of the chemistry of planet-forming disks.

The potential of infrared observations for chemical studies of disks became apparent from pioneering ground-based observations (see Brown et al. 2013; Banzatti et al. 2022, 2023b, for recent summaries) as well as initial space-based data from the *Infrared*

Space Observatory (e.g., Waelkens et al. 1996) and the *Spitzer Space Telescope* (see Pontoppidan et al. 2014, for review). This brief paper summarizes results obtained with the Mid InfraRed Instrument (MIRI) on JWST (Wright et al. 2023) in the context of the Mid INfrared Disk Survey (MINDS) guaranteed time program (PI: Th. Henning, co-PI: I. Kamp). This 120 hr program observes ∼30 disks around solar mass T Tauri stars, ∼10 disks around very low-mass stars and brown dwarfs and a few Herbig and debris disks with the Medium Resolution Spectrometer (MRS) at 5–28 μm (Kamp et al. 2023). Comparison with young disks in the earlier protostellar stages is done in the JWST Observations of Young protoStars (JOYS) MIRI-MRS GTO program (PI: E.F. van Dishoeck, co-PI: H. Beuther) (van Dishoeck et al. 2023; van Gelder et al. 2024). There is also a rapidly growing JWST-MIRI data set on protoplanetary disks based on GO programs, with an additional ∼150 disks observed in Cycles 1–3 by other teams, most of them part of the JDISCS collaboration (Banzatti et al. 2023a; Pontoppidan et al. 2024).

The goals of the MINDS program are to use JWST to (1) investigate the chemical inventory of the terrestrial planet forming zone, (2) follow the gas evolution into the disk dispersal stage, and (3) study the structure of protoplanetary and debris disks in the thermal mid-IR. The program builds a bridge between the chemical inventory of planet-forming disks and the properties of exoplanets. Here only some highlights related to the first goal will be presented. See Henning et al. (2024) and `minds.cab.inta-csic.es` for an overview and early summary.

2. What sets inner disk abundances?

Three main scenarios affect the inner disk chemistry: (1) high-temperature gas-phase chemistry; (2) drifting icy pebbles from the outer disk followed by sublimation once they cross their snowlines; and (3) dust traps locking up icy pebbles in the outer disk. In addition, time-variable accretion and ejection processes (e.g., winds) play a role in all three cases.

Regarding scenario (1): gas temperatures in the inner disk are high, ranging from 200 K to >1500 K, even exceeding the dust sublimation temperature (Najita et al. 2003; Woitke et al. 2018). UV radiation from the young stars is also enhanced by several orders of magnitude, photodissociating and photoionizing molecules and atoms. Thanks to the high temperatures, reaction barriers can readily be overcome and drive a rapid chemistry that is different from that taking place in the cold outer part (Walsh et al. 2015). One important case is the balance between H_2O and CO_2 through the intermediate OH: at 100–250 K, OH reacts primarily with CO to form CO_2, but at higher temperatures it reacts with H_2 to form H_2O (van Dishoeck et al. 2023). Because of continuum optical depth at mid-IR wavelengths, JWST observations are limited to the upper layers of the disk (e.g., Bruderer et al. 2015), although grain growth and settling to the midplane make it possible to look deeper into the disk. Indeed, effective gas/(small) dust ratios in the infrared emitting layers are typically 10000 rather than 100 (Meijerink et al. 2009).

The C/O ratio is thought to be an important diagnostic for linking exoplanets with their formation location (Öberg et al. 2011). In the inner disk, dust temperatures are high enough that all ices have sublimated so the C/O ratio in the gas should be close to the stellar value: the chemistry reshuffles the individual molecular abundances but does not affect the overall C/O.

Scenarios (2) and (3) alter the C/O (and C/H and O/H) abundances ratios in the inner region (see Figure 1). In scenario (2), water (and more generally oxygen, O/H) is enhanced by the drifting of water-rich icy pebbles from the outer to the inner disk. Once icy grains cross the disk mid-plane snowline around ∼160 K, the water ice sublimates into the gas and becomes visible to JWST if mixed to high enough altitudes. Enhancements

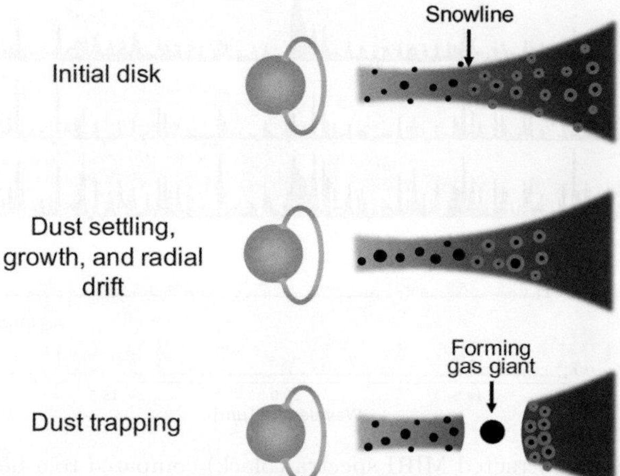

Figure 1. Cartoon illustrating the three scenarios of a full initial disk (top), a disk with significant grain growth, settling and radial drift of icy pebbles (middle) and a disk with a gap, locking up icy pebbles in a dust trap. Figure by Sierra Grant.

in O/H can be more than an order of magnitude compared to the interstellar value (e.g., Bosman et al. 2018; Kalyaan et al. 2021; Mah et al. 2023). Indeed, several JWST spectra show evidence for enhanced "cold" ($\sim$200 K) water at the water snowline (Banzatti et al. 2023a, 2024; Temmink et al. 2024a). This enhancement is expected to be particularly strong for disks that show compact dust millimeter continuum emission with ALMA and large ratios of their gas/dust radii, indicative of pebbles that have drifted inwards (Banzatti et al. 2020; Trapman et al. 2019). Similarly, CO and carbon-containing icy grains can drift in and enhance C/H (e.g., Zhang et al. 2020).

The opposite case holds for disks with gaps and cavities that show prominent dust rings ("substructures") with ALMA (Andrews 2020). In scenario (3), these rings can trap the icy dust grains and prevent them from moving inwards, thereby lowering the inner disk O/H and C/H ratios significantly. Much depends on the timing of the formation of these dust traps (Mah et al. 2024; Sellek et al. 2024): if they form early (<1 Myr), the initial inner disk water rapidly drains onto the star and the disk becomes very dry.

By comparing inner disk abundances found with JWST with disk structures imaged by ALMA, scenarios (2) and (3) can be tested. One caveat is that ALMA can only image cavities at best down to a few au for nearby disks. This is also the region where the millimeter continuum becomes optically thick making it more difficult to see small gaps. The structure of disks inside a few au is therefore still largely unknown territory, although near-IR interferometry and spectrally resolved CO data are providing some constraints.

3. Results

3.1. *T Tauri disks*

Figure 2 summarizes the 13-17 μm spectrum of three disks around T Tauri stars: GW Lup, DR Tau and DF Tau (Grant et al. 2023, 2024; Temmink et al. 2024a,b). The bottom part shows the simulated spectra of the various molecules that contribute to the emission in this region. These so-called slab models assume that the level populations of each molecule are in LTE at a single temperature T with the column density N and emitting area Ω determining the strength of the lines. Typical temperatures are 600–800 K for C_2H_2 and HCN and 300–400 K for CO_2, with emission coming from the

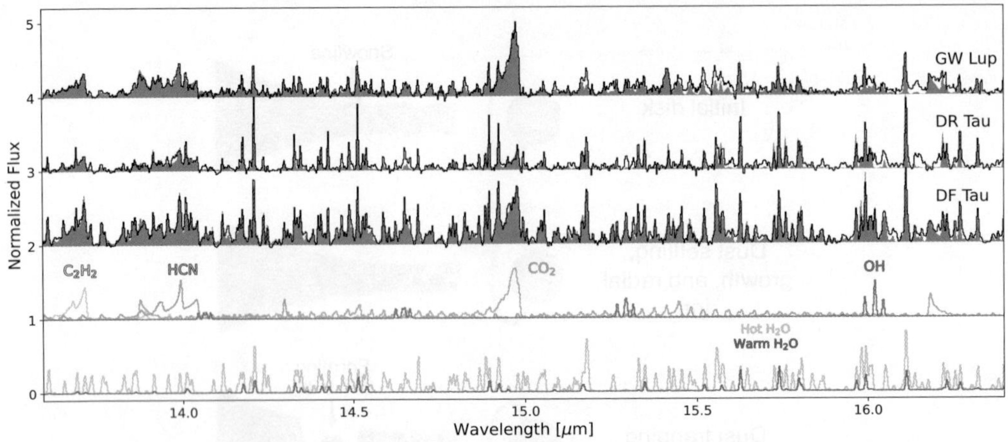

Figure 2. Continuum-subtracted MIRI spectra (black) compared to a best-fitting total model in red, for the disks around GW Lup, a large dust disk with rings (top); DR Tau, a compact dust disk (second row); and DF Tau, a close binary (third row). The spectra are normalized to the peak emission in this wavelength range, 0.035 Jy for GW Lup, 0.52 Jy for DR Tau, and 0.27 Jy for DF Tau (note the much weaker lines of GW Lup). Model emission from C_2H_2 (yellow), HCN (orange), CO_2 (green) and OH (pink) is presented in the fourth row for parameters that best fit the DF Tau disk emission. The bottom row shows two H_2O models in cyan (920 K) and blue (490 K). Data and models taken from Grant et al. (2023); Temmink et al. (2024b,a); Grant et al. (2024), see those papers for best fitting values. Figure by Sierra Grant.

inner ∼0.3 au. H_2O has many lines throughout the MIRI wavelength range indicating a temperature gradient roughly as $R^{-0.5}$ from ∼1200 K down to ∼180 K over the 0.1 – few au range (Gasman et al. 2023; Temmink et al. 2024a), as also found in JDISCS (Romero-Mirza et al. 2024).

It is clear that there is significant diversity among disks: many T Tauri disks are bright in H_2O lines (e.g., DR Tau, DF Tau) whereas others show particularly strong CO_2 features (e.g., GW Lup), including $^{13}CO_2$ (Grant et al. 2023) and even $CO^{18}O$ (e.g., CX Tau, Vlasblom et al. subm.). The detection of isotopologs indicates that the main $^{12}CO_2$ band is optically thick and that the CO_2/H_2O column density ratio is much higher than would be suggested by the relative strengths of the lines. The CO_2 isotopologs are also fitted with colder temperatures suggesting that they are located deeper in the disk (Bosman et al. 2022).

What causes this diversity? DR Tau has a compact millimeter dust disk, whereas DF Tau is a 9 au separation binary system with very small truncated dust disks around both components. Thus, radial drift of icy grains followed by water sublimation in the inner disk could play a role for these disks (§ 2). In contrast, GW Lup has a large dust disk with dust rings that can trap the icy pebbles at large radii well beyond the snowlines.

However, not all compact dust disks are bright in water lines. CX Tau is an example of a drift-dominated disk with brighter CO_2 than H_2O lines (Vlasblom et al., subm.). Thermochemical models suggest a number of factors that can enhance one species over the other (Vlasblom et al. 2024). For example, a small cavity between the H_2O and CO_2 snowlines can selectively remove H_2O. Also, temperature plays a role (§ 2): at low T, OH will react with CO to form CO_2 rather than with H_2 to form H_2O. Finally, the age of the disk can affect relative abundances (Sellek et al. 2024): CO_2-rich ices drift in later than H_2O-rich ices.

Another sign that the situation may be more complex than sketched in § 2 is the detection of water in the inner region of the transitional disks. PDS 70 is a well-known disk with a large cavity in which two young giant planets have been detected (Keppler et al.

2018; Haffert et al. 2019). The prominent dust ring at 45 au is thought to trap all the drifting icy pebbles and prevent them from entering the small (few au) inner disk. It therefore came as a surprise when JWST clearly detected water at $T \approx 600$ K from the inner disk (Perotti et al. 2023). Similarly, emission of water and other molecules is detected from the inner region of the transitional disk SY Cha with a large 70 au cavity (Schwarz et al. 2024). A systematic study of several gapped T Tauri disks in the MINDS sample reveals emission of water and other species from the inner disk in all cases (Gasman et al., subm.). This suggests that gaps do not fully block the delivery of oxygen- and carbon-containing gas and (small, icy) dust to the inner disk: dust traps are clearly "leaky" even if the dust and gas cavities are broad and deep.

Taken together, the data indicate that the gas in the inner disks of T Tauri stars is generally oxygen rich, with C/O<1: H_2O is usually the most abundant molecule, followed closely by CO_2. C_2H_2 and HCN typically have about two orders of magnitude lower column densities. CH_4 is generally not detected (Temmink, priv. comm.). OH is never a significant reservoir of oxygen, although it is a unique probe of the UV field and high-temperature chemistry (Tabone et al. 2024). Note that the various molecules are not necessarily co-located, neither radially nor vertically in the disk, so any column density ratios should be interpreted with caution. More generally, inner disk chemistry may be less strongly related to outer disk substructures than previously thought.

3.2. *Disks around very low-mass stars and brown dwarfs*

The JWST spectra of disks around very low mass stars (VLMS, <0.3 $M_\odot$) and brown dwarfs (<0.08 $M_\odot$) are generally very different from those around T Tauri stars. VLMS are of particular interest since late M-type stars (later than M4) are the most common stars in the Galaxy, and many rocky exoplanet studies focus on them, with the Trappist-1 system being the most famous example. Early hints for bright C_2H_2 emission from VLMS disks came from *Spitzer* spectra (Pascucci et al. 2013), but JWST reveals their rich hydrocarbon emission in unprecedented detail.

The first JWST example was the J16053215-1933159 (J1605) disk which shows two prominent broad bumps at 7.7 and 13.7 μm (Tabone et al. 2023). These can be ascribed to highly optically thick C_2H_2 emission in its $\nu_4+\nu_5$ and ν_5 bands, with orders of magnitude higher column densities ($> 10^{20}$ cm^{-2}) at 525 K coming from the inner 0.033 au. Larger hydrocarbons such as C_4H_2 and C_6H_6 (benzene) are also prominent and detected for the first time.

Subsequent JWST studies have revealed other VLMS disk examples with very rich hydrocarbon spectra, although not with such prominent C_2H_2 bumps as J1605. Rather, the many *P*- and *R*-branch lines of hydrocarbon molecules like CH_4 and C_2H_4 form an underlying pseudo-continuum on which their *Q*-branches are detected (Arabhavi et al. 2024; Kanwar et al. 2024a). C_2H_6 is also detected in a disk for the first time. Of the oxygen-bearing species, CO_2 is detected, but CO only sporadically and H_2O is weak, if detected at all (Tabone et al. 2023; Arabhavi et al. 2024). There appear to be a few exceptions: the disk around the VLMS Sz 114 (0.17 $M_\odot$) shows prominent H_2O lines resembling more that of T Tauri stars (Xie et al. 2023).

Chemical modeling shows that such high abundances of hydrocarbon molecules can only be achieved if C/O>1 in the emitting layers (Kanwar et al. 2024b). This means that either carbon is enhanced, or oxygen is depleted, or both. Mechanisms for enhancing carbon include the destruction of hydrocarbon grains at ~500 K (a so-called "soot" line, Li et al. 2021) or the drifting in of hydrocarbon-rich ices that are subsequently eroded (Tabone et al. 2023; Mah et al. 2024). Options for oxygen depletion focus on locking up most of the oxygen in H_2O ice in a dust trap outside the water snowline. Quantification of these models requires better constraints on the actual H_2O emission which is hidden below the forest of hydrocarbon lines in VLMS disks (Arabhavi et al., subm.). If C/O is indeed >1, this may have consequences for the chemical composition of (terrestrial) planets that are forming in the inner disk. As for Earth, the planets may be poor in carbon if most of the solid carbon has been removed from the disk in the form of carbon-rich gas.

4. Conclusions

The first two years of JWST data have revealed a large diversity in the spectra of protoplanetary disks. It is clear that stellar mass, the presence of dust traps in the inner and/or outer disk, radial drift of icy pebbles, and the age and history of the system all play a role in setting inner disk abundances where planets are formed. Much larger samples are needed to reveal systematic trends, if any, and linking them to exoplanet composition statistics. The community can look forward to many more JWST disk spectra in the coming years.

Acknowledgments

The author is grateful to the entire MINDS team and the MIRI consortium for many years of fruitful collaborations. Special thanks to Sierra Grant for providing Figures 1 and 2. This work is supported by A-ERC grant 101019751 MOLDISK.

References

Andrews, S. M. 2020, *ARAA*, 58, 483
Arabhavi, A. M., Kamp, I., Henning, T., et al. 2024, *Science*, 384, 1086
Banzatti, A., Abernathy, K. M., Brittain, S., et al. 2022, *AJ*, 163, 174
Banzatti, A., Pascucci, I., Bosman, A. D., et al. 2020, *ApJ*, 903, 124
Banzatti, A., Pontoppidan, K. M., Carr, J. S., et al. 2023a, *ApJ*, 957, L22
Banzatti, A., Pontoppidan, K. M., Pére Chávez, J., et al. 2023b, *AJ*, 165, 72
Banzatti, A., Salyk, C., Pontoppidan, K. M., et al. 2024, arXiv e-prints, arXiv:2409.16255
Booth, A. S., Temmink, M., van Dishoeck, E. F., et al. 2024, *AJ*, 167, 165
Bosman, A. D., Bergin, E. A., Calahan, J. K., & Duval, S. E. 2022, *ApJl*933, L40
Bosman, A. D., Walsh, C., & van Dishoeck, E. F. 2018, *A&A*, 618, A182
Brown, J. M., Pontoppidan, K. M., van Dishoeck, E. F., et al. 2013, *ApJ*, 770, 94
Bruderer, S., Harsono, D., & van Dishoeck, E. F. 2015, *A&A*, 575, A94
Gasman, D., van Dishoeck, E. F., Grant, S. L., et al. 2023, *A&A*, 679, A117
Grant, S. L., Kurtovic, N. T., van Dishoeck, E. F., et al. 2024, *A&A*, 689, A85
Grant, S. L., van Dishoeck, E. F., Tabone, B., et al. 2023, *ApJ*, 947, L6
Haffert, S. Y., Bohn, A. J., de Boer, J., et al. 2019, *Nature Astronomy*, 3, 749
Henning, T., Kamp, I., Samland, M., et al. 2024, *PASP*, 136, 054302
Kalyaan, A., Pinilla, P., Krijt, S., Mulders, G. D., & Banzatti, A. 2021, *ApJ*, 921, 84
Kamp, I., Henning, T., Arabhavi, A. M., et al. 2023, *Faraday Discussions*, 245, 112
Kanwar, J., Kamp, I., Jang, H., et al. 2024a, *A&A*, 689, A231
Kanwar, J., Kamp, I., Woitke, P., et al. 2024b, *A&A*, 681, A22
Keppler, M., Benisty, M., Müller, A., et al. 2018, *A&A*, 617, A44
Li, J., Bergin, E. A., Blake, G. A., Ciesla, F. J., & Hirschmann, M. M. 2021, *Science Advances*, 7, eabd3632
Mah, J., Bitsch, B., Pascucci, I., & Henning, T. 2023, *A&A*, 677, L7
Mah, J., Savvidou, S., & Bitsch, B. 2024, *A&A*, 686, L17
Meijerink, R., Pontoppidan, K. M., Blake, G. A., Poelman, D. R., & Dullemond, C. P. 2009, *ApJ*, 704, 1471
Najita, J., Carr, J. S., & Mathieu, R. D. 2003, *ApJ*, 589, 931
Öberg, K. I. & Bergin, E. A. 2021, *Phys. Rep.*, 893, 1

Öberg, K. I., Boogert, A. C. A., Pontoppidan, K. M., et al. 2011, *ApJ*, 740, 109
Öberg, K. I., Facchini, S., & Anderson, D. E. 2023, *ARAA*, 61, 287
Pascucci, I., Herczeg, G., Carr, J. S., & Bruderer, S. 2013, *ApJ*, 779, 178
Perotti, G., Christiaens, V., Henning, T., et al. 2023, *Nature*, 620, 516
Pontoppidan, K. M., Salyk, C., Banzatti, A., et al. 2024, *ApJ*, 963, 158
Pontoppidan, K. M., Salyk, C., Bergin, E. A., et al. 2014, Protostars & Planets VI, ed. Beuther, H., Klessen, R., Dullemond, K., Henning, Th. (Tucson: Univ. Arizona Press), 363–385
Romero-Mirza, C. E., Banzatti, A., Öberg, K. I., et al. 2024, *ApJ*, 975, 78
Schwarz, K. R., Henning, T., Christiaens, V., et al. 2024, *ApJ*, 962, 8
Sellek, A. D., Vlasblom, M., & van Dishoeck, E. F. 2024, arXiv e-prints, arXiv:2412.01895
Tabone, B., Bettoni, G., van Dishoeck, E. F., et al. 2023, *Nature Astronomy*, 7, 805
Tabone, B., van Dishoeck, E. F., & Black, J. H. 2024, *A&A*, 691, A11
Temmink, M., van Dishoeck, E. F., Gasman, D., et al. 2024a, *A&A*, 689, A330
Temmink, M., van Dishoeck, E. F., Grant, S. L., et al. 2024b, *A&A*, 686, A117
Trapman, L., Facchini, S., Hogerheijde, M. R., van Dishoeck, E. F., & Bruderer, S. 2019, *A&A*, 629, A79
van Dishoeck, E. F. 2004, *ARAA*, 42, 119
van Dishoeck, E. F., Grant, S., Tabone, B., et al. 2023, *Faraday Discussions*, 245, 52
van Gelder, M. L., Francis, L., van Dishoeck, E. F., et al. 2024, arXiv e-prints, arXiv:2410.01636
Vlasblom, M., van Dishoeck, E. F., Tabone, B., & Bruderer, S. 2024, *A&A*, 682, A91
Waelkens, C., Waters, L. B. F. M., de Graauw, M. S., et al. 1996, *A&A*, 315, L245
Walsh, C., Nomura, H., & van Dishoeck, E. 2015, *A&A*, 582, A88
Woitke, P., Min, M., Thi, W. F., et al. 2018, *A&A*, 618, A57
Wright, G. S., Rieke, G. H., Glasse, A., et al. 2023, *PASP*, 135, 048003
Xie, C., Pascucci, I., Long, F., et al. 2023, *ApJ*, 959, L25
Zhang, K., Bosman, A. D., & Bergin, E. A. 2020, *ApJ*, 891, L16

Planetary Science and Exoplanets in the Era of JWST
Proceedings IAU Symposium No. 393, 2026
Z. Benkhaldoun, Y. Moulane & A. Szentgyorgyi, eds.
doi:10.1017/S1743921324001868

Implementing an AstroLab Project in Africa

Youssef Moulane and Abdelouahed El Fatimy

Institute of Applied Physics, Mohammed VI Polytechnic University, BenGuerir, 43150, Morocco
email: youssef.moulane@um6p.ma

Abstract. We propose the implementation of an AstroLab project in Africa, dedicated to Astronomy and Planetary Science Education. Our goal is to contribute significantly to the development of digital education in astronomy across universities and schools on the African continent. The AstroLab project seeks to inspire and engage students from diverse cultural backgrounds by providing them with a unique opportunity to conduct astronomical research and education through remotely accessed telescopes. This initiative is particularly crucial in regions where the lack of astronomy infrastructure impedes students from hands-on experience in this field. In addition, we will implement this program in three distinct languages to ensure accessibility and inclusivity for the diverse communities of the continent.

Keywords. AstroLab, Africa, Astronomy, etc.

1. Introduction

The development of astronomy in Africa has seen significant progress in recent years, characterized by increased investments in infrastructure, educational programs, and research initiatives. Historically, African nations faced numerous challenges in building a robust astronomical community, such as limited resources, infrastructure, and access to advanced technology. However, the continent is now experiencing a transformative change as governments, academic institutions, and international collaborations work to elevate the status of astronomy across the region (Povic et al. 2018). In this paper, we propose to implement the AstroLab project in Africa to contribute to the development of astronomy in the continent. This project will tackle the issue of limited access to resources for astronomy education, particularly in regions lacking infrastructure for astronomy learning. Through a fruitful collaboration with the Las Cumbres Observatory (Brown et al. 2013), we have secured access to their 40-cm network of optical telescopes. This partnership provides invaluable resources, allowing students to actively engage in practical astronomical research and education. Las Cumbres Observatory (LCO) operates a global network of 25 telescopes across seven sites, renowned for its excellence in both science and Education and Outreach. Using the LCO flagship education program, *Global Sky Partners*†, AstroLab aims to promote astronomy education by granting remote telescope access, ensuring students from all backgrounds participate in hands-on learning experiences.

In addition, the project addresses the need for innovative educational approaches in Africa, particularly within astronomy. By integrating online platforms such as Zoom and Google Meet, alongside real astronomy data, into the curriculum, AstroLab not only enriches the educational quality, but also cultivates critical thinking, problem-solving, and scientific inquiry skills among students. In addition, AstroLab champions inclusivity

† https://lco.global/education/partners/

Figure 1. Overview of AstroLab strategy.

by offering the program in multiple languages that accommodate Morocco's diverse linguistic and cultural landscape. This commitment ensures full participation of students in different regions and communities, regardless of their linguistic background, fostering an equitable learning environment.

2. Overview

Figure 1 provides an overview of the AstroLab strategy, which is designed to improve astronomy education and research through four key components. The first element, Observations, enables students to gain hands-on experience by using remote telescopes for real astronomical observations and data collection. Training is a core focus, with sessions and workshops aimed at equipping students with essential skills in observational techniques, data analysis, and software usage. The program also offers comprehensive educational resources, such as online courses and tutorials, to support participants in this project. In addition, AstroLab encourages students to contribute to science by publishing research papers in regular journals addressing different projects, such as exoplanet detection and variable star, allowing them to produce valuable data and findings for the scientific community. This strategic approach aims to provide an inclusive and enriching educational experience, empowering students to actively participate in and contribute to the field of astronomy.

2.1. *Projects*

The collaboration with LCO holds immense promise for the use of its unique capabilities within the AstroLab project. Specifically, the project intends to capitalize on LCO expertise and resources in several educational and research projects:

• **Detection of exoplanet using Transit method**: Using LCO telescopes for observing exoplanets in transit will allow students to engage in the detection and analysis of exoplanetary systems. This hands-on experience enables learners to understand the transit methods used in discovering exoplanets and studying their characteristics. The transit method is amongst the most widely employed techniques used to detect and characterize exoplanets orbiting distant stars. Using the subtle dimming of the brightness of a host star as an exoplanet passes in front of it, astronomers can obtain information about the exoplanet such as its size and orbital characteristics. The aim of this computational laboratory is to familiarize students with the basics of the transit method by making use of data from the Kepler Telescope to redetect exoplanets that have already been discovered (Yang et al. 2023).

• **Detection of new Asteroids and Comets**: We will be using LCO's capabilities to conduct systematic searches for asteroids and comets within the solar system (Sarath et al. 2021). This involves engaging students in tracking and characterizing these celestial bodies, contributing to the understanding of their orbits, compositions, and potential impacts. Figure 2 (right panel) shows an example of the reconstructed 3D model of 337 Devosa (top) in projection from five angles, together with the light curve data, and the best-fit light cuvre (Vernazza et al. 2009).

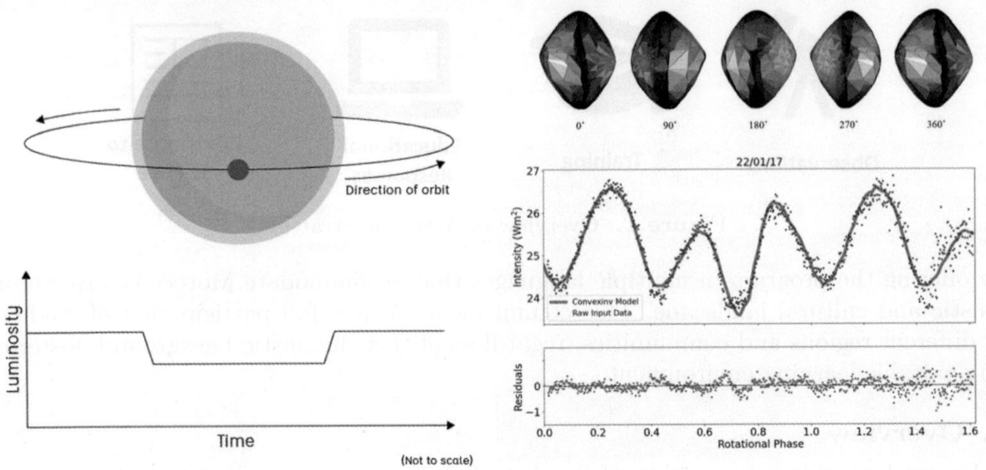

Figure 2. Examples of exoplanet transit (left) and asteroids light-curves (right).

• **Variable Star**: Using LCO 40-cm telescopes to monitor and study variable stars. Students can actively participate in observing these stars, analyzing their light curves, and understanding the underlying physical processes causing their variability (Parrott et al. 2020).

• **Transients and Supernovae Monitoring**: Engage in real-time monitoring of transient events and supernovae using LCO telescopes. This provides an opportunity for students to witness and study these astronomical phenomena, fostering a deeper understanding of stellar evolution and dynamics (Gomez et al. 2021).

• **Long-Term Observational Campaigns**: Initiating long-term observational campaigns for specific objects or phenomena, enabling students to gather data over extended periods, contributing to comprehensive studies and scientific inquiries (Lister et al. 2024).

2.2. *Software*

Using free software such as AstroImageJ (Collins et al. 2017), Astrometrica, and Tycho in conjunction with LCO telescopes can greatly enhance the educational value of the AstroLab projects. These software tools are vital for processing and analyzing the data collected from telescopic observations.

• **AstroImageJ**†: This software is particularly useful for photometric analysis, which is essential in detecting exoplanets via the transit method. Students can use AstroImageJ to analyze the light curves obtained from LCO telescopes, observing the subtle dimming in a star's brightness as an exoplanet transit. By plotting the light curves, they can determine the period, size, and orbital characteristics of the exoplanet. They can also compare their findings with known exoplanet data from sources such as the Kepler Telescope, reinforcing their understanding of exoplanet detection. We can also use AstroImageJ to monitor transient events and to process images and detect changes in brightness over short and large timescales (Collins et al. 2017).

• **Astrometrica**‡: This software is designed for astrometric analysis, which is crucial for detecting and tracking comets and asteroids (Sarath et al. 2021). It allows students to measure the precise positions of these objects against the background stars in sequential images taken by the LCO telescopes. Students can use Astrometrica to search for moving objects in the images, confirm their discoveries against databases such as the Minor

† `https://www.astro.louisville.edu/software/astroimagej/`
‡ `http://www.astrometrica.at/`

Planet Center, and calculate orbital parameters. This process gives them hands-on experience in contributing to the identification and study of new solar system bodies. It can also be used for precise astrometry for different targets (stars, clusters, etc.).

• **Tycho Tracker**†: Tycho is an advanced tool for analyzing variable stars, capable of handling large datasets and producing detailed light curves. Students can use this software to monitor the brightness variations of variable stars over time and identify patterns that indicate different types of stellar variability. Students can analyze the physical processes driving these variations, such as pulsations, eclipses, or rotational modulation. They can also compare their observational data with established catalogs to classify the observed variables. Tycho can then be used to perform more detailed light curve analysis and modeling of these events (Parrott et al. 2020).

3. Impacts and Output Expectations

The AstroLab project is poised to have a transformative impact on African students, particularly those in underserved regions with limited access to astronomy education resources. The project will provide students with access to a network of 40-cm optical telescopes, a unique opportunity to learn how to use remote telescopes. This exposure to practical astronomical research not only enhances their understanding of the universe but also inspires a passion for science and exploration. The expected output of the AstroLab project includes the development of robust scientific and research skills among students. Students will gain first-hand experience in data collection, analysis, and interpretation, skills that are vital in many scientific fields. This experience will likely lead to a deeper interest in Science, Technology, Engineering, and Mathematics (STEM) careers, contributing to the growth of a new generation of African scientists and researchers.

4. Conclusion

The AstroLab project has the potential to create a lasting legacy in Africa by bridging educational gaps, promoting scientific literacy, and inspiring a new generation of African astronomers and scientists. Through this initiative, students will not only gain valuable knowledge and skills, but also the confidence and motivation to pursue their passions and contribute meaningfully to the global scientific community.

References

Brown, T. M., Baliber, N., Bianco, F. B., Bowman, M., Burleson, B., Conway, P., Crellin, M., et al. 2013, *PASP*, 125, 1031. doi:10.1086/673168

K. A. Collins, J. F. Kielkopf, K. G. Stassun, and F. V. Hessman, "ASTROIMAGEJ: Image processing and photometric extraction for ultra-precise astronomical light curves," *The Astronomical Journal*, vol. 153, no. 2, p. 77, Jan. 2017.

Gomez, S. (2021). *Exotic Transients and How to Find Them* (Doctoral dissertation, Harvard University, Massachusetts).

Lister, T., Constantinescu, C., Ryan, W., Ryan, E., Gomez, E., Phillips, L., Rożek, A., Usher, H., Murphy, B. P., Chatelain, J., & Greenstreet, S. (2024). Long-term Monitoring of Didymos with the LCOGT Network and MRO after the DART Impact. *The Planetary Science Journal*, 5(5), 127.

Parrott, D. (2020). Tycho Tracker: A New Tool to Facilitate the Discovery and Recovery of Asteroids Using Synthetic Tracking and Modern GPU Hardware (Abstract). *Journal of the American Association of Variable Star Observers*, 48(2), 262.

Pović, M., Backes, M., Baki, P., et al.(2018). Development in astronomy and space science in Africa. *Nature Astronomy*, 2, 507–510.

† `https://www.tycho-tracker.com/`

Sarath, S, Santhosh, A., & Bhatkar, R. R. (2021). Moving Object Detection in Space using Astrometrica Software. *2021 12th International Conference on Computing Communication and Networking Technologies (ICCCNT)*, 1-6. IEEE.

Vernazza, P., Brunetto, R., Binzel, R. P., Perron, C., Fulvio, D., Strazzulla, G., and Fulchignoni, M. 2009, *Icarus*, 202, 477. doi:10.1016/j.icarus.2009.03.016

Yang, G. (2023). Detection of Exoplanets based on the Transit Method. *Highlights in Science, Engineering and Technology*, 31, 196–203.

Planetary Science and Exoplanets in the Era of JWST
Proceedings IAU Symposium No. 393, 2026
Z. Benkhaldoun, Y. Moulane & A. Szentgyorgyi, eds.
doi:10.1017/S1743921325000328

Exoplanets in the Classroom: A Bilingual K12 Educational Suite for Exploring Exoplanet Science

Marie-Eve Naud[1], **Julie Bolduc-Duval**[2], **Heidi White**[1], **Frédérique Baron**[1] **and Nathalie Nguyen-Quoc Ouellette**[1]

[1]Trottier Institute for Research on Exoplanets and Department of Physics, Université de Montréal, 1375 Ave. Thérèse-Lavoie-Roux, Montréal, QC H2V 0B3, Canada
email: marie-eve.naud@umontreal.ca

[2]Discover the Universe
email: julie@discovertheuniverse.ca

Abstract. A groundbreaking collaboration in Canada has united astrophysicists, science educators, and teachers to create *Exoplanets in the Classroom* – a dynamic suite of K-12 resources designed to inspire the next generation of scientists. Featuring hands-on activities, practical slide decks, engaging videos, and profiles of trailblazing Canadian astronomers, this comprehensive collection of resources is freely accessible online in both French and English. Since 2021, the Trottier Institute for Research on Exoplanets (IREx) at the Université de Montréal, in partnership with Discover the Universe and other educational innovators, has crafted these resources with input from over one hundred Canadian teachers. This rigorous, iterative process ensures seamless integration into a wide range of subjects, from science to the arts, all while meeting Canadian K-12 curriculum standards.

These innovative resources provide educators with the tools to captivate students with the wonders of exoplanet research and the stories of diverse, local scientists at the forefront of discovery. Already tested and embraced by students and teachers from diverse backgrounds, these materials are now poised to inspire a global audience, offering astrophysicists and educators a powerful way to ignite curiosity and engage learners in classrooms and beyond.

Keywords. Exoplanets, Outreach, K12, Astronomy education

1. Introduction

Exoplanets have captivated the imagination of both the scientific community and the general public over the past few decades. Their study not only leads to astonishing discoveries but also opens up enriching pedagogical opportunities. Integrating the study of exoplanets into K-12 curricula has the potential to ignite curiosity in students and teachers alike, deepening their appreciation of science and its vital role in unravelling mysteries in our Universe.

Although this rapidly evolving field remains largely absent from formal elementary and secondary education, it offers a thrilling pathway for exploring key curriculum topics. In particular, students can compare our Solar System with distant planetary systems, contemplate Earth's unique position, and reflect on the possibility of life beyond Earth. These explorations not only align with curriculum goals but also foster critical thinking and encourage a scientific mindset.

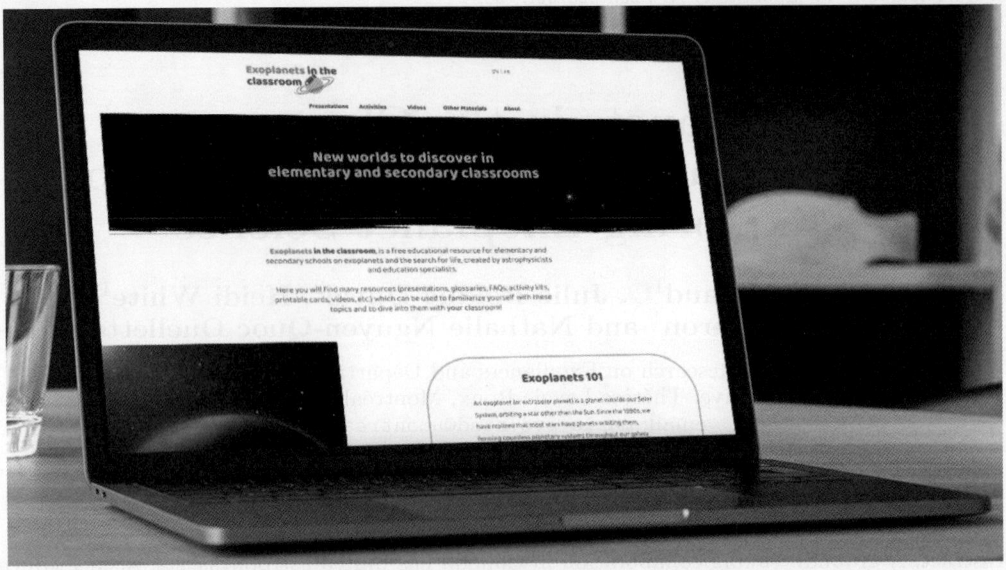

Figure 1. All the resources are available in French and English, free of charge, on the website, www.exoplanetsintheclassroom.ca.

Despite the vibrant research community in Canada dedicated to exoplanets, particularly within the Trottier Institute for Research on Exoplanets (IREx)†, a significant gap has persisted in the availability of French-language educational materials and training for educators. This lack of access has posed challenges for francophone educators eager to bring the wonders of exoplanet science into their classrooms.

Exoplanets in the Classroom, a portal of resources available online (see figure 1) was developed to bridge this gap, offering a wide variety of bilingual astronomy resources that inspire students in both French and English, the two official languages in Canada. By connecting learners with local scientists, these materials not only foster a deeper interest in science but also showcase potential career paths in astronomy, empowering the next generation of Canadian scientific minds.

2. About the Project

2.1. *Motivation*

The Trottier Institute for Research on Exoplanets (IREx), based at Université de Montréal in the French-speaking province of Quebec, Canada, has emerged as a leading force in exoplanet science since its founding in 2014. With a team of 70 researchers, IREx plays a pivotal role in major international projects focused on the discovery and characterization of exoplanets. These efforts include groundbreaking work with the James Webb Space Telescope and the ground-based SPIRou and NIRPS spectrographs, providing the institute with privileged access to cutting-edge data that has led to numerous breakthrough discoveries and transformative insights into these distant worlds.

Alongside its research achievements, IREx is home to a vibrant science education and outreach program, led by astrophysicists who specialise in science communication. Researchers at the institute are deeply committed to engaging the public, actively participating in a wide range of outreach efforts such as innovative science fair exhibits, dynamic classroom and library workshops, impactful career days, captivating video content, and engaging social media campaigns.

† https://exoplanetes.umontreal.ca/en

In 2019, Discover the Universe†, a Canadian program that promotes astronomy education and trains educators, noticed an increased interest in educators for teaching exoplanets and the search for life. Realising the great educational potential of including these topics in the classroom, they approached IREx to collaborate on creating the resources. IREx embraced the opportunity, recognizing the potential to amplify its educational impact.

A core team was formed, consisting of IREx astrophysicists with extensive expertise in education and outreach, along with the director of Discover the Universe. They secured funding through the NovaScience Program of Quebec's ministère de l'Économie, de l'Innovation, et de l'Énergie. Their mandate was specifically to train elementary and secondary school educators, empowering them to introduce exoplanet science and enrich students' learning experiences with the latest astronomical discoveries.

2.2. *Key Aspects of the Resources*

Drawing on years of experience working with teachers and classrooms, the core team developed the Exoplanets in the Classroom initiative based on several foundational principles. Together, these principles enhance the educational impact of the resources and their relevance in today's learning environments.

Central to the development process was the commitment to actively **integrate teacher feedback** at every stage. That team collaborated with over a hundred teachers to refine the materials, ensuring they were responsive to real classroom needs and seamlessly aligned with the curriculum. This process first ensured that the resources would **support educators of all backgrounds and confidence levels**. Whether teachers are seasoned science experts or just beginning their journey into astronomy, the resources are tailored to meet varying levels of knowledge and motivation. This guarantees that every educator can find materials that align with their goals and ignite their students' curiosity about the fascinating world of exoplanets. Through this consulting process, it also became clear that establishing **strong connections to the school curriculum** was essential for providing teachers with innovative strategies to engage students in the scientific topics they are required to teach.

Educators participated in brainstorming sessions, tested the resources, and provided feedback, mostly online, due to the pandemic. The virtual nature of the process ended up being a huge asset, enabling the core team to connect with a broader range of educators, including those in rural or remote areas. Additionally, the pandemic's challenges prompted focusing heavily on creating **digital resources** that are fully **accessible and compatible with remote and asynchronous learning**.

Another key aspect is the **multidisciplinary approach** adopted, linking exoplanet studies to various branches of science — such as geology, biology, and environmental science — but also to subjects like French, English, mathematics, and the arts. This not only allows teachers to use the resources in more varied educational contexts but also encourages learners to appreciate the interconnectedness of knowledge and the relevance of science in their daily lives. It can also be a strong motivator in some cases to tackle challenging concepts in core disciplines through the captivating lens of exoplanet exploration.

A key feature that distinguishes these resources is the connection to local scientists, providing teachers with confidence in accessing **up-to-date and validated content**. It also allows students to **get to know local experts** and get insights into breakthroughs happening in Quebec. The collaborative nature of scientific discovery is highlighted,

† `https://www.discovertheuniverse.ca/`

showcasing how **researchers with diverse qualities and strengths** can unite to push the boundaries of knowledge.

2.3. *Development Process*

The Exoplanets in the Classroom project has undergone significant development since its inception, marked by collaborative efforts from a diverse group of stakeholders. The journey began in Febuary 2021, at the **reception of the grant** from the ministère de l'Économie, de l'Innovation et de l'Énergie of the Quebec provincial government.

In Spring 2021, following several meetings of the core team, the **project was publicly announced** and a **community for teachers was established** to foster discussion and gather insights from educational professionals. This Facebook group, however, turned out to be less effective than anticipated. It proved challenging to generate sufficient content to stay visible in teachers' feeds, resulting in inconsistent engagement. However, existing teachers' groups on Facebook provided a very efficient way to recruit teachers for virtual brainstorming sessions and workshops. The **first series of workshops** aimed at understanding the needs of the community brought together approximately thirty school staff members, including teachers and pedagogical advisors of elementary and secondary schools, allowing for the collection of valuable feedback that would inform future work.

These collaborative efforts led to the **drafting of educational activities by the core team** for both elementary and secondary schools during Summer 2021. A **second series of workshops** was organised in Fall 2021 and included a dozen educators, half from elementary and half from secondary. The feedback from the elementary teachers was extremely positive, and by Winter 2022, the first activity from the elementary school kit, **the Challenge Explore the sky, was successfully deployed**, providing additional testing across multiple classrooms in Quebec. The secondary school activities needed much more modifications, to meet with the various constraints teachers face (see section 2.4). A **second iteration was thus presented to a focus group of 8 teachers** in the Summer 2022.

A major milestone was achieved in Summer-Fall 2022 with the **launch of the French-language website, Exoplanètes à l'école**, making all resources easily accessible to educators. During this time, the video series **"Exoplanets and Us"** was developed and the **ExoBites series** – where scientists of IREx break down complex scientific concepts into engaging, student-friendly videos – was expanded. In Fall 2022, the core team further extended its outreach by leading a **"Community of Practice"** on exoplanets with their partner, École en réseau, consisting of regular meetings with a group of motivated teachers who were interested in learning how to use the resources and testing. **Interactive workshops and booths** at various teachers' conferences allowed to inform teachers, train some of them and get more feedback.

The **English translations** of the resources began in Summer 2023 and were completed by Summer 2024, with the **launch of the English-language website, Exoplanets in the classroom**, expanding the reach of the initiative to a broader audience in Canada. Throughout this period, the project gained visibility with **presentations at key events in the scientific community**, such as the CASCA 2023 meeting, the AstroEDU conference in 2023, and the International Astronomical Union's General Assembly in 2024.

Looking ahead, the collaboration will continue with the aim of not only educating but also empowering both educators and scientists to become ambassadors of exoplanetary science within their communities.

2.4. *Lessons Learned Working with Canadian Teachers*

The adopted approach emphasized intensive collaboration with a diverse group of school teachers from Quebec, Canada's French-speaking province. Two series of virtual workshops were organized, a Community of Practice was facilitated, various classroom engagements were conducted, and the core team participated in multiple teachers' conferences. The insights gained from these interactions have been instrumental in shaping the Exoplanets in the Classroom resources and will continue to inform future initiatives.

A recurring theme among ***all teachers*** was their **appreciation for resources developed by credible local experts**. They especially valued the inclusion of "real" scientists in the materials through videos and biographies, as this brings the subject to life and provides role models for students aspiring to pursue careers in science. Feedback also highlighted their strong appreciation for **online, copyright-free activities that are easy to implement**, enabling them to enhance their lessons seamlessly.

Teachers expressed varying degrees of attachment to their school curricula, with some adhering strictly to prescribed content – that does not explicitly include exoplanets. However, many others were open to integrating innovative subjects, especially when they could connect these topics to established areas like the Earth and the Solar System. To support this, **explicit connections to the curriculum** were made, not only with the science curricula but also with other subjects.

During the second series of workshops, many proposed activities resonated with ***elementary school teachers***, who deemed them relevant and engaging. They expressed **enthusiasm for the multidisciplinary nature** of the activities, particularly those that integrated French and mathematics. Activities designed to **span multiple class periods** were seen as feasible, reflecting their commitment to thorough exploration when they could address various concepts through them. Some elementary school educators mentioned a lack of confidence in teaching scientific content, and they **appreciated the comprehensive instructions and additional materials** provided.

Engaging with ***secondary school teachers*** revealed further learning opportunities. While they expressed interest in incorporating exoplanets into their curriculum, **time constraints emerged as a significant barrier**. Many secondary teachers have **limited opportunities for enrichment activities**, often confining themselves to those that can be completed within a single 60 to 75-minute session. Although they were theoretically enthusiastic about multidisciplinary activities, logistical challenges – such as scheduling conflicts and varying subject expertise — hampered their adoption. Secondary school teachers voiced a strong desire for **adaptable activities** that could easily align with their instructional styles and classroom dynamics. To address this, significant adjustments were made to the secondary school activities. Notably, they were shortened, and the multidisciplinary elements were placed in optional sections rather than in the core of the activities.

Overall, the consultation process was met with enthusiasm, and **many teachers expressed gratitude for the opportunity to provide input**, reinforcing the collaborative spirit that underpins this initiative.

2.5. *Reception of the Material*

The reception of the Exoplanets in the Classroom materials from both the educational and scientific communities has been overwhelmingly positive, reflecting their **effectiveness and relevance** across various educational settings. Educators have praised the resources for their **accessibility and ability to engage students**.

For instance, Sandy, an elementary school teacher, remarked on the turnkey nature of the materials, stating, "I found it so informative and interesting." Similarly, Marie-Eve,

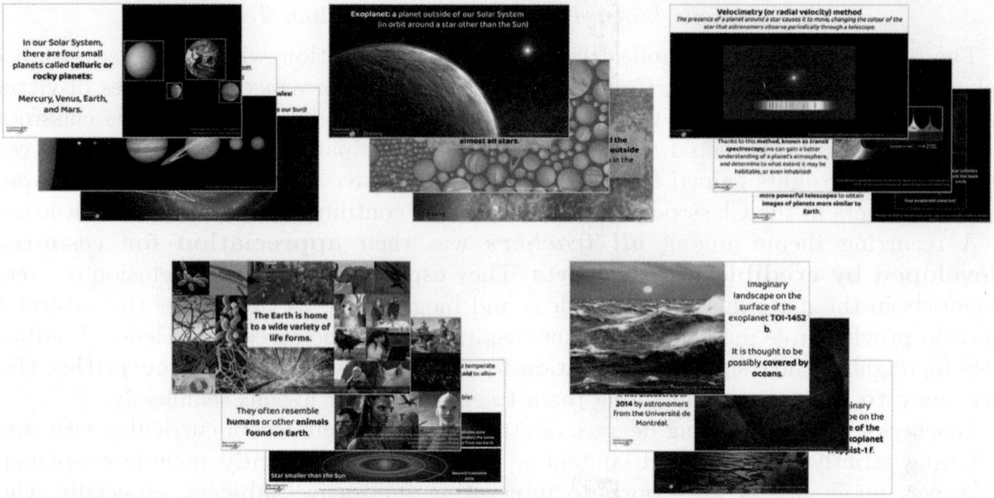

Figure 2. The slide decks are Google Slides presentations featuring 20-25 slides, designed for teachers and students to explore various topics, whether at home or in the classroom. The colorful and abundant illustrations are all carefully identified and their source is given.

a secondary school teacher, highlighted the presentations as "very well done, clear, and complete."

Beyond the classroom, the resources have received high accolades from astrophysicists both within IREx, among Discover the Universe team members and globally. Members of IREx, including educators and graduate students, have actively embraced these materials, integrating them into diverse settings such as career days and science fairs.

For example, Alexandrine, an IREx PhD student who utilized the resources while teaching at the Pan-African School for Emerging Astronomers (PASEA) in Monastir, Tunisia in 2024, shared her enthusiasm: "I really appreciate that the resources are accompanied by detailed documentation and can easily be adapted to any audience."

Furthermore, because these resources are fully bilingual, they are exceptionally well-positioned for use by a global audience; the participants at PASEA 2024 were very diverse, arriving from many countries in Africa with English and/or French as common languages among them.

3. Resources

This section introduces the main resources available through the Exoplanets in the Classroom website (see figure 1), designed to enrich the educational experience of both students and teachers. The resources cover a wide range of topics related to exoplanets, astronomy, and their connections to various disciplines.

3.1. *Slide Decks, Glossary and FAQs*

The **five Slide Decks** (Google Slides presentations) (see figure 2) offer a visual and informative way to explain fundamental concepts about the planets of our Solar System, exoplanets, and the search for extraterrestrial life. Educators can utilize these decks as-is or modify them to suit better their students' needs, making them versatile teaching tools. Students can also use them for independent learning. Complementing these are the **Glossary** and **Frequently Asked Questions** sections, which provide definitions of key scientific terms and answers to over 25 common queries answered by astrophysicists, catering to the curiosity of learners of all ages.

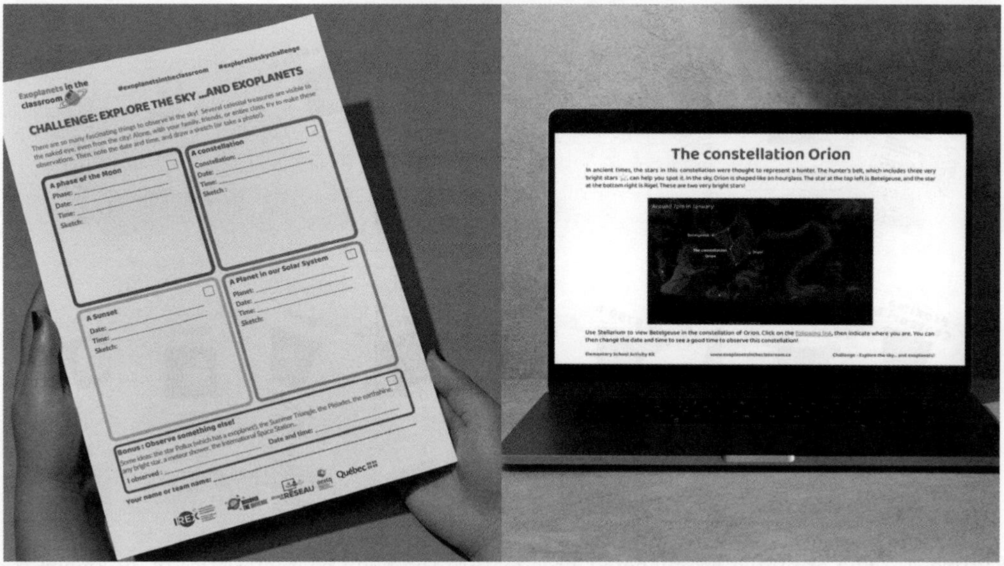

Figure 3. The first Elementary School Activity is a Challenge that invites students to observe various celestial objects and fill out their own observer's log (left). This activity is supported by tip sheets for each potential observation (right) and encourages the use of the Stellarium web tool.

3.2. *Elementary School Activity Kit*

The **Elementary School Activity Kit** contains four interactive activities specifically designed for grades 3-6 (9-12 years old) but easily adaptable for younger learners. The activities encourage hands-on exploration of astronomical concepts.

(a) **The Challenge - Explore the Sky... and Exoplanets!** activity (see figure 3) encourages students to observe celestial objects and astronomical phenomena, deepening their connection to the night sky while teaching them the fundamentals of maintaining an observer's log.
(b) In **Extraterrestrial Planetary Systems**, learners create scale models of our Solar System and other planetary systems, fostering comprehension of relative sizes and distances.
(c) The **My Planetary System Adventure** allows students to invent a planetary system and compose a story taking place in it, helping them understand the unique nature of our own Solar System and the diversity of planetary systems.
(d) The **Interstellar Travel Agency** activity engages them in researching a given exoplanet, learning their main characteristics and how they compare to Earth and Solar System planets, helped by the Exoplanets Cards (see figure 4). They can then present these exoplanets creatively to their peers.

3.3. *Secondary School Activity Kit*

The **Secondary School Activity Kit** offers four short activities designed for students aged 12 to 17, suitable for grades 7-12.

(a) **Eyes on Exoplanets** encourages the discovery of various exoplanets and introduces students to a NASA's web-based tool of the same name.

Figure 4. The printable Exoplanets Cards feature the main characteristics of 30 known exoplanets and the Solar System cards the 8 planets of our Solar System. They are used in one elementary and one secondary school activity, but can also be used on their own, for example, to distribute in booths at science fairs.

(b) **The Exoplanet Zoo** encourages them to classify and compare known exoplanets, using the Exoplanet Cards (see figure 4), learning more about their characteristics and the subjective nature of classifying.
(c) **In an Astronomer's Shoes**, learners select an exoplanet to investigate using the James Webb Space Telescope, developing their understanding of the process astronomers use to learn more about the questions they want to solve.
(d) **Where are the Aliens?** prompts discussions about the factors influencing the possibility of extraterrestrial life, encouraging critical thinking and argumentative discussion (see figure 5).

3.4. *Videos Series and supporting material*

The **ExoBites series**, short 4 to 7-minute videos featuring IREx researchers (see figure 6), provide an engaging overview of current exoplanet research and scientists' insights, supported by a **Viewing Guide** with discussion prompts, and quizzes. The topics are:

(a) "An exoplanet zoo",
(b) "Earth 2.0",
(c) "Alien Life",
(d) "The stars that dance",
(e) "Telescopes of the Future",
(f) "The Modern Astronomer", and
(g) "Cosmic Wink"

The second video series, **"Exoplanets and Us"**, allows students to listen to scientists answering three popular questions:

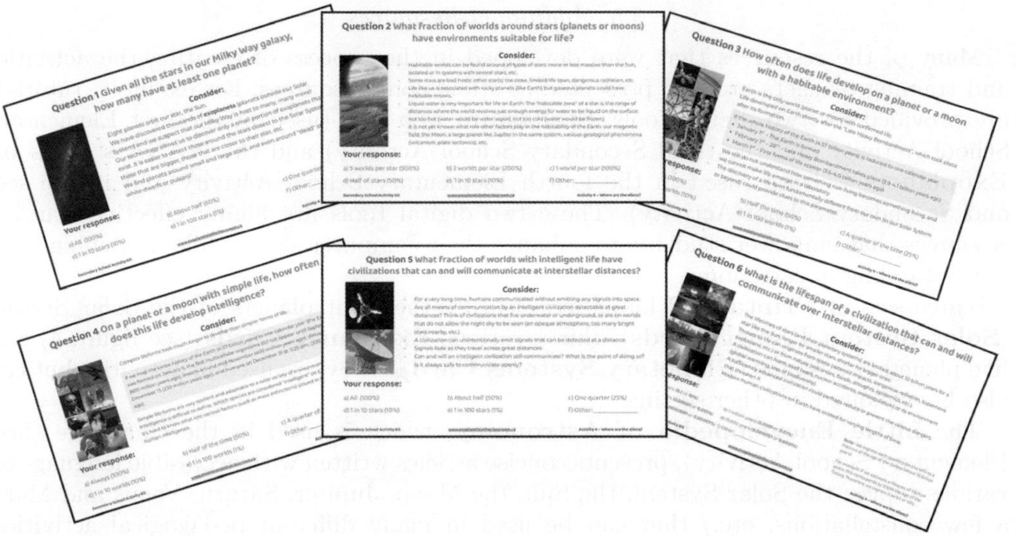

Figure 5. The last Secondary School Activity is a discussion in small teams about what factors need to be considered to evaluate the presence of life outside of Earth. The cards provided help them learn and articulate various relevant discussion points.

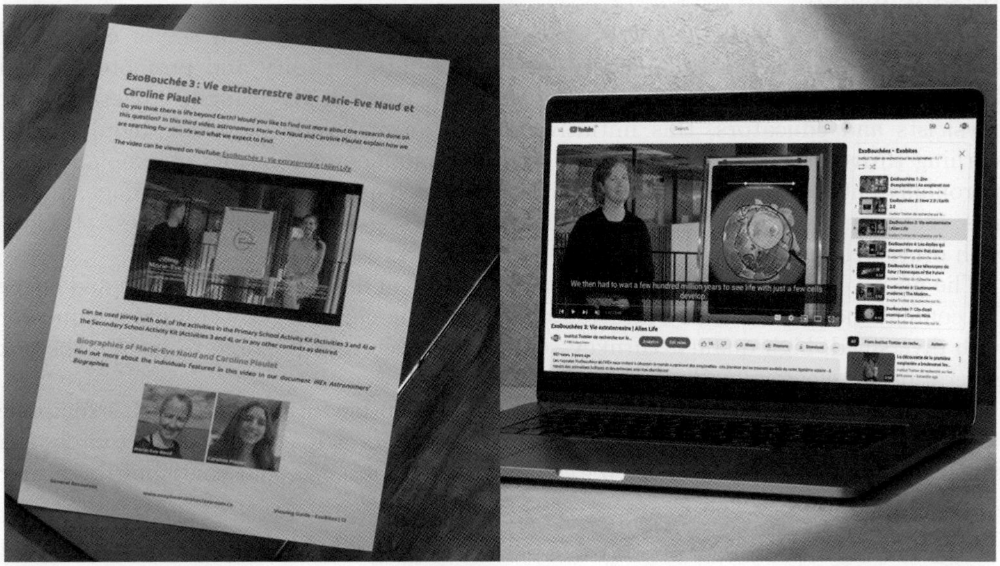

Figure 6. The Exobites series was integrated into Exoplanets in the Classroom and was expanded as part of this project because it is very well suited to the classroom. Teachers can use these videos (right) to introduce their students to different topics and can use the supporting material included in the Viewing Guide (left) to support their teaching.

(a) "What is your favourite exoplanet?",
(b) "Do you think there life elsewhere?", and
(c) "Tell me about the history of exoplanets".

All researchers featured in videos are also presented in the **Astronomer Biographies**. Highlighting the journeys of diverse Canadian astronomers from IREx allows for promoting connections between students and local scientific expertise.

3.5. *Additional Resources*

Many of the resources that were developed in the process of creating the activities and training teachers are also presented as standalone resources. For example, tutorials are provided for two digital tools: **Stellarium** web tool (used in the first Elementary School Activity and the third Secondary School Activity) and the **NASA's Eyes on Exoplanets** website (used in the fourth Elementary School Activity and in the second Secondary School Activity). These two digital tools are highly effective learning resources that empower students to enhance their computer skills while deepening their understanding of astronomy.

Three sets of **Printable Flashcards** feature the eight planets of the Solar System (**Solar System Planet Cards**), thirty various **Exoplanet Cards** (see figure 4) and five planetary systems (**Planetary Systems Card**). They are used in activities but can also be exploited in other settings.

The **Little Encyclopedia of Astronomy**, which is used in the Challenge (first Elementary School Activity), presents concise articles written with accessible language on various topics: the Solar System, the Sun, the Moon, Jupiter, Saturn, Venus and Mars, a few constellations, etc.) that can be used in many different pedagogical activities. Finally, the **Exoplanet Posters** offer visually appealing materials to enhance classroom environments.

4. Conclusion

In summary, Exoplanets in the Classroom is more than just a set of educational materials; it embodies a collaborative effort to ignite curiosity and promote scientific literacy among students in Quebec, Canada, and beyond. By bringing together astrophysicists and educators, this initiative fosters an environment where inquiry-driven learning thrives. The resources not only cover fundamental concepts of exoplanets and astronomy but also connect to various disciplines such as mathematics, language, and the arts, motivating students and encouraging them to see the interconnectedness of science with other fields.

The positive reception from teachers indicates a real demand for accessible and engaging content that can be easily integrated into existing curricula. By equipping educators with diversified, high-quality, turnkey resources, it becomes possible to inspire the next generation of scientists and astronomers.

The success of this initiative lays a solid foundation for future projects. It exemplifies how the collaborating between educators and scientists helps to develop resources that captivate students, inspire them to explore careers in science, and contribute to a more scientifically literate society.

Planetary Science and Exoplanets in the Era of JWST
Proceedings IAU Symposium No. 393, 2026
Z. Benkhaldoun, Y. Moulane & A. Szentgyorgyi, eds.
doi:10.1017/S1743921324001893

Bringing the sky to the living room: Reflections on more than three years of live streaming astronomical events

Claudia Mignone[1], Federica Duras[2], Giulia Mantovani[2], Federico Di Giacomo[3], Gianluigi Filippelli[4], Livia Giacomini[2], Sandro Bardelli[5] and Emanuele Scalise[2]

[1]INAF - Istituto Nazionale di Astrofisica, Viale del Parco Mellini 84 - 00136 Roma, Italy
email: claudia.mignone@inaf.it

[2]INAF - IAPS, Via del Fosso del Cavaliere 100 - 00133 Roma, Italy

[3]INAF - Osservatorio Astronomico di Padova, Vicolo Osservatorio 5 - 35122 Padova, Italy

[4]INAF - Osservatorio Astronomico di Brera, Via Brera 28 - 20121 Milano, Italy

[5]INAF - Osservatorio di Astrofisica e Scienza dello Spazio, Via Piero Gobetti 93/3 - 40129 Bologna, Italy

Abstract. "Il cielo in salotto" (in English, "The sky in your living room") is a format for live streaming astronomical observations created by the Italian National Institute for Astrophysics (INAF). The project started in late 2020 in the midst of the Covid-19 pandemic to engage the general public and students with astronomy and space science remotely, when observatory visits were not possible. The format later evolved, in the "new normal" scenario, as a complementary activity to in-person events, featuring astronomical phenomena - such as planet alignments, eclipses, different Moon phases, comets, sunspots, exoplanet-hosting stars - observed live by the INAF network of telescopes all around the country. These events are enriched by live interviews with INAF researchers and supported by large-scale public engagement campaigns conducted together with space enthusiasts, amateur astronomy associations and other partners. Viewers can ask questions live to the experts and also select some of the targets to observe. We present lessons learnt and best practices from three and a half years running the project, along with some of the technical and logistics, content and communication solutions adopted in the format of potential interest to colleagues organising similar events in other countries. We also discuss the results of a focus group conducted for three years in a row during the summer "SuperMoon" broadcasts, a regular programme taking advantage of the popularity of this event in the news to cover current topics in planetary science.

Keywords. Astronomy outreach, Public engagement, Live streaming, Planets, Moon, Comets

---

1. Introduction

INAF, the Italian National Institute for Astrophysics, is Italy's public research institute dedicated to the study of the Universe. With 16 premises in 12 cities across Italy plus its headquarters in Rome and the National Galileo Telescope in the Canary Islands, it comprises over 1000 researchers investigating topics from the Solar System to stars in the Milky Way, from galaxies to cosmology, from high-energy astrophysics to astronomical instrumentation. As mandated by its statute, the institute implements a wide public engagement programme including outreach, education and informal science learning activities.

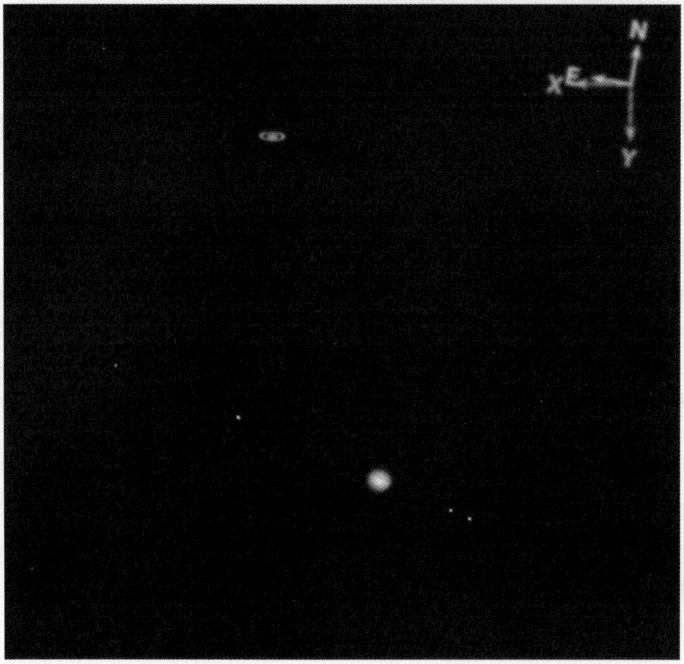

Figure 1. Saturn (top) and Jupiter (bottom, with its major moons) as imaged during the 2020 conjunction by INAF's Rapid Eye Mount (REM) telescope in Chile. Credit: INAF/REM.

"Il cielo in salotto" (Mantovani et al. (2021), Duras et al. (2022)) is a live streaming format for astronomical events run by EduINAF (edu.inaf.it), the institute's online magazine dedicated to education and public outreach. Aimed at the general public as well as teachers and students, it has broadcast 16 programmes (available here) between December 2020 and September 2024, an average of roughly 4 per year, mostly featuring live telescope observations of astronomical events from INAF observatories across the country. The title, meaning "The sky in your living room", was inspired by the comment from a young student, who contacted the magazine newsroom because there are no INAF premises in their region and suggested to create one in their parents' living room.

The project was born in late 2020, when most people in Italy and around the world were sheltering in place due to the Covid-19 pandemic. On 21 December, a rare Jupiter-Saturn conjunction occurred: the two giant planets had not been that close in the sky for centuries, so public interest was huge, boosted by the very unique emotional moment the world was experiencing, as described by Walkowicz (2020). Observatories, however, were closed to the public due to national health regulations. Therefore, INAF astronomy communicators decided to embark on a format that was new to them at the time: live streaming astronomical observations. The initiative, sharing the feed from several telescopes in real time, proved very successful, with over 100,000 views on YouTube (Fig. 1).

After this successful first experiment, a newly formed team (consisting of the paper authors) started working on a dedicated format for astronomical live streaming. Uncertainty about the pandemic evolution, still very high at the time, prompted astronomy communicators to seek new ways to engage with the public remotely while observatories were still closed to visitors (see e.g. Bolduc-Duval, Mann & Baron (2021), Christoph & Hüttemeister (2021), Monfardini Penteado et al. (2021) and Monteiro et al. (2023)). At the same time, the team saw an opportunity to reach new pockets of the public - individuals, families and schools who do not (or cannot) visit our observatories.

Table 1. Viewer statistics of "Il cielo in salotto".

Date	Topic	Live views	Total views	Live obs.
September 2024	SuperMoon	650	5,000	YES
April 2024	Total solar eclipse (in North America)	74,000	104,000	YES
March 2024	The Sun	360	1,300	YES
July 2023	SuperMoon	4,800	9,200	YES
April 2023	Spring alignment, incl. stars, exoplanets and galaxies	200	840	YES
January 2023	A comet among the winter stars	4,600	10,500	YES
December 2022	Mars vs. Moon	300	2,700	YES
October 2022	Partial solar eclipse (also visible from Italy)	650	5,000	YES
July 2022	SuperMoon	5,700	8,200	YES
June 2022	The sky of Margherita Hack	460	2,600	NO
March 2022	Equinox at the museum	300	730	NO
February 2022	Winter stars	1,500	3,200	YES
December 2021	Planetary alignment	3,300	6,900	YES
November 2021	Happy hour with a comet	960	2,700	NO
May 2021	SuperMoon	21,000	28,000	YES
December 2020	Saturn-Jupiter great conjunction	92,000	109,000	YES

An online, video-based format would increase the reach of our activities even in the "new normal", once the pandemic would be over (Impey et al. (2023)).

2. The format

A major question loomed above the team: would the effort be worth it? After all, looking at planets, stars, nebulae and other celestial bodies through (a telescope mediated by) a screen - possibly a small, smartphone screen - and an internet connection - possibly a lousy one - is certainly not comparable to the unique experience of looking through a telescope's eyepiece with one's own eyes (Marques et al. (2020), Marques et al. (2024)). Besides, through the same screen, people have access to majestic astronomical images from world-class ground- and space-based telescopes - most notably including JWST - that are masterfully processed by astronomy visualisation experts (Chung (2015)).

To tackle this question, we adopted an "evidence-based" approach to science communication (Jensen & Gerber (2020)). We tested several options over the past three years and a half, perfecting the format each time to keep it entertaining and appeal to a broad public, while at the same time featuring engaging imagery accompanied by accurate scientific content. In particular, for three years in a row we ran a dedicated focus group during the "SuperMoon" live stream (further details are provided in Sect. 4).

As summarised in Table 1, live and cumulative views of each episode vary depending on several factors, some of which are not under our control. However, some overall trends can be identified and fed into the programming of the next season. The following lessons learnt, based on over three years of "Il cielo in salotto", might be of interest to astronomy communicators around the world who wish to adapt this format in their country.

- **A newsworthy topic**. When the live stream covers current astronomical events, mainstream media tend to promote our programme, boosting the audience (as discussed also by Mignone et al. (2023)). Favourite events by the media are: eclipses, comets and the so-called SuperMoon (i.e. the full Moon at perigee).
- **A host (or two) in a virtual studio**. A regular host helps growing an audience online (Wellbourne & Grant (2016)). Occasionally, we also resort to a two-host conduction to provide extra spark. An example of the look and feel of the virtual studio is provided

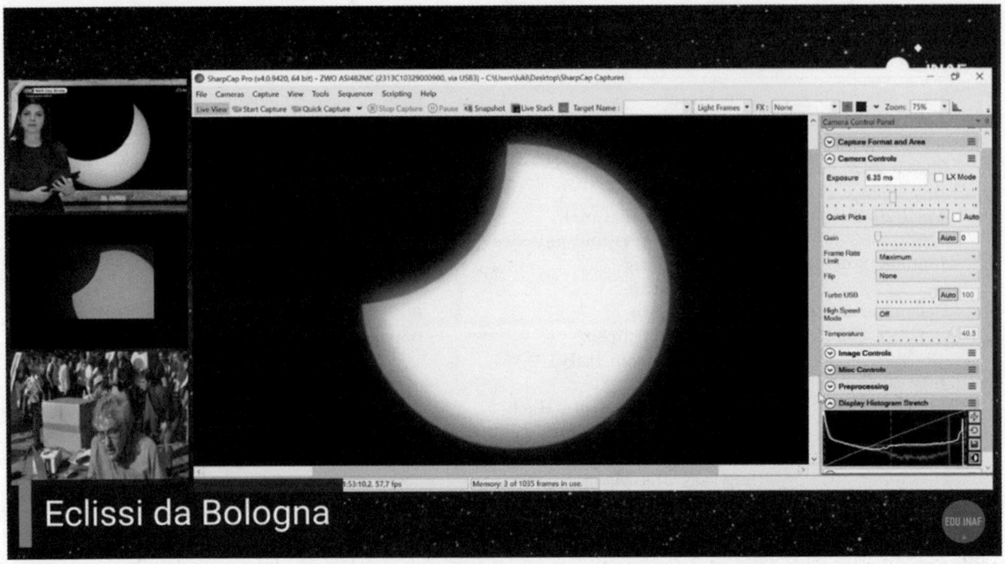

Figure 2. A frame from the live stream about the 2022 partial solar eclipse. Credit: INAF

in Fig. 2. The virtual studio (with lights, a green screen and 3D graphics) is nice but not necessary: a sparkling host works great also with a virtual background on their webcam.

- **Two (or more) observing stations**. Astronomical observations are subject to the weather, which may change even at short notice. Relying only on one observing station at a single geographical locations may put the programme in jeopardy. We usually work with at least two stations, leveraging INAF's widespread presence across the Italian territory.
- **Backup observations from the previous days or weeks**. For the same reason, we also make sure to have back-up observations (images and, if possible, also videos) collected during the previous days or weeks, so that even in the worst weather scenario, we still have something to show and comment on.
- **Engaging astronomers**. Astronomers and astronomy communicators running the observations present what they are showing through the telescope in simple and accessible terms. Depending on the subject, we feature a variety of different telescopes: small, portable telescopes; medium-size telescopes usually employed for education and public outreach; and also large professional telescopes. In most cases, these are not tools used for astronomical research: it is important that astronomers participating in the programme are familiar with smaller telescopes and with the principles of astrophotography. Further details on instrumentation are provided in section 3.
- **A "special guest"**. Besides the astronomers conducting live observations, who are recurring guests of the series, we usually invite a researcher who is an expert on the particular topic covered that night; on two occasions, we also invited best-selling authors who are popular with our audience.
- **Always show what is going on in the sky**. Even when switching different guests on the various windows, it is essential to always keep a window visible with the celestial phenomenon (see e.g. Fig. 3, Fig. 4). Viewers might come and go throughout the programme, in this way so this helps grabbing the attention of newcomers.
- **Maximum length of 1 hour**. The first two episodes lasted up to two hours. However, video consumption habits during the Covid-19 pandemic were different. Today, the attention span of viewers is rather short, so we try to avoid exceeding 1 hour, with the exception of long-duration events like eclipses.

Figure 3. Frames from the live stream dedicated to the SuperMoon in 2022, observed here with a 0.2-m telescope in Cagliari (left) and from the live stream dedicated to comet C/2022 E3 (ZTF), observed with the 0.92-m Schmidt telescope in Asiago (right). Credit: INAF

Figure 4. Frame from a live stream dedicated to the 2022 Mars opposition and other Solar System planets, featuring Jupiter as observed with a 0.2-m telescope in Palermo. Credit: INAF.

• **No talk show without live observations**. On a few occasions, we broadcast a talk-show format without live observations of the sky, around e.g. special anniversaries. It was not as popular as streaming live observations of the sky.

• **No Power Point style slides**. Experts and guests are welcome to show images and videos, but not in a Power Point style presentation. The backstage team shows them individually, so they appeal to a large audience. We also discourage the use of plots.

• **Storyboard scripted in advance**. A storyboard, even a rough one, with planned transitions according to the observability of different phenomena at different locations, helps keeping the programme flow smoothly, coordinating among different parties and sticking to the allotted time.

• **Social media platform**. So far, "Il cielo in salotto" is live streamed on the YouTube and Facebook channels of EduINAF. We are considering Instagram because of its fast growing audience in Italy (We Are Social & Meltwater (2024)).

• **Live Q&A with the experts**. The audience can ask questions to the experts via the YouTube or Facebook chat. Experts reply either on video or in the chat. Outstanding questions become the topic of a dedicated column on the magazine in the future.

• **Live comment moderation**. Especially when thousands of people are connected, the chat can become heated, with off-topic comments interspersed between genuine messages and questions. Moderation is essential to hide inappropriate or offensive comments and keep trolls under control, providing a safe space for viewers to engage via the chat.

• **Live surveys**. Since 2024, we have been experimenting with live surveys, asking the audience where they are based, what they see in the sky, and other questions through the online platform Mentimeter.

• **Pre-recorded videos**. We usually prepare one or two short pre-recorded videos, e.g. to explain the celestial phenomenon with animated simulations using the open source Stellarium software. The video also helps the backstage production, e.g. providing a break to switch between different instruments or cameras.

• **The more the merrier**. The team behind the scenes makes sure all transitions are smooth, identifies relevant questions, solves last-minute problems, moderates and answers questions in the chat.

3. Equipment

A key element to broadcast astronomical observations is a **telescope** equipped with a **digital detector** (a CCD or a digital camera mounted on the eyepiece, with a T-piece adapter). The detector should be connected to a computer with an **astronomy camera capture tool** (an example is the SharpCap software). For special events, such as the Moon rising above the horizon, or planetary alignments, we also feature a **digital camera with a wide-angle lens** (on a tripod) to provide a view of the sky embedded in the landscape, similar to what viewers might be experiencing from their own window. A sensitive camera is a versatile investment for astronomy public engagement: it can also be used for public outreach events in person, mounted on a modest-sized telescope, to live stream real-time images on a screen that several people can view, e.g. while waiting in line to look through a telescope (Burcher & McDonald (2021), Carlin et al. (2021)).

We feature a wide range of telescopes (see examples in Fig. 3 and 4), from portable ones (12,5-20 cm in diameter) to larger professional telescopes dedicated to education, public outreach and research (e.g. 60 cm in diameter). On particular occasions, we also take advantage of the network of professional telescopes managed by INAF, such as the robotic REM 0.6-m diameter telescope at La Silla, Chile; the 0.92-m Schmidt telescope in Asiago; the 1.52-m Cassini telescope in Loiano; and the National Galileo Telescope (TNG) in the Canary Islands. On one occasion, we also featured radio observations of the Sun with the 32-m Gavril Grueff dish at the radio astronomical station in Medicina (similar to what Altunkaynak & Batishchev (2024) did during a solar eclipse).

A **stable internet connection** at all involved sites is required, as well as **live streaming software** (e.g. Streamyard or the Open Broadcaster Software, OBS). Furthermore, a **video camera** (on a tripod) or a **computer webcam** is recommended to broadcast the astronomers performing observations at each location. In addition, **microphones** are a must for online video streaming, to ensure good audio quality. To film the host(s), we also experimented with a virtual studio equipped with a 4k video camera, **lights** and a **green screen** (Fig. 5); this is by no means necessary, but can be handy to explain some complex topics with the help of 3D animations.

Depending on the subject, real-time observations could be perceived as "disappointing" with respect to astronomical imagery coming from cutting-edge ground- and space-based telescopes, processed for public consumption. This is especially true for deep sky objects (nebulae, galaxies). In that case, besides real-time, monochrome images, we also show a stacked image collected over a longer exposure time (either during the live stream or in advance), possibly in more than one filter and quickly post-processed to create a colour image. This provides an opportunity to explain how astronomical images are made.

Figure 5. Backstage view at INAF's "Il cielo in salotto" live stream. Credit: INAF.

4. Focus group results

On the occasion of the second episode, dedicated to the SuperMoon in May 2021, we conducted a focus group to evaluate the newly developed format. The event, which lasted about 2 hours, proved quite successful, with over 20,000 live views, prompting the team to dedicate an episode to the same astronomical phenomenon - and more broadly, to the Moon, including lunar geology and exploration - every year. Having implemented lessons learnt from the 2021 focus group, with a shorter programme, higher-paced rhythm and improved visuals and graphics, the team repeated the focus group on the occasion of the SuperMoon live stream also in 2022 and 2023 to further improve the format. Highlights of what participants liked best/least are shown in Tab. 2.

4.1. *Focus group 2021*

During the first focus group, viewers were asked to evaluate the technical aspects of the live stream as well as some content related aspects. The focus group (40 people) included a sample of the target audience: 38.6% students, 7% parents, 13.6% teachers, 13.6% researchers, 6.8% amateur astronomers, and 20.5% others.

In particular, we asked participants their expectation (before watching the live stream) and then their assessment (during and after the live stream) of the following questions: **Did you have fun? Did you see the sky? Did you learn interesting stuff? Did you learn useful stuff?** and **Was the content suitable for you?** Participants had to grade their experience on a Likert scale from 0 (very bad) to 5 (very good). Considering audience satisfaction with a grade greater or equal to 4, 64.1% of the participants reported satisfaction with respect to "seeing the sky", 79.5% with respect to "learning interesting stuff" and 64.1% with respect to "learning useful stuff". Furthermore, 61.6% of the participants found the "content suitable for them" and only 42.5% had fun. In some cases, expectations were higher than the final assessment. We also asked participants whether they would watch again a similar live stream, and 75.4% answered "yes".

Table 2. Outstanding comments by focus group participants evaluating the SuperMoon live stream for three consecutive seasons.

What people liked most		
2021	2022	2023
Live astronomical images Participation from many experts from across Italy, each one contributing different information	Live astronomical images Availability of experts answering questions live	Zoomed-in view of the Moon Dialogue among different experts, including the special guest
Clear language	In-depth explanations	Explanation of the SuperMoon
	Discovering new information	Virtual "tour" of the Moon surface
	Visual layout, featuring the experts, the host and the sky, all at the same time	Double host
What people liked least		
2021	2022	2023
Too long / too late at night	Video quality	Video extract from previous year
Slow rhythm at times	Too much detail in some explanations	Too much detail in some explanations
Did not see much of the Moon	Virtual spacecraft-style studio	Visual layout
Distracting comments in the chat	Silly comments in the chat	Off-topic comments in the chat
Audio not always great	Intro theme	Digression about the guest's books

4.2. *Focus group 2022*

During the second focus group, we focused on the content related aspects of the live stream. The focus group (48 people) had a somewhat different composition from the previous year, with a much larger presence of teachers and a reduced presence of students, including: 8.4% students, 8.4% parents, 63.6% teachers, 4.2% researchers, 4.2% amateur astronomers, and 10.4% others.

We asked participants their expectation (before watching the live stream) and then their assessment (after the live stream) about the same five key aspects. In most cases, the final assessment was higher than the expectations, indicating an improvement with respect to the previous year. In line with the trend of the previous year, 66.7% of the participants reported satisfaction with respect to "seeing the sky", whereas all other indicators are higher: 86.7% reported satisfaction with respect to "learning interesting stuff" and 73.3% with respect to "learning useful stuff". Furthermore, 76.6% of the participants found the "content suitable for them" and 60% had fun. We also asked participants whether they would watch again a similar live stream, and 96.7% answered "yes".

4.3. *Focus group 2023*

The third and final focus group also focused on the content related aspects of the live stream. The slightly smaller focus group (30 people) had yet again a different composition from the previous years, including: 18.5% students, 3.7% parents, 29.6% teachers, 14.8% researchers, 3.7% amateur astronomers, and 29.6% others.

Like in 2022, we asked participants their expectation (before watching the live stream) and then their assessment (after the live stream) about the same five key aspects. In nearly all cases, the final assessment was higher than the expectations, and all indicators score in line with or better than in 2022: 89.5% of the participants reported satisfaction with respect to "seeing the sky", whereas 89.5% reported satisfaction with respect to "learning interesting stuff" and 73.7% with respect to "learning useful stuff". Furthermore, 78.9% of the participants found the "content suitable for them" and the entertainment aspect was also higher: 89.5% of the group had fun. We also asked participants whether they would watch again a similar live stream, and 94.7% answered "yes".

5. Engagement opportunities

Each episode of the live stream series is launched with a news article on the EduINAF magazine and social media channels. In most cases, the INAF press office announces the event to national media, resulting in a massive promotion especially for the most newsworthy events (e.g. eclipses). To further engage the public, an astrophotography or drawing contest is often launched at the same time, inviting members of the public to share their images of e.g. a comet or the Moon, or encouraging students to share their drawings of that particular celestial phenomenon.

While the live stream is aimed at the general public, also considering the broadcast time (usually in the evening), teachers and schools are also part of the target audience. For evening broadcast, teachers often show the replay in the classroom the morning after. On the occasion of the October 2022 partial eclipse, the live stream conveniently happened during school time, and we recorded the participation of hundreds of schools (primary and secondary alike) from the entire national territory. This prompted the team to organise another morning live stream, dedicated to observations of the Sun, in 2024, and to plan a streaming of the March 2025 partial eclipse. The North-American eclipse of 2024 was also extremely successful: in that case, we made use of third-party observations (provided by TimeAndDate.com) which were commented in real time by Italian researchers. The broadcast reached around a hundred thousand views and was featured during the prime-time news on Italy's first national channel.

The live streaming boosts traffic towards both EduINAF's website and YouTube channel, bringing a few to several thousand extra page visualisations above a daily average of around one thousand. A survey conducted with teachers reading EduINAF in the 2022/2023 school year indicated that the live streaming series is the single most appreciated topic among teachers from the magazine offer (by 32% of the surveyed sample) and that it is one of the most popular educational resources from the magazine to be employed in the classroom.

6. Conclusions

After almost four years since the first broadcast, "Il cielo in salotto" has become a staple product of INAF's education and public outreach programme. According to views and engagement of the various episodes, as well as the focus group results, the series continues to be an event appreciated by the public. Varying levels of views and engagement depend on several factors, chiefly the newsworthiness of the topic and the ensuing media promotion. The possibility to watch live telescope images, together with the experts' explanations and the clarity of the presentation, are among the most appreciated elements; the live chat continues to be popular, even if not everyone appreciates the profusion of messages, many of them off-topic, which requires continuous monitoring throughout the broadcast.

Having improved the format on the basis of the focus group results, measurements of overall public appreciation and engagement from each season inform the team while programming the calendar of the following season. We recommend live streaming newsworthy events such as solar eclipses (even if partial or not occurring on the national territory) and comets, as well as the SuperMoon, which continues attracting public interest even after several years. Alternatively, we recommend to feature "the Moon and...", combining first-quarter or full-Moon observations with planets and other deep-sky objects (nebulae, galaxies) for evening events and the Sun for a morning live stream dedicated to schools.

Acknowledgements

We are grateful to our INAF colleagues Conrad Boehm, Albino Carbognani, Silvia Casu, Francesco D'Alessio, Gian Luigi Deiana, Mario Guarcello, Giulia Iafrate, Riccardo Leoni, Maria Teresa Menna, Paolo Ochner (UniPD), Maurizio Oliviero, Simona Righini, Paolo Romano, Lorenzo Terranegra and Simone Zaggia for sharing their live observations, to Laura Barbalini for the graphics and to everyone else who has contributed to "Il cielo in salotto" over the past years.

References

Altunkaynak, B. & Batishchev, O. 2024, *Physics Today*, 77 (7), 10–11
Bolduc-Duval, J., Mann, L. P. & Baron, F. 2021, *CAP Journal*, 30, 33
Burcher, S. & McDonald, B. 2021, *CAP Journal*, 29, 22
Carlin, A. P., Marques, J. B. V. & Moutinho, R. 2021, *Physics Education*, 56, 5
Christoph, J., & Hüttemeister, S. 2021, *CAP Journal*, 30, 38
Chung, J. 2015, *Public Outreach Applications. In: Astro-Imaging Projects for Amateur Astronomers. The Patrick Moore Practical Astronomy Series. Springer, Cham*, p. 79
Duras, F., Mantovani, G., Mignone, C., Giacomini, L., Bardelli, S., Di Giacomo, F. & Filippelli, G. 2022, *EPSC Abstracts*, 16, 907
Impey, C., Pereira, V., Danehy, A. & Wenger, M. 2023, *Astronomy Education Journal*, 03, 1
Jensen, E. & Gerber, A. 2020, *Frontiers in Communication*, 4
Mantovani, G., Mignone, C., Di Giacomo, F., Duras, F., Giacomini, L., Bardelli, S. & Filippelli, G. 2021, *EPSC Abstracts*, 15, 179
Marques, J. B. V., Carlin, A. P. & Moutinho, R. 2020, *Revista Brasileira de Ensino de Física*, 42
Marques, J. B. V., Carlin, A. P. & Gomes, M. 2024, *Astronomy Education Journal*, 56, 5
Mignone, C., Buckle, A., Jones, G. & Usher, H. 2024, *Europlanet Magazine*, 5, 25
Monfardini Penteado, E., Figueiró Spinelli, P., Silva do Nascimento, W. R., Kelly de Lima, G., Oliveira do Nascimento, J., Carrelli, F. & Cortesi, A. 2021, *CAP Journal*, 30, 6
Monteiro, S., Salcides, P. F., Mello, D. R. C., de Mello, A. B., de Nader, R. V., Abreu, E. F., Liberato, G. & Loureno, M. C. H. 2023, *Boletim da Sociedade Astronômica Brasileira*, 34, 1, 233
Walkowicz, L. 2020, *Slate* (Future Tense), The Great Conjunction Will Give Us a Once-in-a-Lifetime Night Sky, 21 December 2020
We Are Social & Meltwater 2024, Digital 2024 Report
Wellbourne, D. J. & Grant, W. J. 2023, *Public Understanding of Science*, 25(6), 706–718

Planetary Science and Exoplanets in the Era of JWST
Proceedings IAU Symposium No. 393, 2026
Z. Benkhaldoun, Y. Moulane & A. Szentgyorgyi, eds.
doi:10.1017/S1743921324001649

Approach of the NGC1977 star cluster to the TOI-2796 host star. II

H. Y. Wakjira[1,2], S. H. Negu[1], M. D. Sizova[3], E. S. Postnikova[3], D. A. Mosunova[3], N. V. Chupina[3], Anna M. Fateeva[3], Alexander M. Fateev[3] and Sergei V. Vereshchagin[3]

[1]Space Science and Geospatial Institute, Entoto Observatory and Research Center, Astronomy and Astrophysics Department, P. O. Box 33679 Addis Ababa, Ethiopia
email: hawiy@essti.gov.et

[2]Finfine Aerospace and Robotics International Institute (FARIS), P.O. Box 12059, Addis Ababa, Ethiopia

[3]Institute of Astronomy of the Russian Academy of Sciences, Moscow, 119017 Russia
email: svvs@ya.ru

Abstract. It was found that the cluster NGC 1977, located in the region of Orion's Sword, in past epochs approached in space the star TOI-2796, which has a planetary system. It should be noted that in the vicinity of NGC 1977, the James Webb Space Telescope observed more than 500 free planets and free binary planets. Numerical calculations of the orbit around the Galactic Center were carried out. It is shown that these objects approached at a distance of ∼7.8 pc about ∼4.4 million years ago. This approach could have a gravitational effect on the Oort cloud analogue TOI-2796, leading to a change in the orbital elements of small bodies. It is important to note that the observed effect could be the loss of objects from the TOI-2796 planetary system into interstellar space.

Keywords. planetary system, exoplanet, TOI-2796, open star cluster, NGC 1977

1. Introduction

One of the best applications of the data obtained by Gaia and the James Webb Space Telescope (JWST) has been the study of objects in the Orion complex. Notably, within just 1.5° of Orion's Sword there are several adjacent and likely overlapping open star clusters. Most notably, NGC 1981, NGC 1977, ONC (Orion Nebula Cluster including Ori Trapezium), NGC 1980. The famous Orion Nebula (M 42) is also located here. One of the recent discoveries of JWSC was the detection of 42 Jupiter-Mass Binary Objects (JuMBOs) in the Trapezium region, Pearson and McCaughrean (2023). Zwart, Hochart (2023) identified a possible mechanism for the formation of the population of JuMBO-type objects. It is proposed that pairs of Jupiter-mass bodies spontaneously formed together with stars in the cluster, rather than being ejected from planetary systems. Limbach et al. (2022) present also about 150 exoplanets transiting young Orion stars, of which $>50\%$ will orbit mid- to late-M dwarfs and about ten will be habitable zone, terrestrial $(0.1M_\oplus - 5M_\oplus)$ exoplanets.

This work is a continuation of our work Wakjira et al. (2023), Paper I on the study of the cluster complex in Orion's Sword. We have already made such estimates (Paper I), but took different initial data and got a new assessment.

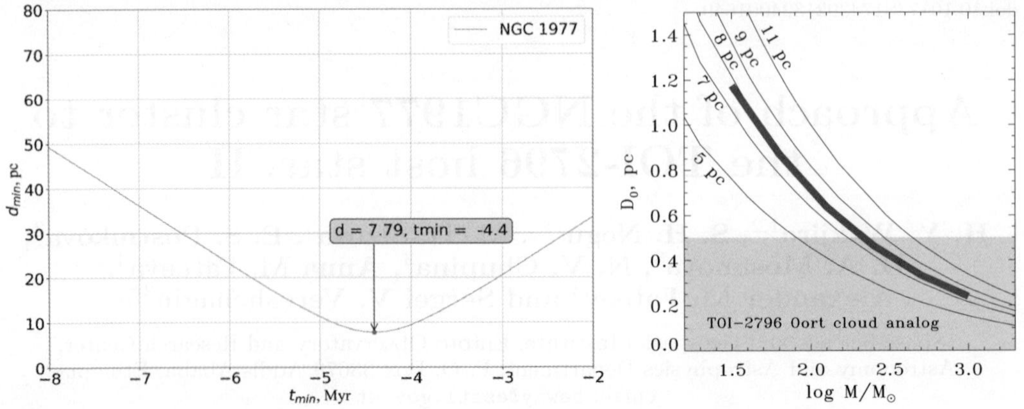

Figure 1. Left panel: The dependance of $d_{min} - t_{min}$. Right panel: Evaluation of the effects of the OSC approach to the planetary system. Dependence of D_0 on the logarithm of the cluster mass for different values of d_{min} (signed in the figure). The Oort cloud region is shown at the bottom in blue. Error $\varepsilon_{d_{min}} \approx 5$ pc.

2. Data

To initialize our calculation in *galpy* Bovy (2015), we used initial values of the spatial coordinate and velocity components. For this, we used SIMBAD with a query for NGC 1977. We took the data for TOI-2796 from the GDR3. We first transformed the astrometry and radial velocity data into a galactocentric Cartesian coordinate system using *astropy* Astropy Collaboration (2018). Note that the sign of the X coordinate in *astropy* is in a left-handed coordinate system.

3. Results of the galpy calculations

A star and a star cluster are represented as point objects and move in space around the Galactic Center. To perform calculations of the orbital motion, we used the *galpy* package. It uses the Milky Way potential MWPotential14, described by an analytical axisymmetric model that contains a spherical halo, bulge and disk.

To estimate the gravitational influence of the NGC 1977 cluster on the planetary system TOI-2796, we applied an approach that allows us to estimate the minimum distance between the cluster and the planetary system d_{min} in the considered time range. We calculated the distance between the NGC 1977 and TOI-2796 at each time step, and selected the minimum distance and the corresponding time.

Fig. 1 (left panel) shows the approach of objects, which made it possible to estimate the parameters of the approach. We correlated the integration time with the age estimates of NGC 1977 and TOI-2796. Approach calculations yielded $d_{min} = 7.8$ pc, $t_{min} = -4.4$ million years (Fig. 1, left panel), that did not exceed the age of NGC 1977 according to Cantat-Gaudin et al. (2020) is 97.7 Myr (lg age = 7.99).

4. The approach effect

Fig. 1 (right panel) shows how d_{min} affects the dependence of the distance from the star to the comet or small planet D_0 on the mass $M_{cluster}$ (Paper I) of the approaching object. In the case of $D_0 \leq R_{Oort}$, the comet will still leave the Oort cloud analogue. Fig. 1 (right panel) shows the dependence for different values of the mass of the star cluster and the minimum distance at which the cluster center and the star are closest. We get $R_{Hill} = R_{Oort} = 0.55$ pc. In case $D_0 \leq R_{Oort}$ the small body does leave the Oort cloud. The area of the Oort cloud analog TOI-2796 according to (Paper I) is shown at the bottom on Fig. 1 (right panel) (the solid area at the bottom of the figure).

In Fig. 1 (right panel) the picture of dependence is presented for various values of the mass of star cluster and the minimum distance, at which the cluster center and the star are as close as possible. The gravitational effect of the NGC 1977 cluster is marked by bold curve at 8 pc. It is clearly seen that if the mass of the cluster exceeds ≈ 200 solar masses, the gravitational effect of the cluster on the body in the Oort cloud exceeds the attraction of the star (the red curve falls into the blue region). As a result, small bodies can leave the parent star.

5. Conclusions

The approach of NGC 1977 and TOI-2796 could have happened ≈ 4.4 million years ago, that is less than the age of the cluster, which is < 10 million years. Approach parameters for NGC 1977 and TOI-2796 are $(d_{min}, t_{min}) = (7.8pc, -4.4Myr)$.

Uncertainties in the calculations may contribute to the need to adjust our results. They arise from the imprecision of the input data, neglect of the peculiar motions of stars within the cluster, the rotation of the cluster around its axes, and treating the cluster as a point mass. In this paper we took the initial data for NGC 1977 from SIMBAD, so we got a different version of the results from (Paper I).

NGC 1977 and the ONC have similar proper motions, Clark, Lada (2023). As mentioned above, the JWST observations of free planets covered the ONC region. This suggests that similar observations in the neighboring region will also help to detect free planets. If JWST had observed not only ONC, but also the NGC 1977 region, it would likely have been able to detect free small bodies, including those detached from TOI-2796.

Acknowledgments

This work used data from the European Space Agency (ESA) Gaia mission (`https://www.cosmos.esa.int/gaia`) processed by the Gaia Data Processing and Analysis Consortium (DPAC, `https://www.cosmos.esa.int/web/gaia/dpac/consortium`). Funding for DPAC was provided by national institutions, in particular institutions participating in the Gaia Multilateral Agreement. Mission Gaia website: `https://www.cosmos.esa.int/gaia`. Gaia Archive website: `https://archives.esac.esa.int/gaia`. This study used the SIMBAD database (`http://cds.u-strasbg.fr`) operated by CDS, Strasbourg, France. In this work we used the galactic dynamics library (galpy) created by Bovy (2015). The authors are grateful to Grigory Tsurikov for useful advice.

References

Astropy Collaboration (Price-Whelan, A.M., et al.) 2018, *AJ*, 156, 123

Bovy, J. 2015, *ApJS*, 216, 29

Cantat-Gaudin, T., Anders, F., Castro-Ginard, A., et al. 2020, *A&A*, 640, 1

Clark, Sophie; Lada, Elizabeth 2023, *Bulletin of the American Astronomical Society*, 55, e-id 2023n6i202p02

Limbach, M.A., Soares-Furtado, M., Vanderburg, A. et al. 2022, *arXiv* astro-ph/2209.12916

Pearson, S.G. and McCaughrean, M.J. 2023 *arXiv* astro-ph/2310.01231v1 URL=`https://archive.stsci.edu/doi/resolve/resolve.html?doi=10.17909/vjys-x251`

Wakjira, H.Y., Negu, S.H., Mosunova, D.A. et al. 2023, *arXiv* astro-ph/2310.14193 (Paper I)

Zwart, S.P., Hochart, E. 2023, *arXiv* astro-ph/2312.04645

Planetary Science and Exoplanets in the Era of JWST
Proceedings IAU Symposium No. 393, 2026
Z. Benkhaldoun, Y. Moulane & A. Szentgyorgyi, eds.
doi:10.1017/S1743921324001650

Stellar variability monitoring and vistas of its support of JWST programs

Alisher S. Hojaev

Ulugh Beg Astronomical Institute, Uzbekistan Academy of Sciences, Tashkent, Uzbekistan
email: hojaev@yahoo.com

Abstract. The vast observational survey project aimed to search and study of variable stars in different regions of the Galaxy is presented. This project is undertaking within the frame of international research collaboration between Indian and Uzbekistan astronomers. The main targets were the open star clusters and stellar associations but we also considered a multitude of known variable stars of various types with prominent activity. The observations were carried out at Maidanak astrophysical observatory (Uzbekistan) and ARIES observatories at Devasthal and Manora Peak as well as Hanle observatory (India). The main results of the projects are presented briefly. The prospects of the project especially in a point of view to support the possible programs and projects using the James Webb Space Telescope are discussed.

Keywords. (Galaxy:) open clusters and associations: general; stars: variables: general; exoplanet detection methods; techniques: photometric

---

1. Introduction

In recent decades, there has been significant exponential progress, especially in observational astronomy, both in the technology of obtaining data and in the amazing results obtained using the latest instruments both on the ground and in space. Certainly one of the unrivalled and very promising telescopes is the James Webb Space Telescope (JWST) (see, e.g., Gardner et al. 2006). However, despite the significant expansion of the capabilities of observational astrophysics represented by JWST, the importance of ground-based observations not only does not lose their relevance, but also becomes an indispensable aid to the space telescope. The need for synchronous ground-based observations of targets observed by JWST is obvious, as well as support for JWST observations using ground-based time domain astronomy and indication for JWST of new targets detected by ground-based telescopes. In this sense, long-term monitoring of a large number of star clusters and associations as well as a number of known variable stars with their wide vicinity carried out in the frame of the international Uzbek-Indian research project, initially implicated by Ulugh Beg Astronomical Institute (UBAI) and Aryabhatta Research Institute of observational sciencES (ARIES), is of particular interest. However, it should be emphasized that, in fact, the cluster monitoring program was started by us back in 2001–2003 (Hojaev 2005).

2. Observational datasets and reduction

Observations were carried out at 1.5 m telescope AZT-22 and two 60 cm telescopes Zeiss-600 of the high-mountain Maidanak Astrophysical Observatory (Uzbekistan, Longitude: *66:53:04*E Latitude: *38:40:22*N, Altitude: *2 593* m), located in the spurs of the Pamir-Alai mountain system, using CCD cameras Andor iKon-XL (4096 × 4112 pixels) [or sometimes SNUCAM (4096 × 4096 pixels)], Andor iKon-L (2048 × 2048 pixels)

Table 1. Observation statistics of clusters and associations.

Target	Telescope	Frames
NGC 6611	0.6 m	364
NGC 6705	0.6 m	508
NGC 6791	0.6 m	1085
NGC 6823	1.5 m, 0.6 m	2361
NGC 6871	0.6 m	1513
NGC 7772	1.5 m	250
IC 1396	1.5 m,0.6 m	932
IC4996	1.5 m,0.6 m	1355
Berkeley 55	1.5 m,0.6 m	765
Berkeley 98	1.5 m	402
King 1	0.6 m	689
King 13	1.5 m,0.6 m	821
King 18	0.6 m	833
BRC 5	0.6 m	25
BRC 7	0.6 m	148

and Apogee Aspen CG230 (2048 × 2048 pixels), respectively. Noteworthy that the observatory has one of the highest astroclimate indicators worldwide (median seeing is about 0.7 seconds of arc). The exposure time was selected every night according to the observed brightness of the object and the S/N ratio. Processing was carried out using standard IRAF/NOAO packages however we used some original pipelines as well.

3. Targeted open clusters and stellar associations

Dense photometric time series of 15 galactic open clusters of different ages and stellar associations located in various distances from the Sun were obtained over 3 years (2021–2023). The data on these observations are presented in Table 1. In total, 12 051 frames were obtained with observational images of areas of the sky covering the clusters and their surroundings. Moreover, the extensive time series were also obtained for 7 clusters (NGC 7801, IC 4996, Berkeley 55, King 1, King 13, King 18, King 20) in 2003 at 60 cm telescope Zeiss-600 of the Maidanak Observatory, using CCD camera Apogee 10. Some of the clusters (IC 4996, Berkeley 55, King 1, King 13, King 18) were observed in both monitoring campaigns, which will provide data on variability on a long time scale spanning 20 years.

4. Variables in clusters

The combined analysis, together with the available data obtained by TESS, made it possible to discover a significant number of new variable stars and study the few known variables in these clusters. We have carried out a meticulous study of the young cluster NGC 6823, associated with a star-forming complex that includes molecular clouds, OB-association and the bright nebula NGC 6820. The identification and characterization of 88 variable stars were made, of which only 14 have been previously recognized; 72 variable stars are found to be cluster members, of which 25 Main-Sequence stars (β Cep, δ Scuti, slowly pulsating B type, and new class variables) and 48 Pre-Main-Sequence (8 Classic T Tauri Stars, 4 HAe/Be Stars, 36 Weak-lined T Tauri Stars) (Lata et al. 2023). The Figure 1 combined from this our paper shows the two-color (TCD), color-magnitude (CMD) and luminosity-temperature (LTD) diagrams for cluster variables. Preliminary analysis of open star cluster NGC 6791 data obtained at Maidanak Observatory revealed more than 430 variables in the region, covered by our observations: there characterization and preparation for publication are in progress. Beside the analyzing our own observation databases, we processed and analyzed photometric monitoring databases, in particular, TESS data bank. We have detected 65 variable stars in open cluster Gulliver 35, 136 in

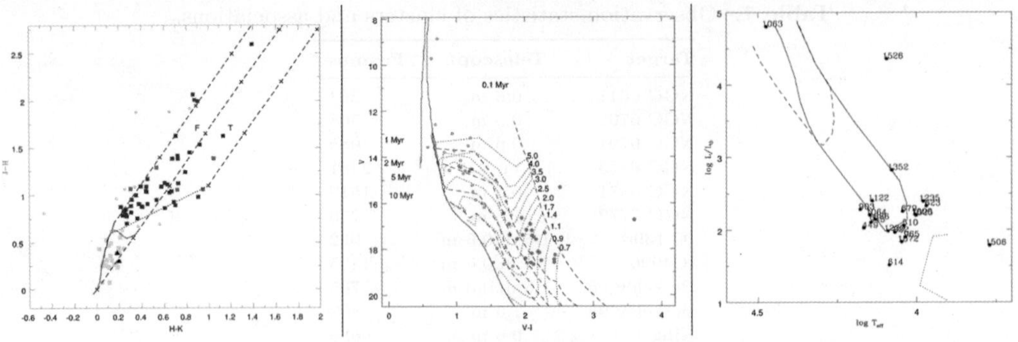

Figure 1. TCD, CMD and LTD for NGC 6823 cluster variables.

Table 2. Observations of cataclysmic variables.

MCV	Telescope	Frames
V533 Her	0.6 m	3301
RX J0153.3+7446	1.50 m	108
RX J0953-1458	0.6 m	282
1RXSJ1743-0429	0.6 m	363
Swift J2241+5644	1.5 m	43
Swift J2237.2+6324	1.5 m	92
SDSS J204827.91+005008.9	1.5 m	90
1RXSJ174320.1-042953	1.5 m,0.6 m	418
RXJ0953-1458	1.5 m	46
1RXSJ163100.5+695000	1.5 m	108
LAMOST J071553.88+581606.4	0.6 m	45
2MASS J18560806+4537400	0.6 m	459
SwiftJ2116	0.6m	494
RXJ0153	1.5 m	326

NGC 6871, 43 in King 18 and 26 in Berkeley 98 (see, for instance, Hojaev & Karimov 2024). It is also noteworthy that part of these clusters were photometrically monitored at Maidanak Observatory, and the combined analysis of space and ground-based data sets is planed.

5. Individual variable stars

In addition to searching for and studying variables in clusters, we have observed other individual variables. Particular attention was paid to a large sample of magnetic cataclysmic variables (MCV), consisting of a number of cataclysmic variable stars recently discovered by space telescopes, mainly in X-rays, supplemented by spectral and polarimetric observations from foreign observatories in order to determine the physics of these variable stars, to study the geometry of the accretion flow in these systems and their belonging to the class of *magnetic polars*. In total 14 MCV were patrolled during 170 nights with a cadence more than 4 hours per night. The general statistics of these observations at the Maidanak Observatory, are given in Table 2. A total of 6 539 frames were obtained with observational images of areas of the sky covering MCV and their environs. The R-band light curve of the star 1RXS J174320.1-042953 (J1743), obtained from our observations at Maidanak Observatory in 2021–2022, is presented in Figure 2. It demonstrate the precision of data is sufficiently high to make analysis and modelling of the star. We could confirm two MCV as *polars* (Rawat et al. 2023).

High-quality observations of a number of eclipsing binary systems HS Hya, V699 Cyg, VV Ori, 2MASSJ23194851+3603503, TYC3224-1304-1, the recurrent nova RS Oph, the

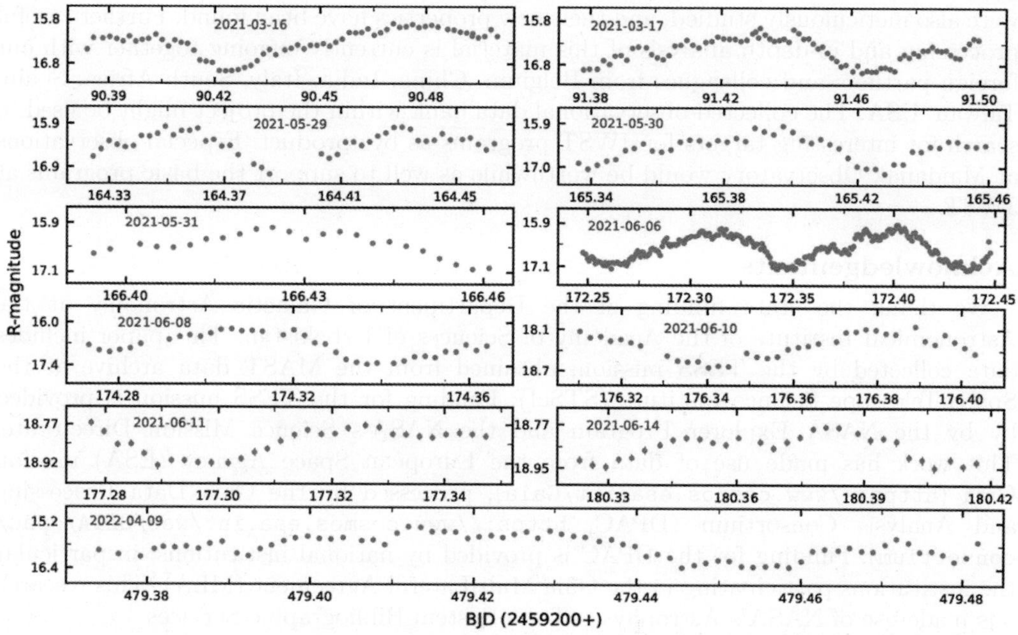

Figure 2. The light curve of the star 1RXS J174320.1-042953.

flare star GJ 3147=LP 245-10 (within an international multi-site multiwavelength synchronous photometric and polarimetric monitoring campaign) and many other variables were also obtained. The results are prepared to be published in the mean time.

6. Discussion and prospects

The combination of excellent astroclimatic properties, which despite global changes in the Earth's climate are still present at the Maidanak Observatory (stable atmosphere above the site, providing good seeing, small background of the night sky, a large number of clear nights, etc.) and the relatively good quality of the used telescopes/CCD cameras allow us to obtain a quite high photometric accuracy, cadence, time resolution and a fairly deep limit on faint stellar magnitudes (see, for example, Hojaev 2005). These conditions are suitable not only for time-domain study of different types of variable objects (NEOs, solar system planets, asteroids and comets, stars and star clusters, galaxies, quasars, transients, etc.) but to detect and characterize of extrasolar planets (exoplanets) too.

The existing extensive observational material (for instance, obtained within the star clusters and variable stars study project descibed earlier) is of sufficiently high accuracy, cadence and FOV of each image frame (about 20×20 *arcmin*) also makes it possible to search for transits of probable new exoplanets which along with various other detected at ground-based observatory unique objects might become as targets for JWST or another telescope facility in the space or on the ground.

7. Conclusions

Resuming above-said we could conclude that the search and study of variable stars in big number of star clusters and beyond were productive. 12,051 image frames were obtained covering the clusters with their surroundings, and 6,539 image frames were obtained covering MCV with their surroundings. Many new variable stars were discovered, studied and characterized together with few known ones in star clusters. The selected peculiar variable stars beyond the clusters or in the general field of the Galaxy were also meticulously studied, and their new properties have been found. Further careful processing and in-depth analysis of this material is currently ongoing together with our foreign partners and colleagues from Belgium, China, India, Italy, South Africa, Spain, Taiwan, USA. The collected observational data bank within the project might be used to search for interesting targets for JWST programs as bye-product. Especial observations at Maidanak Observatory would be worthwhile as well to support the basic programs at JWST.

Acknowledgements

We thank the state funding of the Department of Galactic Astronomy at the Astronomical Institute of the Academy of Sciences of Uzbekistan. This paper includes data collected by the TESS mission, obtained from the MAST data archive at the Space Telescope Science Institute (STScI). Funding for the TESS mission is provided by by the NASA Explorer Program and the NASA's Science Mission Directorate. This work has made use of data from the European Space Agency (ESA) mission Gaia (`https://www.cosmos.esa.int/Gaia`), processed by the Gaia Data Processing and Analysis Consortium (DPAC, `https://www.cosmos.esa.int/web/GAIA/dpac/consortium`). Funding for the DPAC is provided by national institutions, in particular the institutions participating in the Gaia MultiLateral Agreement (MLA). This research has made use of NASA's Astrophysics Data System Bibliographic Services.

References

Gardner, J. P., Mather, J. C., Clampin, M., Doyon, R., Greenhouse, M. A., Hammel, H. B., Hu. J., Mountain, M., Nella, J., Rieke, G. H., Rieke, M. J., Rix, H-W. Smith, E. P., Sonneborn, G., Stiavelli, M., Stockman, H. S., Windhorst, R. A. & Wright, G. S. 1995, *Space Sci Rev*, 123, 485

Hojaev, A. S. 2005, Collaborative Research of Open Star Clusters *JAA*, 26, 293

Hojaev, A. S. & Karimov, R. G. 2024, Study for variable stars in King 18 *Bulletin of Liège Royal Society of Sciences*, 93, 236

Lata, S., Chen, W. P., Pandey, J. C., Dileep, A., Ai, Z., Hojaev, A. S., Panwar, N., Joshi, S., Mondal, S., Biswas, S., & Bhatt, B. C. 2023, Confirmation of two magnetic cataclysmic variables as polars: 1RXS J174320.1-042953 and YY Sex *MNRAS*, 520, 1092

Rawat, N., Pandey, J. C., Joshi, A., Potter, S. B., Hojaev, A. S., De Becker, M., Rao, S. M.& Yadava, U. 2023, Photometric variable stars in the young open cluster NGC 6823 *MNRAS*, 521, 2729

Planetary Science and Exoplanets in the Era of JWST
Proceedings IAU Symposium No. 393, 2026
Z. Benkhaldoun, Y. Moulane & A. Szentgyorgyi, eds.
doi:10.1017/S174392132500047X

Instability of In-plane Equilibrium Points of the Restricted Three-Body Problem with P-R Drag and Variable Masses

Tajudeen Oluwafemi Amuda[1] and **Leke Oni**[2]

[1]Department of Mathematics, Faculty of Physical Science, Air Force Institute of Technology
Kaduna, Kaduna State, Nigeria
email: teejaymath@gmail.com

[2]Department of Mathematics, College of Physical Science, Joseph Sarwuan Tarka University
P.M.B. 2373, Makurdi, Benue-State, Nigeria
email: leke.o@uam.edu.ng

Abstract. This paper studies instability of the in-plane equilibrium points (EPs) which lies on the line joining the main bodies in the frame of the restricted three-body problem with Poynting-Robertson (P-R) drag force and variable masses. The dynamical structures of the primaries and their mass variation are governed by the Gylden-Mestschersky problem (GMP) and the Mestschersky combined law (MCL), respectively, under the conditions that both primaries are radiation emitters with the Doppler shift and the absorption and subsequent re-emission of the incident radiation, considered. It is seen that there can be three, four and even up to six Collinear In-plane Equilibrium Points (CIEPs). Some are independent of the P-R drag while others depend on the mass parameter and P-R drag of the primaries. The stability of the CIEPs is investigated analytically and numerically for a dust grain particle in the gravitational field of all binary systems which covers most astronomical systems, and it is seen that all the CIEPs are unstable. Additionally, it is seen that the zero velocity curves (ZVC) around the CIEPL_0, for all mass parameter do not exist as the P-R drag effects are annulled in the force function. Consequently, the resulting CIEP will be located on same point as the bigger primary, which in turns means that the force function will be undefined in this case. The stability of the CIEPs of the non-autonomous system which differ from those of the autonomized system by a function of time, are unstable as they do not converge as time tends to infinity. Our problem can be applied to the description of a dust grain in the gravitational field of stellar systems with radiation force and variable masses.

Keywords. R3BP; Stability; P-R Drag; Variable Masses

1. Introduction

The Restricted Three-Body Problem (R3BP) constitutes one of the most important problems in dynamical astronomy. This formulation defines the motion of an infinitesimal mass moving under the gravitational influence of two primary bodies which move in circular orbits around their common center of mass on account of their mutual attraction. To tackle the problem of classical R3BP, Lagrange considered the behavior of the distances between the bodies without finding a general solution. He discovered two distinct classes of constant-pattern solutions in a rotating frame of reference where gravitational equilibrium can be maintained. These points are called equilibrium points $L_i (i = 1, 2..5)$ and are very important in space missions (Szebehely 1967).

The classical problem of Szebehely (1967) did not consider the primaries as sources of radiation pressure. Given this, several studies when one or both primaries are sources of radiation pressure have been examined. Notable among these are Singh and Ishwar (1999); Singh and Leke (2010, 2012). Further research which has to do with radiation force, is the perturbing effect of Poynting–Robertson (P–R) drag were investigated. The radiation force can sweep small particles of the solar system into the Sun at a cosmically rapid rate. Authors have conducted researches on the R3BP with P-R drag, amongst them are Singh and Abdulkarim (2014); Singh and Amuda (2019); Amuda and Singh (2022).

The classical R3BP assumes that the masses of primaries are constant. However, the occurrence of absorption in stars motivated scientists to formulate the R3BP with variable mass(es). Several authors such as Luk'yanov (1989); Singh and Leke (2010, 2012, 2013); Gao and Wang (2020); Leke and Singh (2023), and, Leke and Orum (2024) have carried out investigations of the variable mass R3BP under diverse characterizations.

Inspired by the vast applications of the R3BP with variable masses, we thought it timely to examine in this paper, the instability around collinear in-plane equilibrium points (CIEPs) of the R3BP when both primaries have variable masses and are sources of radiation under effects of P-R drag forces. The importance of the study of the R3BP with variable mass is that the model provides a more accurate representation of the real world dynamics of celestial bodies and allows for more accurate dynamical predictions of their impending behaviors. For instance, the motion around the EPs of the photogravitational R3BP studied by Singh and Amuda (2014, 2019), may be inappropriate to discuss the model in the gravitational environment of radiating stars with varying masses. This study will provide better insights into the dynamical environment of a test particle around radiating-variable stars with P-R drag forces.

The paper is organized in the following order: Section 2, contains the description of the equations of motion and Section 3 deals with the locations of the CIEPs. The stability of the CIEPs is investigated in Section 4 while the numerical analysis and discussion are given in Section 5. The conclusion is drawn in Section 6.

2. Dynamical Equations

We follow the methodology of Singh and Leke (2010), where the equations of motion of the photogravitational R3BP with variable masses have the form:

$$\begin{aligned}
\ddot{x} - 2\omega\dot{y} &= \omega^2 x + \dot{\omega} y - \frac{\mu_1 q_1 (x - x_1)}{r_1^3} - \frac{\mu_2 q_2 (x - x_2)}{r_2^3} \\
\ddot{y} + 2\omega\dot{x} &= \omega^2 y - \dot{\omega} x - \frac{\mu_1 q_1 y}{r_1^3} - \frac{\mu_2 q_2 y}{r_2^3} \\
\ddot{z} &= -\frac{\mu_1 q_1 z}{r_1^3} - \frac{\mu_2 q_2 z}{r_2^3}
\end{aligned} \tag{1}$$

where (x, y, z) are the coordinates of the test particle; $r_i^2 = (x - x_i)^2 + y^2 + z^2$ and the distances of m_1 and m_2 from m_3, respectively, while q_i $(i = 1, 2)$ are the radiation pressure of the primaries. $\omega(t)$ is the angular velocity of the primaries while μ_i is the product of the masses of the primaries with the gravitational constant G.

In estimating the light radiation force of the photogravitational R3BP with variable masses, Singh and Leke (2010) took into account just one of the three components of the light pressure field, which is due to the central force: the gravitational and the radiation pressure. The other two components arising from the Doppler shift and the absorption and subsequent re-emission of the incident radiation, was not considered. In this study, the last two components that constitute the Poynting-Robertson (P-R) effect will treated in the equations of motion to have the form:

$$\ddot{x} - 2\omega\dot{y} = \omega^2 x + \dot{\omega} y - \frac{\mu_1 q_1 (x - x_1)}{r_1^3} - \frac{\mu_2 q_2 (x - x_2)}{r_2^3} - \frac{W_1}{r_1^2}\left[\frac{(x - x_1)}{r_1^2}\{(x - x_1)\dot{x} + y\dot{y} + z\dot{z}\}\right.$$
$$\left. + \dot{x} - \omega y\right] - \frac{W_2}{r_2^2}\left[\frac{(x - x_2)}{r_2^2}\{(x - x_2)\,\dot{x} + y\dot{y} + z\dot{z}\} + \dot{x} - \omega y\right],$$
$$\ddot{y} + 2\omega\dot{x} = \omega^2 y - \dot{\omega} x - \frac{\mu_1 q_1 y}{r_1^3} - \frac{\mu_2 q_2 y}{r_2^3} - \frac{W_1}{r_1^2}\left[\frac{y}{r_1^2}\{(x - x_1)\,\dot{x} + y\dot{y} + z\dot{z}\} + \dot{y} + \omega\,(x - x_1)\right]$$
$$- \frac{W_2}{r_2^2}\left[\frac{(x - x_2)}{r_2^2}\{(x - x_2)\,\dot{x} + y\dot{y} + z\dot{z}\} + \dot{y} + \omega\,(x - x_2)\right],$$
$$\ddot{z} = -\frac{\mu_1 q_1 z}{r_1^3} - \frac{\mu_2 q_2 z}{r_2^3} - \frac{W_1}{r_1^2}\left[\frac{z}{r_1^2}\{(x - x_1)\,\dot{x} + y\dot{y} + z\dot{z}\} + \dot{z}\right]$$
$$- \frac{W_2}{r_2^2}\left[\frac{(x - x_2)}{r_2^2}\{(x - x_2)\,\dot{x} + y\dot{y} + z\dot{z}\} + \dot{z}\right], \quad (2)$$

where $W_1 = \frac{\mu_1(1-q_1)}{c_d}$ and $W_2 = \frac{\mu_2(1-q_2)}{c_d}$ represent the P-R drag effect of the primaries while c_d is the velocity of light.

Following the lead of Singh and Leke (2010), the autonomized equations with constant coefficients are:

$$\xi'' - 2\eta' = \kappa\xi - \frac{\kappa q_1\,(1-\upsilon)\,(\xi+\upsilon)}{\rho_1^3} - \frac{\kappa q_2 \upsilon\,(\xi+\upsilon-1)}{\rho_2^3}$$
$$- \frac{W_{10}}{\rho_1^2}\left[\frac{(\xi+\upsilon)}{\rho_1^2}\{(\xi+\upsilon)\,\xi' + \eta\eta' + \zeta\zeta'\} + \xi' - \eta\right]$$
$$- \frac{W_{20}}{\rho_2^2}\left[\frac{(\xi+\upsilon-1)}{\rho_2^2}\{(\xi+\upsilon-1)\,\xi' + \eta\eta' + \zeta\zeta'\} + \xi' - \eta\right]$$
$$\eta'' + 2\xi' = \kappa\eta - \frac{\kappa q_1\,(1-\upsilon)\,\eta}{\rho_1^3} - \frac{\kappa q_2 \upsilon\eta}{\rho_2^3} - \frac{W_{10}}{\rho_1^2}\left[\frac{\eta}{\rho_1^2}\{(\xi+\upsilon)\,\xi' + \eta\eta' + \zeta\zeta'\} + \eta' + (\xi+\upsilon)\right]$$
$$- \frac{W_{20}}{\rho_2^2}\left[\frac{\eta}{\rho_2^2}\{(\xi+\upsilon-1)\,\xi' + \eta\eta' + \zeta\zeta'\} + \eta' + (\xi+\upsilon-1)\right]$$
$$\zeta'' = (\kappa-1)\,\zeta - \frac{\kappa q_1\,(1-\upsilon)\,\zeta}{\rho_1^3} - \frac{\kappa q_2 \upsilon\zeta}{\rho_2^3} - \frac{W_{10}}{\rho_1^2}\left[\frac{\zeta}{\rho_1^2}\{(\xi+\upsilon)\,\xi' + \eta\eta' + \zeta\zeta'\} + \zeta'\right]$$
$$- \frac{W_{20}}{\rho_2^2}\left[\frac{\zeta}{\rho_2^2}\{(\xi+\upsilon-1)\,\xi' + \eta\eta' + \zeta\zeta'\} + \zeta'\right] \quad (3)$$

where

$$W_{10} = \frac{\kappa\,(1-\upsilon)\,(1-q_1)}{c_d} \text{ and } W_{20} = \frac{\kappa\upsilon\,(1-q_2)}{c_d} 0 < \kappa < \infty \quad (4)$$

$$\rho_1^2 = (\xi+\upsilon)^2 + \eta^2 + \zeta^2,\ \rho_2^2 = (\xi+\upsilon-1)^2 + \eta^2 + \zeta^2 \text{ and } 0 < \upsilon \le 0.5 \quad (5)$$

υ is the mass parameter and defined as the ratio of the mass of the smaller primary to the sum of the masses of the primaries and is such that $0 < \upsilon \le 0.5$.

Equations (3) has the Jacobi integral

$$2\Omega\,(\xi,\eta,\zeta) - \left(\xi'^2 + \eta'^2 + \zeta'^2\right) = C \quad (6)$$

where C is the Jacobi or energy constant.

3. Locations of collinear in-plane equilibruim points

The CIEPs are the solutions of equations (7), when the LHS is set to zero and when $\xi \neq 0,\ \eta = \zeta = 0$. That is

$$\xi - \frac{q_1 (1-\upsilon)(\xi+\upsilon)}{\rho_1^3} - \frac{q_2 \upsilon (\xi+\upsilon-1)}{\rho_2^3} = 0 \tag{7}$$

$$-\frac{(\xi+\upsilon) W_{10}}{\rho_1^2} - \frac{(\xi+\upsilon-1) W_{20}}{\rho_2^2} = 0 \tag{8}$$

$$\rho_1 = (\xi+\upsilon) \text{ and } \rho_2 = (\xi+\upsilon-1) \tag{9}$$

Solving (8), we get

$$\xi = -\upsilon + \frac{W_{10}}{W_{10}+W_{20}} \tag{10}$$

The point defined by equation (10) depends on the mass parameter and the P-R drag of both primaries. We shall denote this point by L_0.

Following Singh and Leke (2010), we solve equation (7) to get the following three polynomials of degree five, each of which corresponds to the CIEPs L_i $(i = 1, 2, 3)$:

$$\begin{aligned}
&\varepsilon_1^5 + (2+\upsilon)\varepsilon_1^4 + (1+2\upsilon)\varepsilon_1^3 - (q_2\upsilon - \upsilon + q_1 - q_1\upsilon)\varepsilon_1^2 - 2q_1(1-\upsilon)\varepsilon_1 - q_1(1-\upsilon) = 0\\
&\varepsilon_2^5 - (3-\upsilon)\varepsilon_2^4 + (3-2\upsilon)\varepsilon_2^3 - (1-\upsilon+q_2\upsilon - q_1 + q_1\upsilon)\varepsilon_2^2 + 2q_2\upsilon\varepsilon_2 - q_2\upsilon = 0\\
&\varepsilon_3^5 + (3+\upsilon)\varepsilon_3^4 + (3-2\upsilon)\varepsilon_3^3 + (1-\upsilon-q_2\upsilon - q_1 + q_1\upsilon)\varepsilon_3^2 - 2q_2\upsilon\varepsilon_3 - q_2\upsilon = 0
\end{aligned} \tag{11}$$

Equations (11) are equations of degree five in ε_i $(i = 1, 2, 3)$, and positions of the CIEPs L_i $(i = 1, 2, 3)$ are respectively given as

$$\xi^1 = -\upsilon - \varepsilon_1, \xi^2 = 1 - \upsilon - \varepsilon_2, \xi^3 = 1 - \upsilon + \varepsilon_3, \tag{12}$$

The first CIEP lies on the left of the bigger primary, while the second and third are located in between the primaries and the right of the smaller primary, respectively.

The CIEPs L_i $(x_i,\ 0 : i = 0, 1, 2, 3)$ of the system of equations of the non-autonomous system (2) are sought in the form (Luk'yanov 1990):

$$x_i(t) = \xi_i R(t) \tag{13}$$

where ξ_i $(i = 0, 1, 2, 3)$ are the CIEPs of autonomized system obtained in equations (10) and (12). All the CIEPs in this case are function of time $\rho_1 = \varepsilon_2 < 1$ and always change with time.

4. Stability of collinear in-plane equilibrium points

To examine the linear stability of the CIEPs (Szebehely 1967), we use a small displacement method by shifting the coordinates of the third body and linearize the equations of motion (3) around the CIEPs with coordinates $(\xi_0, 0, 0)$. The variational equations are obtained and the corresponding characteristic equation in this case is

$$\lambda^4 - \left(4 - \Omega_{\eta\eta}^0 - \Omega_{\xi\xi}^0\right)\lambda^2 + \Omega_{\xi\xi}^0\Omega_{\eta\eta}^0 - \left(\Omega_{\xi\eta}^0\right)^2 = 0 \tag{14}$$

where

$$\begin{aligned}
\Omega_{\xi\xi}^0 &= \kappa\left[1 - \frac{q_1(1-\upsilon)}{|\xi_0+\upsilon|^3} + \frac{3q_1(1-\upsilon)}{|\xi_0+\upsilon|^3} + \frac{2q_2\upsilon}{|\xi_0+\upsilon-1|^3}\right]\\
\Omega_{\eta\eta}^0 &= \kappa\left[1 - \frac{q_1(1-\upsilon)}{|\xi_0+\upsilon|^3} - \frac{q_2\upsilon}{|\xi_0+\upsilon-1|^3}\right], \Omega_{\xi\eta}^0 = 0
\end{aligned} \tag{15}$$

Table 1. Locations of Collinear In-plane EPs for $0 < \upsilon \leq 0.5$.

Mass parameter	L_1	L_2	L_3	L_0
0.000000001	−0.99997	0.99930	1.00068	0.754902
0.000001	−0.99997	0.99308	1.00694	0.754901
0.00001	−0.99998	0.98514	1.01499	0.754892
0.0001	−1.00002	0.96817	1.03242	0.754802
0.01	−1.00414	0.85036	1.14671	0.744902
0.1	−1.04158	0.61760	1.25740	0.654902
0.2	−1.08281	0.44914	1.26364	0.554902
0.3	−1.12318	0.29711	1.24189	0.454902
			−1.48122	
			−0.38351	
0.4	−1.16202	0.149555	1.20637	0.354902
			−1.89527	
			−0.27495	
0.49	−1.19490	0.015242	1.16696	0.264902
			−2.21122	
			−0.25344	
0.49999	−1.19838	0.000011	1.16225	0.254912
			−2.24441	
			−0.25402	
0.5	−1.19839	−0.000005	1.16225	0.254902
			−2.24444	
			−0.58351	

The stability of the CIEPs L_i $(i = 1, 2, 3)$ have already been discussed by Singh and Leke (2010, 2012, 2013a). Therefore, we shall only discuss the stability of the CIEP L_0.

Now, substituting equation (10) in the partial derivatives (15), we get

$$\Omega^0_{\xi\xi} = \kappa\left[1 + 2q_2\upsilon\left(1 + \frac{3W_{10}}{W_{10}+W_{20}}\right)\right], \Omega^0_{\eta\eta} = \kappa\left[1 - q_2\upsilon\left(1 + \frac{3W_{10}}{W_{10}+W_{20}}\right)\right] \quad (16)$$

Next, we substitute (16) in the characteristic equation (14), to get

$$\lambda^4 - \left[4 + 3\kappa\upsilon\left\{1 - (1-q_2) + \frac{3W_{10}}{W_{10}+W_{20}}\right\}\right]\lambda^2 + \kappa^2\left[1 + \upsilon\left\{1 - (1-q_2) + \frac{3W_{10}}{W_{10}+W_{20}}\right\}\right] = 0 \quad (17)$$

We shall numerically deduce the roots of equation (17) in a later subsection of the paper to be able to conclude on the stability outcome.

The stability of the CIEPs (13) are unstable as they do not converge as time increases and tend to infinity.

5. Numerical treatments and discussion

In this section, we devote our efforts to the numerical applications of the obtained analytical results, for a general case in which the test particle is a dust grain in the gravitational field of the primaries having variable masses and emitting radiation coupled with the P-R drag effects. We carry out our numerical exploration for all binaries having mass parameter in the interval $0 < \upsilon \leq 0.5$, while values for the radiation pressure and P-R drag for the bigger and smaller primary, are taken from Amuda et.al (2020): $q_1 = 0.99992$, $q_2 = 0.99996$, and $W_{10} = 1.04548 \times 10^{-9}$, $W_{20} = 3.39442 \times 10^{-10}$, respectively. All numerical computations have been carried out using the software package, Mathematica (Wolfram 2017).

The CIEPs have been obtained in equations (10) and (12). In this subsection, we shall evaluate the points numerically for the chosen parametric values in Table 1.

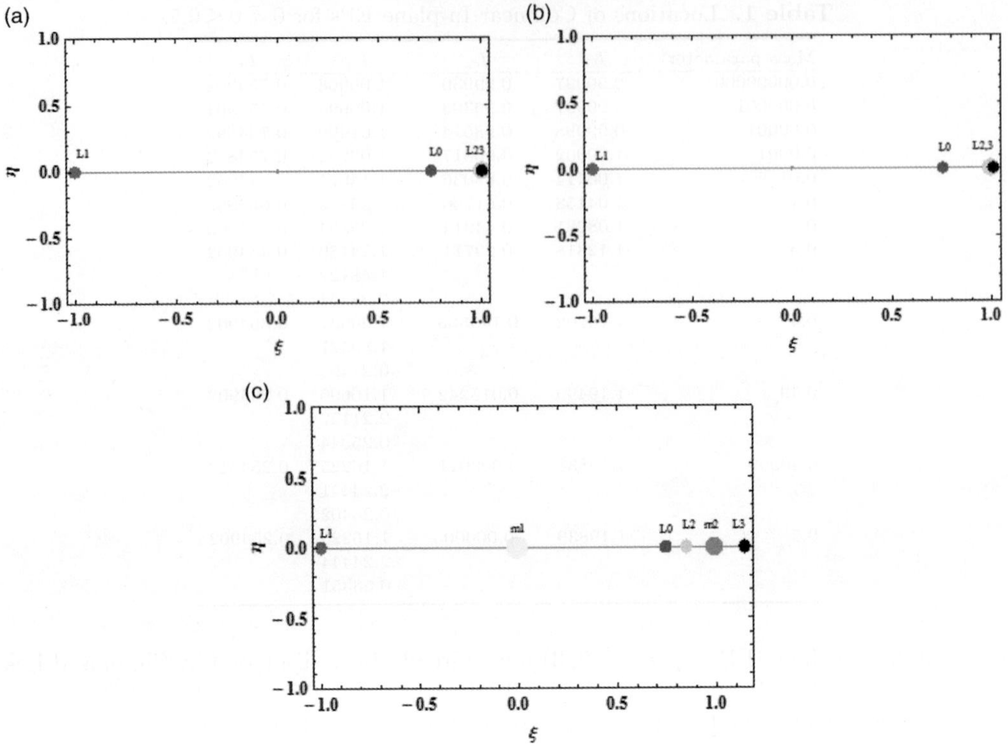

Figure 1. Locations of CIEPs, L_0, L_1, L_2, L_3 when (a) $\upsilon = 0.000000001$ (b) $\upsilon = 0.000001$ (c) $\upsilon = 0.01$

It is seen that when $0 < \upsilon \leq 0.2$, there are only four CIEPs. However, when $0.3 \leq \upsilon \leq 0.5$, two additional CIEPs lying on the left of the bigger primary evolves, and we denote them as L_{11} and L_{12}. Therefore, there are six CIEPs L_0, L_1, L_{11}, L_{12}, L_2 and L_3 of the R3BP with variable masses under radiation pressure and P-R drag effects of the primaries. These are different from the result of Szebehely (1967), where there exists only three CIEPs and from those in Singh and Leke (2013), where there exist five CIEPS except the CIEP which exist due to P-R drag of the primaries. Also results of Singh and Amuda (2016), and Leke and Orum (2024), are different from our results in terms of the numbers of the CIEPs.

Below, we show the positions of the CIEPs graphically in Fig. 1 to Fig. 4, under different characterizations. The CIEP L_1 is depicted in red, L_2 is green, L_3 is black while L_0, L_{11} and L_{12} are depicted in brown, pink and purple, respectively. The bigger and smaller primary are depicted in yellow and orange, respectively.

Fig. 1, gives the plots of CIEPs for different mass parameter. In Fig. 1a and Fig. 1b, the points are plotted when $\upsilon = 0.000000001$ and $\upsilon = 0.000001$, respectively. In this case, the points L_2(green) and L_3 (black) are nearly coincident. In Fig. 1c, it is seen that the CIEP L_1 is located on the left of the bigger primary, while the points L_0 and L_2 are located between the primaries, while, the point L_3, is located on the right of the smaller primary. Fig. 2, gives the location of the six CIEPs when the binaries are $\upsilon = 0.3$ and $\upsilon = 0.5$. It is seen that the points L_1, L_{11} and L_{12} are located on the left of the bigger primary and the points L_0 and L_2 are located between the primaries, while, the point L_3, is located on the right of the smaller primary.

The variations in the locations of the CIEP L_0 is plotted for $0 < \upsilon \leq 0.5$ and P-R drag of the primaries, in Fig. 3. It is seen that for increasing mass parameter, the point L_0

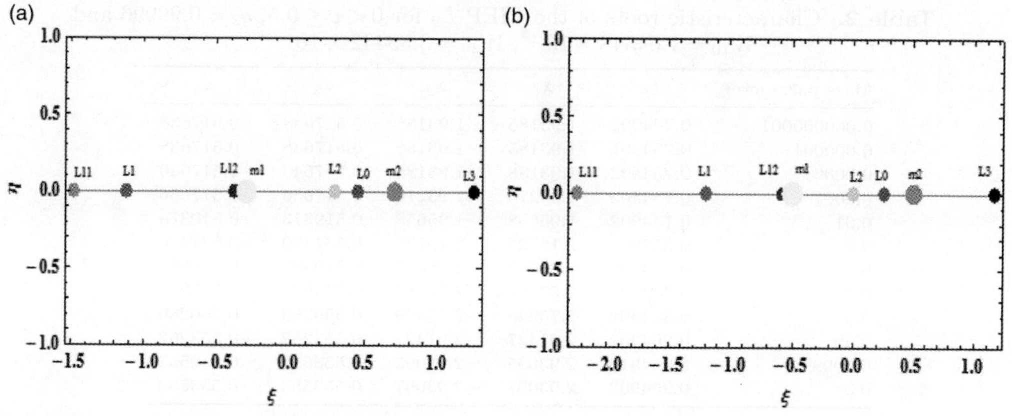

Figure 2. Locations of six CIEPs, $L_0, L_1, L_{11}, L_{12}, L_2, L_3$ when (a) $\upsilon = 0.3$ (b) $\upsilon = 0.5$

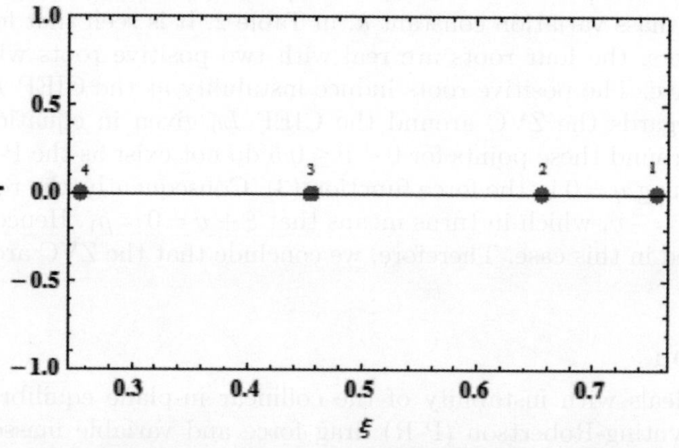

Figure 3. Locations of the CIEP L_0 for (1) $\upsilon = 0.00000001$ (2) $\upsilon = 0.1$ (3) $\upsilon = 0.3$ (4) $\upsilon = 0.5$

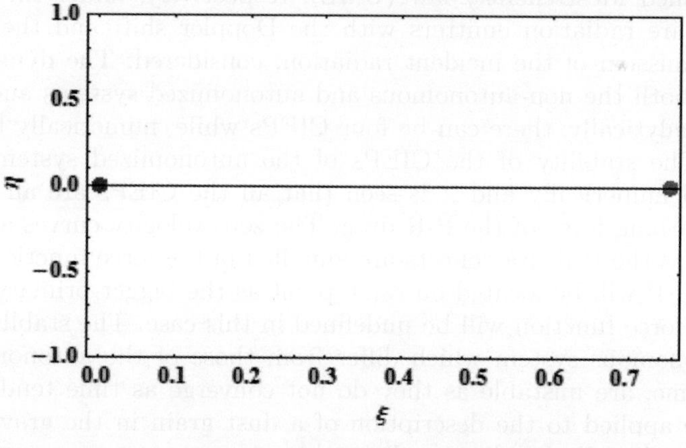

Figure 4. Variations in the CIEP L_0 in the absence of P-R drag (Blue) and in the Presence of P-R drag (Brown), for $\upsilon = 0.000000001$

moves closer to the origin. Fig. 4 shows the variation in the location of the CIEP L_0 in the presence (brown point) and absence (blue point) of P-R drag. It is observed that in the absence of P-R drag, the point L_0 is located close to the origin

Next, to examine the stability of the CIEPs, we numerically evaluate the roots of the characteristic equation (17) in the case of the CIEP L_0 for values of the mass parameter

Table 2. Characteristic roots of the CIEP L_0 for $0 < \upsilon \leq 0.5$, $q_2 = 0.99996$ and $W_{10} = 1.04548 \times 10^{-9}$, $W_{20} = 3.39442 \times 10^{-10}$.

Mass parameter	L_0	λ_1	λ_2	λ_3	λ_4
0.000000001	0.754902	1.93185	−1.93185	0.517638	−0.517638
0.000001	0.754901	1.93185	−1.93185	0.517638	−0.517638
0.00001	0.754892	1.93188	−1.93188	0.517640	−0.517640
0.0001	0.754802	1.93210	−1.93210	0.517656	−0.517656
0.01	0.744902	1.95658	−1.95658	0.519373	−0.519373
0.1	0.654902	2.16725	−2.16725	0.531420	−0.531420
0.2	0.554902	2.38057	−2.38057	0.540065	−0.540065
0.3	0.454902	2.57684	−2.57684	0.545983	−0.545983
0.4	0.354902	2.75949	−2.75949	0.550286	−0.550286
0.49	0.264902	2.91427	−2.91427	0.553262	−0.553262
0.49999	0.254912	2.93095	−2.93095	0.553553	−0.553553
0.5	0.254902	2.93097	−2.93097	0.553553	−0.553553

$0 < \upsilon \leq 0.5$, radiation pressure of the smaller primary and P-R drag of both primaries under effects of mass variation constant κ, in Table 2. It is seen that for all values of the model parameters, the four roots are real with two positive roots while the other two roots are negative. The positive roots induce instability at the CIEP L_0.

Finally, as regards the ZVC around the CIEP L_0 given in equation (10). It is seen that the ZVC around these points for $0 < \upsilon \leq 0.5$ do not exist as the P-R drag effects are annulled by setting $\eta = 0$ in the force function (3). Consequently, the resulting CIEP (10) will reduce to $\xi = -\upsilon$, which in turns means that $\xi + \upsilon = 0 = \rho_1$. Hence, the equation (3) will be undefined in this case. Therefore, we conclude that the ZVC around the CIEP L_0 do not exists.

6. Conclusion

This paper deals with instability of the collinear in-plane equilibrium points of the R3BP with Poynting-Robertson (P-R) drag force and variable masses. The motion of the primaries and their mass variation is governed by the Gylden-Mestschersky problem (GMP) and unified Mestschersky law (UML), respectively, under the conditions that both primaries are radiation emitters with the Doppler shift and the absorption and subsequent re-emission of the incident radiation, considered. The dynamical equations are derived for both the non-autonomous and autonomized systems and the CIEPs are investigated. Analytically, there can be four CIEPs while, numerically the numbers can go up to six. The stability of the CIEPs of the autonomized system is investigated analytically and numerically and it is seen that all the CIEPs are all unstable due to the strong perturbing force of the P-R drag. The zero velocity curves around the CIEP L_0 do not exist as the P-R drag effects are annulled in the force function. Consequently, the resulting CIEP will be located on same point as the bigger primary, which in turns means that the force function will be undefined in this case. The stability of the CIEPs of the non-autonomous system which differ from those of the autonomized system by a function of time, are unstable as they do not converge as time tend to infinity. Our problem can be applied to the description of a dust grain in the gravitational field of stellar systems with radiation force and variable masses.

References

Amuda, T.O. and Singh, J., 2022. Instability of Triangle Equilibrium points in the restricted three-body problem under effects of circumbinary disc, radiation pressure and P-R drag. Earth, Moon and Planets 126: (1).

Gao, F., Wang, Y., 2020. Approximate analytical periodic solutions to the restricted three-body problem with perturbation, oblateness, radiation and varying Mass. *Universe, 6*, 110.

Leke, O., Orum, S., 2024. Motion and zero velocity curves of a dust grain around collinear libration points for the binary IRAS 11472-0800 and G29-38 with a triaxial star and variable masses. New Astronomy, 108, 102177.

Leke, O., Singh, J., 2023. Out-of-plane equilibrium points of extra-solar planets in the central binaries PSR B1620-26 and Kepler-16 with cluster of material points and variable masses. New Astronomy, 99, 101958.

Luk'yanov, L. G., 1989. Particular solutions in the restricted problem of three bodies with variable masses. Astronomical Journal of Academy of Sciences of USSR. 66:180.

Luk'yanov, L. G., 1990. Stability of libration points in the restricted three-body problem of variable masses. Soviet Astronomical Journal, 67, 167–172.

Singh, J. and Abdulkarim, A., 2014. Instability of triangular libration points in the perturbed photogravitational R3BP with Poynting-Robertson (P-R) drag. Ap&SS 351:473-482.

Singh, J and Amuda, T. O., 2014. Poynting-Robertson (P-R) drag and oblateness effects on motion around the triangular equilibrium points in the photogravitational R3BP. Ap&SS 350:119–126.

Singh J, and Amuda T.O., 2019. Stability of triangular equilibrium points in restricted three-body problem under effects of circumbinary disc, radiation and drag forces. Journal Astronomy and Astrophysics. 40:5.

Singh, J. and Amuda, T. O., 2016. Binaries application around the collinear equilibrium points in the photogravitational CR3BP with bigger primary oblate. *European Physical Journal Plus (Springer)*. 131: 137.

Singh, J and Ishwar, B., 1999. Stability of triangle points in the generalized photogravitational restricted three-body problem. Bulletin of Astronomical Society of India. 27, 415–424.

Singh, J., Leke, O., 2010. Stability of the photogravitational restricted three-body problem with variable masses. Ap&SS 326, 305–314.

Singh, J., Leke, O., 2012. Equilibrium points and stability in the restricted three- body problem with oblateness and variable masses. Ap&SS 340: 27–41.

Singh, J., Leke, O., 2013a. Effects of oblateness, perturbations, radiation and varying masses on the stability of equilibrium points in the restricted three-body problem. Ap&SS 344: 51–61.

Szebehely, V.G. 1967: Theory of Orbits Addison Wesley.

Wolfram, S. The Mathematica Book, 10th Ed. Wolfram Media, Campaigns; 2017.

Planetary Science and Exoplanets in the Era of JWST
Proceedings IAU Symposium No. 393, 2026
Z. Benkhaldoun, Y. Moulane & A. Szentgyorgyi, eds.
doi:10.1017/S1743921324002515

Solar Hydrologic Cycle Determines Earth and Venus Atmospheres Mass and Composition

Xuguang Leng

Freelance, Breinigsville, PA 18031, USA
email: xuguangleng@gmail.com

Abstract. Earth's and Venus' atmosphere mass and composition today are accumulated results of the solar hydrologic cycle over billions of years with comet collisions infusing gassy material onto the planets and solar wind blowing away the lighter gasses leaving primarily only carbon dioxide and nitrogen on both Earth and Venus. The comets entering the inner solar system have higher evaporation rate than collision probability, resulting in far less mass infusion onto inner planets than outer planets. The infused gasses onto the Earth are further processed by its chemical processes. Being too close to the Sun to have liquid water, Venus lacks chemical processes. The solar hydrologic cycle theory predicts comet mass infusion ratio between Venus and Earth to be 1:0.76. Comparing carbon dioxide and nitrogen mass in Venus atmosphere with nitrogen mass in Earth atmosphere plus carbon and nitrogen deposits in the crust, the ratio is 1:0.45-0.54.

Keywords. Earth, Venus, comet, planets formation

1. Introduction

The atmospheres of the eight planets in the solar system vary significantly, even though the solar system was born off the same nebula. Two of the inner planets, Mercury and Mars, have almost no atmosphere. Four outer planets, Jupiter, Saturn, Uranus, Neptune, are almost completely made of gasses or frozen gasses. The remaining two inner planets, Venus and Earth, are the closest pair with distance of 0.277 AU, Venus' mass is 82% of Earth's, Venus' diameter of 12,104 km is 95% of Earth's of 12,756 km, yet the atmosphere mass and composition are magnitudes apart. Venus' atmosphere has a mass of 480×10^{18} kg, versus the meager 5.15×10^{18} kg of Earth's atmosphere mass. Venus' atmosphere comprises nearly 0.01% of the total planetary mass, compared to the factor of 8.5×10^{-7} for Earth. The atmosphere compositions are very different between the two planets as well. Venus' atmosphere has 96.5% carbon dioxide and 3.5% nitrogen, Earth's atmosphere has 78% nitrogen and 21% oxygen.

For our closest neighbor, the atmosphere is the easiest part of Venus to study, yet many of its basic properties are still unknown or poorly understood. (O'Rourke et al. 2023) Comets have enough mass and their collisions with planets happen frequent enough that the effect of mass infusion must be considered. The Hale-Bopp comet has an estimated mass of 1.3×10^{19} kg, (Weissman 2006) a single flyby comet has more gas mass than the whole Earth's atmosphere. In our life time, we also witnessed the Shoemaker-Levy comet collided with Jupiter. Opposite to comet infusion, solar wind is strong enough at Mars orbit to overcome Mars gravity and strip Mars barren. (Barabash et al. 2007) The comet infusion and solar wind stripping combined forms the solar hydrologic cycle, (Leng 2021) which could determine the mass and composition of inner planets atmosphere.

In early eons, the gassy material mass in the Kuiper Belt would be magnitudes more than today, before majority of it infused onto outer planets. There were billions of comets, big and small. The network effect causes the probability of close gravitational interaction or collision to increase exponentially, just like probability of two random persons sharing the same birthday is 1/365 yet 23 random people has more than 0.5 probability of two persons sharing a birthday. Such interaction alters velocity and trajectory of comets involved, resulting in some comets falling inside of the Kuiper Belt, as well as stirring up the comet pot, causing even more close gravitational interactions.

For the comets falling inside of the Kuiper Belt, yet staying out of the frost line, the loops continue for millions of years until the comet collides with a planet. For the smaller numbers of comets venturing inside the frost line, solar radiation causes the comet to evaporate. As the loops continue and more gassy material evaporates, the comet disintegrates into smaller pieces. Most of the gassy material evaporated before colliding with an inner planet, and was blown back to the Kuiper Belt as recycled comet building material. The rocky material within the comet scattered in the inner solar system eventually collides with planets or the Sun. Rocky material in the vicinity of Mercury or Mars orbits could fall onto the Sun, or form the asteroid belt, respectively. Earth and Venus collected a large share of rocky material because of their orbits being more remote to a large mass object. Such rocky material supplement could be the reason Earth and Venus have the larger masses amongst inner planets.

Over time, a large portion of comet mass in the Kuiper Belt has been transported and stored in outer planets, like ice stored in Antarctica. The overall comet numbers have diminished, and more massive outer planets intercept more comets, resulting in a sharp reduction of comet infusion to inner planets. (Leng 2023) On the solar wind stripping side, the more atmospheric mass, the lighter the gasses, the lesser planet gravity holds on the outer layer, the easier for solar wind to strip. Thus, solar wind stripping rate is somewhat a time lag of comet infusion rate. The planet mass change timeline shall look like Fig. 1. Most of the infusion and stripping were done in the earlier half of planets' existence.

The mass of water infusion onto Venus should be of the same magnitude as Earth. Such infusion continues periodically to this day, though the intensity has drastically decreased. The fact that water is almost completely absent from the Venus, while the heavier carbon dioxide and nitrogen gasses remain indicates that solar wind is able to strip water vapor fairly fast, yet unable to strip heavier gasses at Venus' gravity.

The constant comet incursion into the inner solar system keeps the asteroid belt replenished. These asteroids have plenty of comet flybys over the eons, which could disturb their orbits and cause them to cross and collide with planets. These asteroids have the same dynamic as comets in the Kuiper Belt. The more asteroids, the more chance of gravitational interaction with the flyby comet, the more chance of orbit change and withdrawal from the asteroid belt. The number of objects in the asteroid belt is dynamically balanced.

The effect of comet evaporation inside the frost line means comet mass infused onto inner planets contains far more rocky material than outer planets. The solar hydrologic cycle is a conveyor belt that transports rocky material from the Kuiper Belt to the inner planets; to a lesser extent, it is also a conveyor belt to concentrate heavier gasses in the inner solar system.

There are three types of comet infusions, comet disintegrated/evaporated completely in the atmosphere; comet remnant collided with the crust; comet remnant penetrated the crust. When comet remnant penetrated crust, it provides gasses and water in the mantle that come up to the surface at a later time through volcano eruptions. Venus' thick

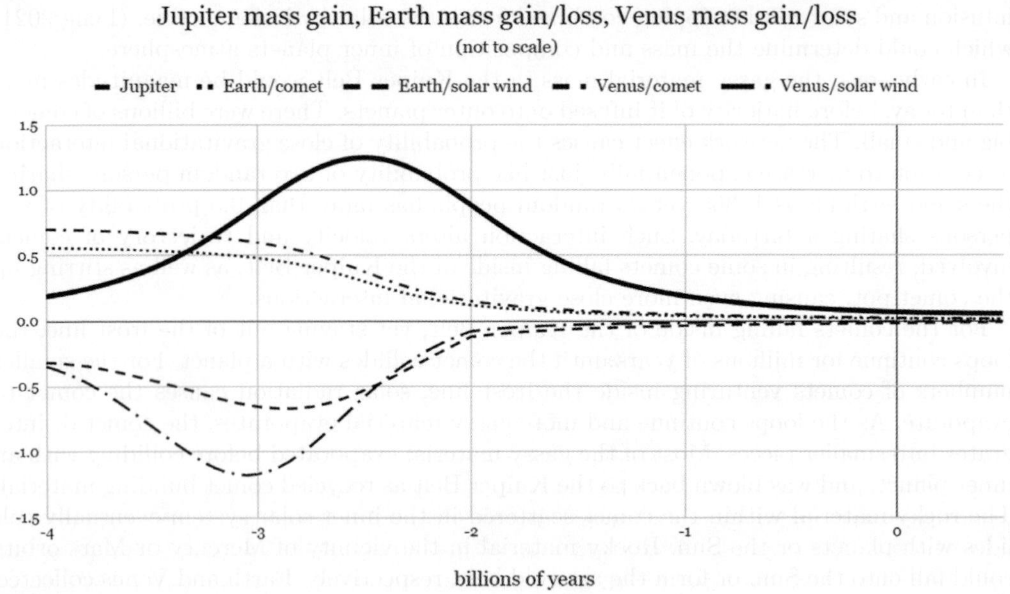

Figure 1. planet mass change timeline.

atmosphere prevented most comet collisions from reaching and penetrating the crust. That Venus has only indications which indirectly suggestive of current or geologically recent volcanism (O'Rourke et al. 2023) is consistent with the theory that volcanoes are time delayed comet infusions that penetrated the crust, (Leng 2023) not an independent source of atmospheric gasses.

Prior to the forming of crust, the tidal force from the Sun acts like constant hammer strikes on planet's molten mantle, squeezed out the gas and water contained within. Mantle degassing was mostly complete at the time of crust formation. Tidal force pounding of molten mantle also sinks heavy metals into planet's core, deprived them in the upper mantle that formed the crust. The heavy metals like gold in Earth's crust are the results of comet infusion.

The solar hydrologic cycle theory provides that comet infusion is the overwhelming input to the planet's atmosphere, and solar wind stripping is the only outlet of gas leaving the planet. For hard to strip gasses like carbon dioxide and nitrogen, all the infused masses shall still remain on Earth and Venus, after other factors, like storage in crust and mantle, and chemical conversion to other gasses, are taken into account. Comet infusion ratio between Earth and Venus calculated by solar hydrologic cycle theory can be used as a proxy for mass quantities; this ratio shall match the carbon dioxide/nitrogen mass ratio.

2. Comet Collision Mass Infusion

The amount of mass comets infused onto a planet is determined by two factors: the average comet collision probability and average comet mass at the time of collision. To accurately determine all comet collisions in the past and future involves a model of billions of loose objects in three dimensional space that have gravitational interaction with each other, all at once. These billions of objects can collide with each other, merge into one, split into many. In case of comets in the inner solar system, the mass and center of gravity are changing every second, randomly, depending on comet's structure and solar radiation at the moment and the place which varies in accordance with Sunspots and

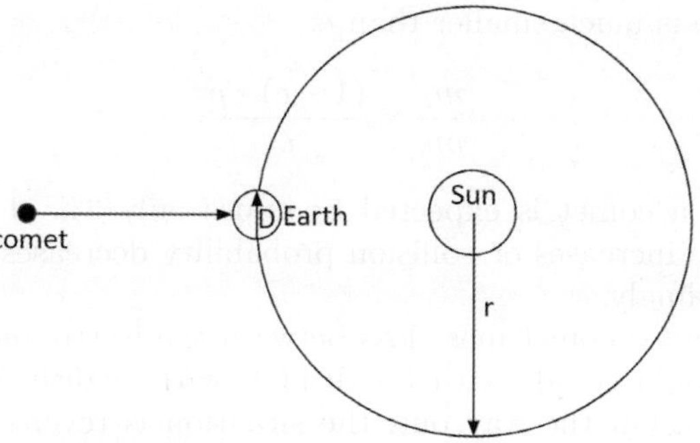

Figure 2. Comet collision probability

other random solar activities. A comet disintegrated versus a comet carries on for more loops yield totally different results afterwards. No amount of modeling or computing power is going to solve the problem of a butterfly flapping its wings causing a hurricane. The true state is unknowable. The best one can hope is a probability of collision, even that may be impossible, and one has to settle for an approximation of collision probability. Such approximation of collision probability fits well with outer planets mass relationship, (Leng 2021) can be useful for inner planets as well.

The eight planets are not on the same orbit plane, nor do the billions of smaller objects including comets. Strictly speaking, the orbit plane is not even a flat plane, rather a hyperbolic paraboloid plane with wrinkles and waves. The planets and smaller objects are constantly changing their orbit shapes and orbit planes. The simplification starts with assuming planets and loose objects loop around the Sun on a flat plane without wrinkles and waves. Since the comet orbit planes are random, they do not favor any particular planet, further simplification that all objects are on the same plane can be made, for the purpose of comparing relative collision probability between two planets. In such two dimensional model, if comet orbit intersects with a planet's orbit then collision is possible. (Fig. 2)

For comets of negligible size that cross Earth's orbit twice (one in and one out), the probability that the orbit is occupied at either crossing moment can be expressed as:

$$p = \frac{D}{\pi r} \tag{1}$$

Earth's distance to the Sun r is 150×10^6 km, thus the collision probability with Earth in a single comet loop around the Sun is 2.7×10^{-5}. For Earth and Venus, their orbit is only 0.277 AU apart, their masses are small and only differ by about 20%, it can be further assumed that all comets come inside Earth's orbit also inside Venus' orbit, and their masses make no difference in attracting comet collisions.

Comet mass at the time of collision is hinged on evaporation loss rate, heated by the solar radiation inside frost line. The evaporation accelerates the closer the comet gets to the Sun. It is estimated that a comet losing 0.1 % of mass for one approach of perihelion of 1.24 AU. (Choukroun et al. 2020) The closer the perihelion is to the Sun, the faster the loss rate will grow. For comet whose perihelion reaches Venus orbit, the loss can be a few percent. The loss rate varies with comet characteristics as well.

The cumulative evaporation of comet material becomes significant after hundreds of loops around the Sun. The expected infusion mass m_c onto planet as percentage of original mass m_0 in the lifespan of a comet is a function of evaporation loss rate e when collision probability p is much smaller than e:

$$\frac{m_c}{m_0} = \frac{(1-e)*p}{e} \tag{2}$$

At 0.1% loss rate, a comet is expected to infuse only 3% of its original mass onto Earth. If the loss rate increases or collision probability decreases, the expected infusion mass decreases accordingly.

The evaporation causes comet mass loss between the Earth and Venus orbits. On the way in, the comet would collide with Earth first, with slightly larger mass than when it collides with Venus. On the way out, the situation is reversed. On the way in, the comet mass is larger, the same solar radiation evaporates more mass. On the way out, the comet is warmer, the same solar radiation evaporates more mass. Assuming the two factors cancel each other, the planet mass gain from comet infusion is in proportion with planet diameter and inverse proportion with distance from the Sun:

$$\frac{\Delta m_1 r_1}{D_1} = \frac{\Delta m_2 r_2}{D_2} \tag{3}$$

Where Δm is the mass infusion to a planet, D is the diameter of a planet, r is the distance the planet is from the Sun.

Mass infusion/collision probabilities ratio for Venus and Earth calculated to be 1:0.76. For comparison, using the same equation, Jupiter and Earth ratio is 1:0.46, as Jupiter has a diameter of 142,800 km, or 11 times Earth's diameter, with an orbit size of 5.2 AU. Jupiter mass of 1.9×10^{27} kg is 10^7 times of Venus' atmosphere mass, both acquired from comet infusion. Evaporation can be a 10^4 reduction factor to mass infusion to Venus. 90% or more comet mass may be composed of lighter gasses, which have been blown away by solar wind, less than 10% is carbon dioxide and nitrogen, which stays on Venus, another 10^1 reduction factor. Jupiter's oversized gravity may attract 10^1 more collision than Venus' meager gravity. Finally, it is possible that for every ten comets ventured into Jupiter's orbit only one ventured further into Venus' orbit. All these factors have a range of possibilities, this is just one combination that adds up to 10^7.

3. Planet Atmosphere Chemical Processes

The solar Wind has stripped most of atmosphere away from smaller mass planets like Mercury and Mars, but only lighter gasses away from Earth and Venus. For heavier gasses, there are many chemical processes on Earth that convert gasses into crust or into lighter gasses to be blown away. The solar wind intertwines with the chemical processes in shaping planet atmosphere.

There are no significant crustal Venusian reservoirs of carbon and nitrogen, (Lécuyer et al. 2000) therefore Venus lacks the chemical processes Earth has to exchange nitrogen and carbon mass between atmosphere and crust. Venus' orbit being closer to the Sun results in surface temperature too hot for liquid water to exist. Today's Venus surface temperature approaches 750 K, (Way et al. 2016), exceeds water critical point of 647 K. In the early eons, when most of the comet infusions occurred, Venus surface temperature could be even higher. The carbon dioxide stays in the atmosphere keeping the temperature high. Over time, water vapor has been all blown away by the solar wind. Venus has the same feed stock as Earth from comet infusion, but evolved to have very different atmosphere mass and composition.

The most significant chemical processes on Earth are related to carbon dioxide. Earth have carbon dioxide infusion 0.76 times the amount in Venus atmosphere according to Eq. 3, which is two magnitudes larger than present day Earth total atmosphere mass.

Most of the carbon dioxide in Earth's atmosphere has been converted into other things stored in the crust; there is only a trace amount left in the atmosphere.

The carbon processes on Earth can be both inorganic and organic. Carbon dioxide can react directly with rocks in the presence of water without involvement of living organisms. Earth orbit is far enough from the Sun that water can stay liquid. The air pressure in early eons could be higher caused by a larger air mass than today, which increases water boiling point, enabling liquid water at a higher surface temperature. The liquid water facilitated the removal of carbon dioxide from the atmosphere, reducing greenhouse effect and cooling down surface temperature further, making life and organic processes possible. The main organic chemical process is photosynthesis, which turns carbon dioxide into oxygen and organic carbon. The total carbon, inorganic and organic, stored in all superficial Earth reservoirs, combined total of 54×10^{18} kg of carbon, while Venus has 120×10^{18} kg of carbon in its atmosphere. (Way et al. 2016) The Venus to Earth atmosphere comet infusion mass ratio calculated by carbon dioxide is about 1:0.45.

From the amount of organic carbon, the oxygen gas converted from carbon dioxide can be estimated. The majority of organic carbon is stored in Earth's lithosphere in the form of kerogen, and is estimated in the amount of 10×10^{18} kg of carbon. (Horsfield et al. 2018) For carbon dioxide to generate this amount of carbon, there would be 26.5×10^{18} kg of oxygen as a byproduct. Earth's atmosphere today has only 1.2×10^{18} kg of oxygen. A large portion of oxygen gas was consumed in weathering the Earth surface. Only after the completion of crust oxidation can oxygen accumulate in Earth's atmosphere in a substantial concentration.

The accumulation of oxygen gas in Earth's atmosphere needs another precondition, the absence of hydrogen gas. The Earth atmospheric oxygen level started to pick up around 2.5 billion years ago (bya), and fluctuated over time. (Holland 2006) Hydrogen gas was constantly infused onto Earth through comet collisions. Solar wind gained upper hand and blew hydrogen away from Earth only after the solar hydrologic cycle lost intensity around 2.5 bya. Still, there are periodic comet collisions that infuse large quantities of hydrogen. Such infusions consume the oxygen in the atmosphere and convert it to water. The fluctuation in atmospheric oxygen level can be caused by large comet infusions of hydrogen or carbon dioxide, or both.

The nitrogen gas mass of 3.9×10^{18} kg dominates the Earth's atmosphere, and 11×10^{18} kg in Venus' atmosphere is the second in abundance. (Way et al. 2016) The chemical processes involving nitrogen are also significant. The lightning on Earth can convert nitrogen gas to nitrogen oxides, which dissolve in moisture to form nitrates that are carried to Earth's surface by precipitation, which then stored in biomass and Earth crust. It is estimated that storage could amount to 2×10^{18} kg. (Lécuyer et al. 2000) Comparing Venus' atmospheric nitrogen mass to Earth's atmospheric nitrogen and crust reservoir, the ratio is about 1:0.54, fairly close to the 1:0.45 ratio calculated using carbon dioxide mass. The deficit of carbon and nitrogen on Earth from solar hydrologic cycle prediction could be due to the production and then blown away by the solar wind of methane and ammonia in early eons when hydrogen was the dominant gas. There was more feed stock of carbon dioxide than nitrogen, and carbon dioxide is less chemically inept than nitrogen, therefore, there was more carbon loss than nitrogen. Given the imprecise nature of both the solar hydrologic cycle model and account of Earth's crust reservoirs, as well as unaccounted for storage in the mantle, the ratio match is satisfactory.

References

Barabash, S., Fedorov, A., Lundin, R., & Sauvaud, J.-A. 2007, Martian atmospheric erosion rates. *Science*, 315(5811), 501–503

Choukroun, M., Altwegg, K., Kuhrt, E., Biver, N., Bockelée-Morvan, D., Drazkowska, J., Hérique, A., Hilchenbach, M., Marschall, R., Patzold, M., Taylor, M. G. G. T., & Thomas, N. 2020, Dust-to-gas and refractory-to-ice mass ratios of comet 67p/churyumov-gerasimenko from rosetta observations. *Space Science Reviews*, 216(3)

Holland, H. D. 2006, The oxygenation of the atmosphere and oceans. *Philosophical Transactions of the Royal Society B: Biological Sciences*, 361(1470), 903–915

Horsfield, B., Schulz, H.-M., Bernard, S., Mahlstedt, N., Han, Y., & Kuske, S. 2018, *Oil and Gas Shales*, 1-34. Springer International Publishing

Lécuyer, C., Simon, L., & Guyot, F. 2000, Comparison of carbon, nitrogen and water budgets on venus and the earth. *Earth and Planetary Science Letters*, 181(1-2), 33–40

Leng, X. 2021, Solar wind and hydrologic cycle. *Proceedings of the International Astronomical Union*, 17(S370), 196–197

Leng, X. 2023, Driver of tectonic plate fracture and movement. *International Journal of Physical and Mathematical Sciences*, 17(10), 149–153

O'Rourke, J. G., Wilson, C. F., Borrelli, M. E., Byrne, P. K., Dumoulin, C., Ghail, R., Gülcher, A. J. P., Jacobson, S. A., Korablev, O., Spohn, T., Way, M. J., Weller, M., & Westall, F. 2023, Venus, the planet: Introduction to the evolution of earth's sister planet. *Space Science Reviews*, 219(1)

Way, M. J., Del Genio, A. D., Kiang, N. Y., Sohl, L. E., Grinspoon, D. H., Aleinov, I., Kelley, M., & Clune, T. 2016, Was venus the first habitable world of our solar system? *Geophysical Research Letters*, 43(16), 8376–8383

Weissman, P. R. 2006, The cometary impactor flux at the earth. *Proceedings of the International Astronomical Union*, 2(S236), 441–450

Planetary Science and Exoplanets in the Era of JWST
Proceedings IAU Symposium No. 393, 2026
Z. Benkhaldoun, Y. Moulane & A. Szentgyorgyi, eds.
doi:10.1017/S1743921325001450

Numerical Geodynamic Simulations of the Thermal Evolution within the Outer Ice Shell of Ganymede

Katherine Villavicencio Valero

Dipartimento di Ingegneria e Geologia, Università degli studi G. d'Annunzio Chieti - Pescara, Italy
email: katherine.villavicencio@unich.it

Abstract. This study presents geodynamic experiments within the outer ice shell of Ganymede to investigate the effects of partial melting into the convective layer, and trying to infer any correlation between internal processes and strain deformations in surface regions such as cryovolcanism and polar deposits. The numerical models were conducted performing a sensitive analysis of the internal dynamic for compressible fluids by varying the viscosity, ice shell thickness, pressure at the melting point of ice, and applying the Anelastic Liquid Approximation (ALA). Results show molten material under restricted geophysical parameters that could allow for emerging material from the subsurface ocean to the surface. These findings will support the objectives of the RIME instrument on board of the JUICE mission, as they point out specific conditions that may, in the future, help to have a better understanding of the internal mechanisms that transfer heat on the surface and modify the geomorphology and atmospheric processes.

Keywords. Ganymede, geodynamic, partial melting, thermal evolution, JUICE mission

1. Introduction

The depth and chemical composition of the outer ice shell of Ganymede have not been well constrained until now. Studies on how tidal dissipation affects Ganymede's thermal evolution during Laplace resonance periods indicate that a minimum ice shell thickness of 10 km is necessary to maintain a passage during a resonance period, a depth of 144 km is required if solid state convection is not present in the layer, and it has been estimated from Galileo magnetometer measurements that a maximum thickness of 200 km is essential for keeping a subsurface ocean beneath the outer ice shell (Vance et al. 2014; Bland et al. 2009; Spohn & Schubert 2003; Schenk 2002).

The chemical composition of this layer is mainly based on pure water ice (Vance et al. 2014). Nonetheless, recent spectral analysis using the data taken by the JIRAM instrument on board of the Juno mission, has found the existence of ammonium chloride (NH_4Cl) and sodium/ammonium carbonate (NH_42CO_3) on the surface (Tosi et al. 2024), giving rise to the possibility of an origin from endogenic reservoirs. The presence of global oceans between two ice layers of different phases on Ganymede support the evidence of large aqueous reservoirs beneath tens of kilometers of ice shells (Prieto-Ballesteros et al. 2015). Internal water-rock interaction may have resulted in the formation of local aqueous chambers or reservoirs, which may have been the source of the current salts found on Ganymede's surface. In addition, these salts could decompose into ammonia (NH_3) as a result of kinetic effect due to freezing processes into subsurface reservoirs (Thomas et al. 2019).

On the other hand, hydrate ammonia has been suggested to be present in the interior of Ganymede in the form of trapped clathrate hydrates (Mousis & Gautierb 2004). Furthermore, it is known that a very small amount of ammonia is sufficient to lower the melting temperature of water ice and to keep the liquid ocean beneath the subsurface outer shell (Grasset et al. 2000). If ammonia acts as an anti-freezer, and some of its compounds detected on surface would have migrated from the interior, this could be linked with the ancient cryovolcanism activity that provoked the global resurfacing on the surface during the early geological history of the moon (Kargel 1994). The existence of volatile molecules like ammonia play an important effect in the generation of melt within icy mantles (Kalousová et al. 2018). However, it is unclear if the potential melt production rates would be sufficiently high to cause isolated eruptions that could reach the surface through the cryovolcano's cavities. Since it is not possible to backtrack this phenomenon, it is also unknown whether the primary melted plumes reached the surface without fractionating the lithosphere and remained trapped in the lower crust. Nevertheless, this study examines the potential circumstances that gave rise to an internal passive production of melt that is ultimately related to cryovolcanism processes. In this proceeding, are summarized the main findings of the simulations based on the assumption that the outer ice shell of Ganymede is composed by water ice and ammonia.

2. Model

The conservation of momentum, mass, and energy equations were solved using the geodynamic modelling software ASPECT (Bangerth et al. 2022), into a 2-d cartesian geometry. A minimum ice shell thickness of 30 km was chosen to initialize convection (McKinnon 2006). Ground-based observations and analysis of Cassini and Voyager data have shown that Ganymede's surface temperature fluctuates between 75 K and 150 K depending on albedo, latitude, and time of day (Squyres 1980). In this study, the simulations were run with the top temperature set at 100 K, which is the average temperature at the ice's surface (Vance et al. 2014), and the bottom temperature set at 273 K, which constitutes the ice's melting point (Jian-Jian et al. 2021). It was adopted the Anelastic Liquid Approximation (ALA), where the density variations are so small that can be neglected. The temperature dependence of the viscosity for water ice was computed setting a value of 273.2 K for the melting temperature, which corresponds to the range of temperatures for ices Ih, and reference viscosities in the range of $\eta_0 = 10^{13}$ Pa s, $\eta_0 = 10^{14}$ Pa s, and $\eta_0 = 10^{15}$ Pa s, for small grain sizes ~1 mm (Showman et al. 1997). It was incorporated the volumetric tidal dissipation for an incompressible Maxwell model, taking into consideration the maximum magnitude of tidal flexing strain of Ganymede during a diurnal orbital frequency. Constitutive equations were included, combining strain rate and various deformation mechanisms such as diffusion, dislocation, and grain boundary sliding. The occurrence of melting was calculated following the approach proposed by Vilella et al. (2020).

3. Results and Conclusions

Figure 1 shows the patterns of the thermal evolution and the melt production for models with different reference viscosities described in the previous section. The upwellings plumes of the temperature models (1a, 1c, 1e) are restricted by the stagnant lid regime at the top of the domains, which are the expected behaviour for celestial bodies strongly dependent on the viscosity (Deschamps & Vilella (2021)). The results of simulations (1b, 1d, 1f) indicate that melt could occur inside an ice crust composed of ammonia and water ice, during 100 million years of model time.

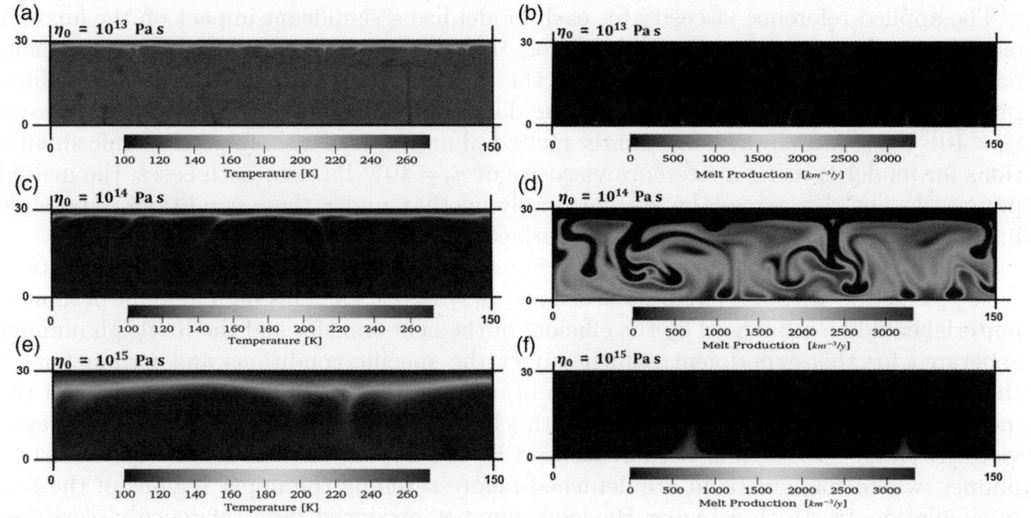

Figure 1. Screenshots of the thermal evolution and the passive production of molten material for compressible models within ice shells thickness of 30 km. Experiments (1a, 1c, 1e) show the thermal evolutionary structures with warm upwellings reaching the upper regions of the domains until the base of the stagnant lid layer. For a reference viscosity of $\eta_0 = 10^{13}$ Pa s, the melt production is very low for model 1b and is concentrated at the bottom of the layer. Melted upwellings plumes are visible on simulation 1d for a reference viscosity of $\eta_0 = 10^{14}$ Pa s, this model shows a highest production of melt among the three models. A reference viscosity of $\eta_0 = 10^{15}$ Pa s for geodynamic simulation 1f displays a visible production of melt at the bottom of the ice shell which, in comparison with model 1b, shows that the melt production is slightly higher but not enough to reach the upper regions of the box.

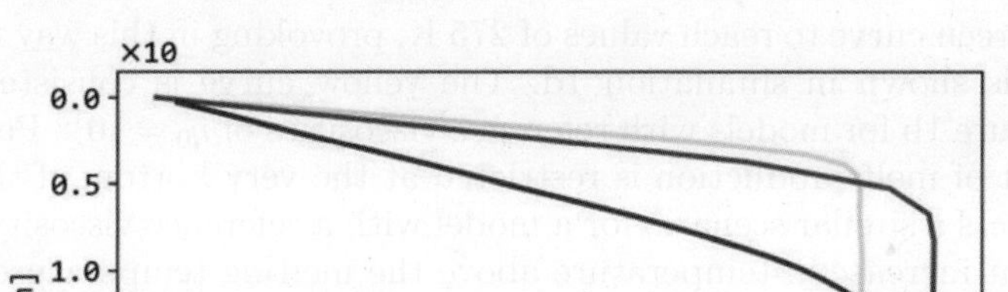

Figure 2. Thermal structures curves for ALA experiments using different reference viscosities into a cartesian domain of 30 km of depth. The jump of the curve for models with viscosities $\eta_0 = 10^{13}$ Pa s and $\eta_0 = 10^{14}$ Pa s takes place at 5 km, while the change in the direction of the curve for a model with reference viscosity of $\eta_0 = 10^{15}$ Pa s occurs at 15 km. For low viscosity (yellow curve), the temperature structure remains below the melting temperature. In contrast, for high viscosity (blue curve), the temperature remains around the melting temperature at 273.2 K. On the other hand, the green curve has an increment of temperature around 275 K, above of the melting temperature.

The applied reference viscosity for each model has a significant impact on the amount of molten material that is produced. Models 1b and 1f show very little production of melt, particularly concentrated at the base of the ice shell. Tiny and small melted upwelling plumes are still present at the base of the domain for model 1b with reference viscosity $\eta_0 = 10^{13}$ Pa s. Moreover, a few faintly melted plumes can be seen in geodynamic simulations for model 1f using a reference viscosity of $\eta_0 = 10^{15}$ Pa s. In both cases, the melted plumes do not rise across the system, implying that under these conditions a possible link between internal mechanisms and surface processes is unlikely.

In contrast to the models discussed previously, simulation 1d displays a high quantity of melt production with a reference viscosity of $\eta_0 = 10^{14}$ Pa s. This high amount of molten material could be influenced by the efficient of the heat transfer and the Rayleigh number generated for this experiment, which, under the specific conditions and parameters of this model, allows for the transportation of more material into the upper regions of the melted upwellings. Convective pattern with a low Rayleigh number generated by reference viscosity of $\eta_0 = 10^{13}$ Pa s, is characterized by the appearance of disconnected thermal plumes, where some of them are detached before reaching the upper regions of the box in simulation 1a. With a higher Rayleigh number produced by a reference viscosity of $\eta_0 = 10^{14}$ Pa s, the collision between hot and cold plumes represents a chaotic convective regime which allows for a more efficient heat transfer and an increase of production of melt as shown in model 1d. Conversely, very high Rayleigh number generated by reference viscosity of $\eta_0 = 10^{15}$ Pa s, makes the system evolving very slowly and producing big convective temperature cells with small amount of melt production, as displayed in simulations 1e and 1f. Although intense melt production is concentrated at the top of the upwellings in model 1d, the stagnant lid layer that restricts the heat transfer to the surface, precludes the melted plumes to reach the surface.

Figure 2 shows the temperature structural curves for the three reference viscosities computed in this study. The presence of ammonia lowers the melting temperature and allows for the green curve to reach values of 275 K, provoking in this way more production of melt, as it is shown in simulation 1d. The yellow curve is consistent with what is displayed in figure 1b for models with reference viscosities of $\eta_0 = 10^{13}$ Pa s, meaning that a small amount of melt production is restricted at the very bottom of the ice layer. The blue curve depicts a similar scenario for a model with a reference viscosity of $\eta_0 = 10^{15}$ Pa s, where a slight increase in temperature above the melting temperature produced some molten material, but not enough to reach the upper regions of the domain. Finally, in light of the complex physics underlying these 2D cartesian simulations, which are intrinsically simpler than 3D models that would have resulted in more thermal anomalies of various sizes, less uniform convection, and more computation resources demanding, as other 3D experiments have demonstrated (Malevsky & Yuen 1993); it was possible to observe thermal variations and diverse amounts of melt production in all 2D simulations.

For the purposes of this study, it was found that it is possible to find pockets of molten material in the interior of the outer ice shell of Ganymede, but they are limited to very restricted geophysical conditions such as a reference viscosity of $\eta_0 = 10^{14}$ Pa s, and the highly necessity of ammonia into the chemical composition of the outer ice shell. Otherwise, it is very difficult to provoke any change in the interior of this layer if it is composed mainly by water ice. The JUICE mission could potentially be able to identify those pockets of molten material inside Ganymede's outer ice shell, which would shed light on their possible link to ancient cryovolcanism activity. If they are present, the 3GM (Gravity and Geophysics of Jupiter and the Galilean Moons) radio science experiment on board the JUICE mission may detect a slight variation in densities within the ice layer due to the assumption of a mixture composition based on water ice-ammonia. Furthermore, subsurface ocean detection will remain unaffected as the measurements made by the 3GM instrument will take into account the moon's gravity field and its tidal response (Di Benedetto et al. 2017). Further analysis of heat transfer by conduction at the stagnant lid regime, along with the influence of faults in cryovolcanism and crater regions, could improve the results obtained in this research.

Supplementary material

To view supplementary material for this article, please visit https://doi.org/10.1017/S1743921325001450

References

Bangerth, W., Dannberg, J., Fraters, M., 2022, *Zenodo*
Bland, M., Showman, A., Tobie, G. 2009, *Icarus*, 200, 207
Deschamps, F., & Vilella, K. 2021, *JGR Planets*, 126
Di Benedetto, M., Ciarcia, S., Iess, L., Kaspi, Y., Mann, R., Shapira, A. 2017, *AGU Fall Meeting*, P12A-03
Grasset, O., Sotin, C., Deschamps, F. 2000, *Planet. Space Sci*, 48, 617
Kalousová, K., Sotin, C., Choblet, G., Tobie, G., Grasset, O. 2018, *Icarus*, 133, 147
Kargel 1994, *Earth, Moons, and Planets*, 67, 101
Malevsky, A., & Yuen, D. 2021, *GRL*, 383, 386
McKinnon, W. 2006, *Icarus*, 183, 432
Mousis, O., & Gautierb., D. 2004, *Planet. Space Sci*, 52, 361
Prieto-Ballesteros, O., Muñoz-Iglesias, V., Bonales, L. 2015, *An Introduction to High-Pressure Science and Technology*, 30
Schenk 2002, *Nature*, 419, 421
Showman, A., Stevenson, D., Malhotra, R. 1997, *Icarus*, 367, 383
Squyres 1980, *Icarus*, 502, 510
Spohn, T., & Schubert, G. 2003, *Icarus*, 161, 456
Thomas, E., Vu, T., Hodyss, R., Johnson, P., Choukroun, M. 2019, *Icarus*, 320, 150
Tosi, F., Mura, A., Cofano, A. 2024, *Nature Astronomy*, 8, 82
Vance, S., Bouffard, M., Choukroun, M., Sotin., C. 2014, *Planet. Space Sci*, 96, 62
Vilella, K., Choblet, G., Tsao, W., Deschamps, F. 2020, *JGR Planets*, 125
Jian-Jian, W., Meng-Qi, K., Feng-Long, H., Yu-Hao, Y., et al. 2020, *Separation and Purification Technology*, 266

Planetary Science and Exoplanets in the Era of JWST
Proceedings IAU Symposium No. 393, 2026
Z. Benkhaldoun, Y. Moulane & A. Szentgyorgyi, eds.
doi:10.1017/S1743921324002606

Observational and Chemical Modelling of Protoplanetary Disks after FU Ori outbursts

Lis Zwicky[1], Tamara Molyarova[2], Vitaly Akimkin[3], Dmitry Semenov[4,5], Ágnes Kóspál[6,7] and Péter Ábrahám[6,7]

[1]Institute of Natural Sciences and Mathematics, Ural Federal University, 19 Mira Str., 620075, Ekaterinburg, Russia
email: lis.zwicky@csfk.org

[2]Research Institute of Physics, Southern Federal University, Rostov-on-Don 344090, Russia

[3]Institute of Astronomy, Russian Academy of Sciences, 48 Pyatnitskaya St., Moscow, 119017, Russia

[4]Max Planck Institute for Astronomy, Königstuhl 17, D-69117 Heidelberg, Germany

[5]Department of Chemistry, Ludwig Maximilian University, Butenandtstr. 5-13, D-81377 Munich, Germany

[6]Konkoly Observatory, HUN-REN Research Centre for Astronomy and Earth Sciences, MTA Centre of Excellence, Konkoly-Thege Miklós út 15-17, 1121 Budapest, Hungary

[7]Institute of Physics and Astronomy, ELTE Eötvös Loránd University, Pázmány Péter sétány 1/A, 1117 Budapest, Hungary

Abstract. FU Ori stars (FUors) are undergoing powerful luminosity outbursts of $\sim$100 $L_\odot$ in magnitude and of several decades in duration. Such outbursts inevitably affect physical and chemical structure of the surrounding protoplanetary disk. Using astrochemical and radiative transfer modelling, we study the lasting impact of the outburst on disks with and without an envelope and how it changes flux in chosen molecular lines. We formulate a number of criteria indicative of a recent outburst activity based on the molecular emission, analyze the chemistry behind the flux change and apply the criteria to available observations of quiescent protoplanetary disks. The latter revealed ten objects with possible outbursts in the past and four of them satisfy multiple proposed criteria.

Keywords. protoplanetary disks, astrochemistry, FU Ori stars

FU Ori stars are young low-mass ($< 2\,M_\odot$) stars that are in a state of a powerful ($\sim$100 $L_\odot$) decades-long luminosity outburst. It has been long established that the reason behind the outburst is temporarily increased accretion rate of material onto the star (Hartmann & Kenyon 1985) but what drives this accretion enhancement is still an open question (Fischer *et al.* 2023). A critical point in investigating any type of objects is good statistics and in case of FUors there is only several dozen objects known, with almost all still being in the outburst state (Audard *et al.* 2014). The goal of this study is to find observational signatures of past FUor outbursts (post-FUors) among the most popular molecular lines.

To tackle this problem we model a set of protoplanetary disks with varying parameters: outburst magnitude, disk size and mass, star mass, inclination and presence of the envelope. For physico-chemical structure of the disk we use the ANDES astrochemical model by Akimkin *et al.* (2013) with modifications from Molyarova *et al.* (2018)

Table 1. A list of post-FUor candidates from observations in several lines. Plus sign means the object fluxes are inside post-FUor area. Minus sign means the object is outside the post-FUor area. Question mark means object is crossing border or criteria cannot be applied. Blank means no data. Multiple signs refer to multiple observations. Objects with asterisks are transitional disks, those without are Class II YSOs.

Object	$CO/^{13}CO$	$CO/C^{18}O$	$H_2CO/C^{18}O$	$HCN/C^{18}O$	$N_2H^+/C^{18}O$
[MGM2012] 307*	+				
[MGM2012] 971*	+				
[MGM2012] 556*	+	+			
CI Tau	++	+		?	
IM Lup	?	?	+	++	
AS 209	−	−	?	+?	
DM Tau*	??	−−	+	+	+
LkCa 15*			+	+	
J1604-2130*				+	
GM Aur*	?	?		?	+

to account for variable accretion rate. Emission in molecular lines is simulated by the radiative transfer code RADMC-3D (Dullemond *et al.* 2012) and further processed by CASA (Bean *et al.* 2022) to imitate ALMA observations. The detailed description of models and their parameters can be found in Zwicky *et al.* (2024).

Radiative transfer modelling is performed in one line per following molecule: CO, ^{13}CO, $C^{18}O$, H_2CO, HCN, CS, HCO^+, N_2H^+. These lines are combined into all possible line pairs ($C_8^2 = 28$) and the pairs showing a systematic difference between their pre- and post-outburst flux ratios are called post-FUor tracers (see poster in Supplementary Material). Overall, we find 10 such tracers for disks with envelope and 21 for disks without envelope. For each tracer we describe an area where only post-outburst values lie and if an observed object falls inside this area, it can indicate that it experienced an outburst in the past. We also quantify the effectiveness of found tracers by measuring how far from the area border post-outburst values are at different times (see Zwicky *et al.* 2024). According to this metric, the most effective tracers are the ones based on H_2CO, HCN, HCO^+ and N_2H^+ lines.

Additionally, we analyze the chemistry behind the flux change and find two main reasons. First is thermal desorption caused by the increased temperature during the outburst and it impacts abundances of CO, HCN and H_2CO directly and of HCO^+ and N_2H^+ indirectly. Second is enhanced photodissociation in the molecular layer that leads to an increased abundance of elementary species which, in turn, reassemble into more complex ones after the outburst. Both processes happen with and without envelope but in the presence of the envelope, light scatters off the dust grains, increasing UV field and temperatures in the upper layers of the disk. Therefore, the effect of the outburst seems weaker in embedded sources.

Finally, we apply our tracers to the existing observations of quiescent disks in CO, ^{13}CO, $C^{18}O$, H_2CO, HCN and N_2H^+ lines. Out of 37 disks we find 10 potential post-FUors, all of them are listed in Table 1 along with tracers that identified them as such. Among them there are four disks that have multiple corraborating identifications: [MGM2012] 556, CI Tau, IM Lup and LkCa 15. We note here that in Zwicky *et al.* (2024) we missed that [MGM2012] 307, 971 and 556 have transitional disks (Grant *et al.* 2018). Moreover, they are the only transitional disks in our [MGM2012] subsample. With that fact, transitional disks now compose a majority of our post-FUor candidates and a half of the best candidates. It is currently unknown why transitional disks are so prevalent among our post-FUor sample. This will be the topic of a follow-up research dedicated to transitional disks with or without outburst activity.

Supplementary material

To view supplementary material for this article, please visit https://doi.org/10.1017/S1743921324002606

References

Akimkin, V., Zhukovska, S., Wiebe, D., *et al.* 2013, *ApJ*, 766, 8
Audard, M., Ábrahám, P., Dunham, M. M., *et al.* 2014, Protostars and Planets VI, 387
Bean, B., Bhatnagar, S., Castro, S., *et al.* 2022, *PASP*, 134, 114501
Dullemond, C. P., Juhasz, A., *et al.* 2012, Astrophysics Source Code Library, record ascl:1202.015
Fischer, W. J., Hillenbrand, L. A., Herczeg, G. J., *et al.* 2023, *ASPCS*, 534, 355
Grant, S. L., Espaillat, C. C., Megeath, S. T., *et al.* 2018, *ApJ*, 836, 13
Hartmann, L. and Kenyon, S. J. 1985, *ApJ*, 299, 462
Molyarova, T., Akimkin, V., Semenov, D., *et al.* 2018, *ApJ*, 866, 46
Zwicky, L., Molyarova, T., Akimkin, V., *et al.* 2024, *MNRAS*, 527, 7652

Planetary Science and Exoplanets in the Era of JWST
Proceedings IAU Symposium No. 393, 2026
Z. Benkhaldoun, Y. Moulane & A. Szentgyorgyi, eds.
doi:10.1017/S174392132400173X

Exploring the collinear Lagrangian points of exoplanet systems with P-R drag and oblateness

S. N. Hasan[1], Priya Hasan[2] and Ibtesam Shaikh[1]

[1]Department of Mathematics, Maulana Azad National Urdu University, Hyderabad 500032, India,
email: hasan.najam@gmail.com

[2]Department of Physics, Maulana Azad National Urdu University, Hyderabad 500032, India

Abstract. We explore the position and stability of the collinear Lagrangian points in the Restricted Three-Body Problem (RTBP) where one primary body is radiative and the other is oblate. We examine the influence of Poynting-Robertson drag and the position and stability of the Lagrangian points which are affected by variations in the radiation parameter and oblateness. We compare our results with ten exoplanet systems, to identify locations in these exoplanet systems where one can detect asteroids, primodial material, or seeds where planet formation can take place. Moreover, for all ten planetary systems examined in this study, the Lagrangian points are unstable and may be possible locations where minor planets, asteroids, or debris can be found. The instability of the Lagrangian points can also be a possible cause of relocation and migration of planetesimals. These could also be used as possible candidates for observations with the James Webb.

Keywords. Three body problem, Lagrange points: collinear, stability, Poynting-Robertson drag, exoplanets

1. Introduction

Research on the three-body problem has been ongoing for almost four centuries. It lacks a universal closed-form solution. An example of this is the restricted three-body problem (RTBP), where one of the bodies has an infinitesimal mass. Khasan (1996) studied the Photogravitational RTBP and obtained the locations and stability of CLEPs. Arantza Jency et al. (2020) studied the effects of its oblateness on the location and stability of the equilibrium points.

In this work, the equations of motion for a special case of RTBP are presented, where the primary with the highest mass radiates while the primary with the lowest mass is oblate. Our analysis considers the impact of radiation pressure as well as the P-R drag and oblateness.

We obtain the Jacobian, plot the zero velocity curves for different values of the parameters *viz.* μ – the mass of the primary, β – the radiation parameter, and a – the oblateness parameter. Find the locations and stability of the Collinear Lagrangian Equilibrium Points (CLEPs). We apply our results to ten exoplanet systems, namely CoRoT-2b, TOI-1278 b, HAT-P-20 b, Kepler-75 b, WASP-89 b, TIC 172900988 b, NGTS 9b, LP 714-47 b, WASP-162 b and XO-3 b data of which have been taken from the NASA exoplanet archives.

Table 1. Location of the Collinear Lagrange Equilibrium points.

LP	X	Y	C
L1	-0.8443	0	2.8575
L2	-1.1256	0	2.8719
L3	0.9696	0	3.5446

2. Formulation of the problem

The equation of motion of a particle with mass m, cross-sectional area A, moving under the gravitational influence of two primary bodies of masses m_1 and m_2, one of which is radiating and the other is an oblate spheroid coupled with the radiation and P-R drag forces in rotating frame is given by:

$$\ddot{x} - 2\dot{y} = x - \frac{(1-\beta)(1-\mu)(x+\mu)}{r_1^3} - \frac{a\mu(x-1+\mu)}{r_2^3} + (1+sw)F_{PR,x}$$

$$\ddot{y} + 2\dot{x} = y - \frac{(1-\beta)(1-\mu)y}{r_1^3} - \frac{a\mu y}{r_2^3} + (1+sw)F_{PR,y}$$

$$\ddot{z} = -\frac{(1-\beta)(1-\mu)z}{r_1^3} - \frac{a\mu z}{r_2^3} + (1+sw)F_{PR,z} \tag{1}$$

where,

$$F_{PR,x} = -\frac{\beta(1-\mu)}{cr_1^2}\left[\frac{[(x+\mu)\dot{x} + (\dot{y}-\mu)y + z\dot{z}](x+\mu)}{r_1^2} + (\dot{x}-y)\right] \tag{2}$$

We get similar expressions for $F_{PR,y}$ and $F_{PR,z}$. We multiply first equation of system of equations(1) by $\dot{x}$ second by $\dot{y}$ and third by $\dot{z}$ and adding and integrating them accordingly we get:

$$\dot{x}^2 + \dot{y}^2 + \dot{z}^2 = 2U - C \tag{3}$$

where

$U = \frac{(1-\beta)(1-\mu)}{r_1} + \frac{a\mu}{r_2} + \frac{1}{2}(x^2+y^2)$, and

$C = \frac{2(1+sw)(1-\mu)\beta}{cr_1^2}\left[\frac{(x\dot{x}+y\dot{y}+z\dot{z})^2}{r_1^2} - [(\dot{x}^2+\dot{y}^2+\dot{z}^2) - (\dot{x}y - x\dot{y})]\right]$

and S, c, Q_{PR} are the solar energy flux density, the speed of light and the radiation pressure force coefficient, respectively, and sw represents the ratio of solar wind drag to PR to drag. We assume sw to be 0.35

3. Collinear Lagrange Equilibrium Points and Zero Velocity Curves

Lagrange equilibrium points are locations in space where the net forces is zero. In the classical case, there are only planar Lagrangian points, which lie in the orbital plane of the primaries mou (1914), but in this case, we have out of plane Lagrangian points as well.

We obtain the location of the CLEPs if $y = 0$ and $z = 0$ we obtain the location of the three collinear Lagrange points, that lie on the axis joining the two primary bodies from:

$$x - \frac{(1-\beta)(1-\mu)}{(x+\mu)^2} - \frac{a\mu}{(x-1+\mu)^2} = 0 \tag{4}$$

For $sw = 0.35$, $\beta = 0.1$, oblateness factor $a = 0.8$, mass parameter $\mu = 0.01$ the positions of the CLEP are given in the Table 1.

Table 2. Characterstic roots for L_3 point

Exoplanet systems	L_3	$\lambda_{1,2}$	$\lambda_{3,4}$
CoRoT-2b	(0.9962,0)	± 4116	± 5836
TOI-1278 b	(1.0125,0)	$\pm 19.037i$	$\pm 27.231i$
HAT-P-20 b	(1.0021,0)	$\pm 78.975i$	$\pm 111.76i$
Kepler-75 b	(1.0003,0)	$\pm 82.518i$	$\pm 116.771i$
WASP-89 b	(0.9992,0)	$\pm 196.43i$	$\pm 289.32i$
TIC 172900988 b	(0.9880,0)	± 44.539	± 62.852
NGTS 9b	(0.9866,0)	± 37.681	± 53.128
LP 714-47 b	(0.9992,0)	1110.13	784.261
WASP-162 b	(0.9958,0)	$\pm 0.00111i$	$\pm 3721.4i$
XO-3 b	(0.9870,0)	± 414.1784	± 585.51

4. Lagrangian points in Exoplanetary Systems

We explore the stability of CLEPs for the sample of ten exoplanet systems for which data was taken from the NASA Exoplanet Data Archive as shown in Table 2. To obtain the Lagrangian points, we selected systems which have atleast one planet around a star. We also needed to find the radiation parameter, i.e., the luminosity of the star as well as oblateness and masses. The sample was selected for exoplanets with varying value of radiation pressure from host stars with temperatures ranging between 3500K to 6500K (K, G and F-type stars).

From Table 2 note that for some of the exoplanetary systems, the eigenvalues of the characteristic equation for L_3 have purely imaginary roots for $0.002 < \beta < 0.02$ and $0.005 < \mu < 0.03$ which makes L_3 conditionally stable.

Zero velocity curves are plotted for different values of mass ratio, radiation pressure parameter, and oblateness factor in Figure 1.

5. Conclusions

- We observe that the location of CLEPs vary by altering the parameters *viz.* – μ mass of the primary, β– the radiation parameter, and a –the oblateness parameter.
- The effect of variation of the radiation pressure is an order of magnitude greater than the effect of variation of the oblateness on the location of the Lagrange points.
- The location of the Lagrange point shifts away from the radiating body as the radiation increases and shift towards the oblate body as the oblateness increases.
- For the ten exoplanet systems we have investigated, we note that L_1 and L_2 are always unstable while L_3 becomes stable if the value of β is in between 0.002 to 0.02 and μ is between 0.005 to 0.03.

From theories of star formation and evolution, we know that in the initial stages, a proto-star as it moves along the Henyey and Hayashi tracks till the Main sequence, goes through variation in its brightness. Due to the change in radiation pressure, there is a change in the location of the Lagrange Points hence this could be one of the mechanisms that leads to migration of planets to other locations from the point of formation. The complete paper is available at Shaikh et al. (2024).

Detailed study of protoplanetary disks is needed to future explore details of the locations of planet formation.

Acknowledgement

This research has made use of the NASA Exoplanet Archive, which is operated by the California Institute of Technology, under contract with the National Aeronautics and Space Administration under the Exoplanet Exploration Program.

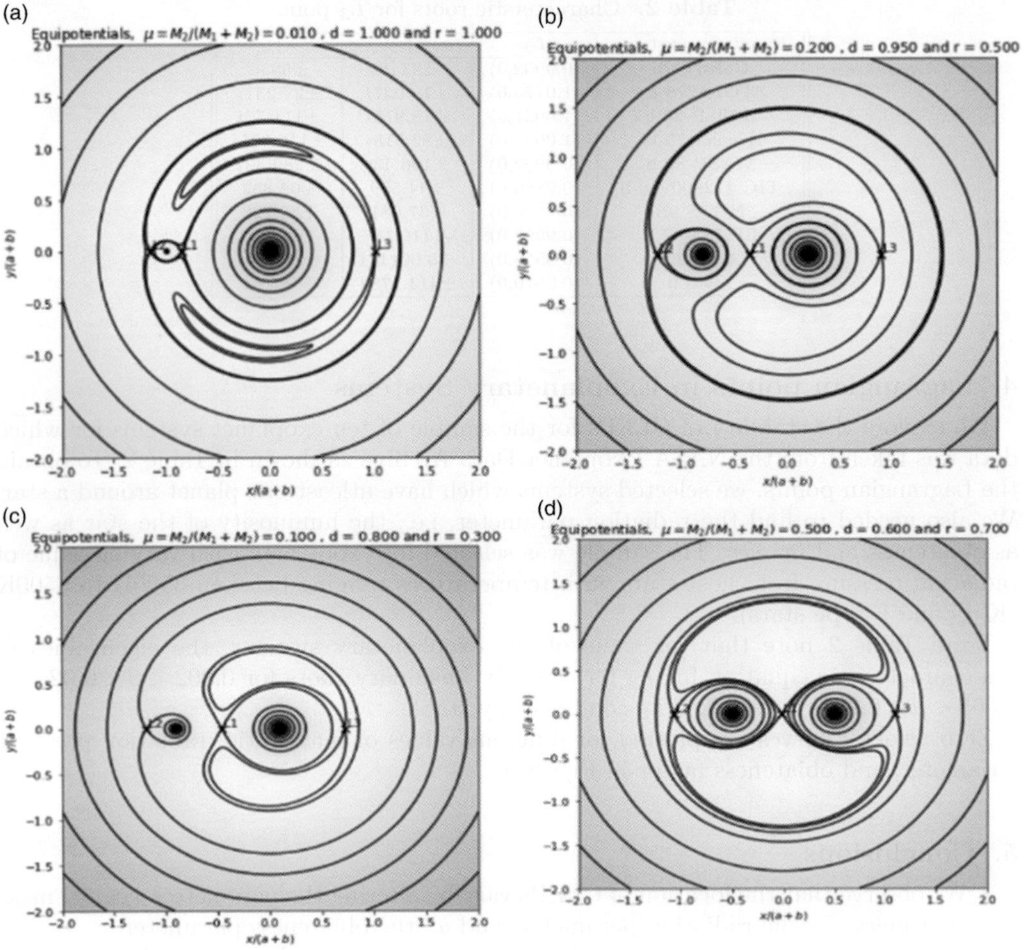

Figure 1. Detailed study of protoplanetary disks is needed to future explore details of the locations of planet formation.

References

A. Arantza Jency, R. K. Sharma, and G. Singh. Stationary solutions, critical mass, Tadpole orbits in the circular restricted three-body problem with the more massive primary as an oblate spheroid. *Indian Journal of Science and Technology* 13.39 (2020), pp. 4168–4188

S. N. Khasan. Librational solutions to the photogravitational restricted three-body problem. *Kosmicheskie Issledovaniia* 34.2 (Apr. 1996), pp. 159–165

F. R. Moulton. An Introduction to Celestial Mechanics. Macmillan, 1914

Ibtisam Shaikh and Priya Hasan and S. N. Hasan. Exploring the collinear Lagrangian points of exoplanet systems with P-R drag and oblateness. *J. Astrophys. Astr*, 45, 6, 2024

Planetary Science and Exoplanets in the Era of JWST
Proceedings IAU Symposium No. 393, 2026
Z. Benkhaldoun, Y. Moulane & A. Szentgyorgyi, eds.
doi:10.1017/S1743921324002618

Search for new members and age estimation of the young Emilkowalski asteroid family

E. D. Kuznetsov, A. S. Perminov, V. S. Safronova and M. A. Vasileva

Ural Federal University, Yekaterinburg, Russia
email: eduard.kuznetsov@urfu.ru

Abstract. We have searched for new members of the young Emilkowalski asteroid family. The number of family members reached 13. We determined proper synthetic orbital elements of family members using Orbit9 software. We investigated the dynamical evolution of the young Emilkowalski asteroid family over a 5 Myr with Orbit9 software. To estimate the ages of the pairs, we used the method of simultaneous convergence of the nodes and perihelion orbits. We considered all possible pairs between asteroids of the family. In the case of the zero-drift rate of the semi-major axis due to the influence of the Yarkovsky effect, the ages of the pairs do not exceed 2 Myr. The obtained estimates of the age of pairs allow us to limit the integration interval in the study of the probabilistic evolution of asteroid family orbits to the range of 2 Myr.

Keywords. Emilkowalski asteroid family, proper elements, pairs, age

1. Introduction

The young Emilkowalski asteroid family is located in the central Main Belt: semi-major axis $a = 2.60$ a.u., eccentricity $e = 0.15$, inclination $i = 17.75°$. Nesvorný and Vokrouhlický (2006) discovered Emilkowalski family as a cluster of three asteroids in close heliocentric orbits: (14627) Emilkowalski, (126761) 2002 DW10, and (224559) 2005 WU179. The first age estimate for the family was 220 ± 30 kyr obtained by Nesvorný and Vokrouhlický (2006) showed that it is a young family. Pravec et al. (2018) estimated the taxonomic type of asteroid (14627) Emilkowalski based on photometric observations as type D.

New family members were discovered by Pravec et al. (2018); Fatka et al. (2020), bringing the number of asteroids in the family to nine. Age estimates of the "main asteroid (14627) Emilkowalski – secondary asteroid" pairs were obtained in Pravec et al. (2018); Fatka et al. (2020). It was shown that at least two decay events of the parent asteroid of this family occurred in the last 5 Myr. This aligns with the scenario of the cascading disintegration of the parent body.

In this paper, we consider the young Emilkowalski asteroid family from the point of view of studying the formation process and attempt to reconstruct the chronological sequence of the parent body breakup. For this purpose, we: 1) search for new family members, 2) numerically investigate the dynamical evolution of asteroids based on known nominal orbits, considering the influence of the Yarkovsky effect, 3) estimate the ages of all possible pairs among the asteroids of the family and based on these estimates 4) construct a basic scenario of the family formation.

2. Search for new family members

The search for new members of young asteroid families is of interest for studying the history of formation of these families. Young families are characterized by strong clustering of both proper and osculating orbital elements. Osculating orbital elements were analyzed when searching for candidates for new members of young families. We evaluated the Kholshevnikov metrics ρ_2 and ρ_5 (see Kholshevnikov et al. (2016)), analyzed the behavior of nodes and pericenters, and searched for low relative-velocity close encounters. Synthetic proper orbital elements (see Knežević and Milani (2000); Knežević et al. (2002); Knežević and Milani (2003)) were calculated for all selected candidates using the OrbFit† software, on the basis of which a conclusion was made about the asteroid's belonging to the family.

The Kholshevnikov metrics are a tool for measuring distances in Keplerian orbital space and its factor spaces, so they are a reliable tool for identifying members of asteroid families. The metric ρ_2 is the distance between two orbits in the five-dimensional space of Keplerian orbits orbital elements a, e, i, Ω, ω (the position of the body in the orbit is not considered). The metric ρ_5 is a distance in three-dimensional factor-space of positional orbital elements a, e, i (at all possible orientations of orbital planes and apsidal lines). Osculating elements of orbits were used to calculate the metric ρ_2 and proper elements of orbits were used to calculate the metric ρ_5. The metric ρ_5 is defined in the factor space of positional orbit elements. The final decision on the membership of candidates to families was made after comparing the values of the metric ρ_5 calculated on the basis of the proper orbital elements. For family members asteroids, the ρ_5 metrics for all possible asteroid pairs were calculated and two maximum values were found: K_1 is among the pairs that include the main asteroid of the family, and K_2 is among all pairs. For candidate asteroids, the metrics ρ_5 were calculated for all possible pairs with asteroids of the family and two maximum values were found: L_1 is among pairs that include the main asteroid of the family, and L_2 is among all pairs. For new family members, one of two conditions must be satisfied:

$$L_1 \leq K_1 \quad \text{or} \quad L_2 \leq K_2. \tag{1}$$

Based on the proper orbital elements of nine known members of the Emilkowalski family, the values of $K_1 = 0.00050$ au$^{1/2}$ and $K_2 = 0.00059$ au$^{1/2}$ were obtained (see Table 4 in Perminov (2024)). For eleven family candidates, the maximum values of L_1 and L_2 of the ρ_5 metric were obtained. Four asteroids satisfy condition (1): 2015 WH29, 2016 CS377, 2017 UY114, and 2022 SA160. The number of members of the Emilkowalski asteroid family has increased to 13 (see Table 4 in Perminov (2024)). After the new asteroids were included in the family, the criteria values changed: $K_1 = 0.00050$ au$^{1/2}$ and $K_2 = 0.00062$ au$^{1/2}$, but there are no other candidates satisfying condition (1) with the new criteria values. Thus, 13 asteroids that are members of the Emilkowalski family are currently known.

3. Modelling the dynamic evolution of asteroids

Numerical modelling of the motion of asteroids of the Emilkowalski family was performed using the Orbit9 software included in the OrbFit software package. When modelling the dynamical evolution, perturbations from eight major planets and the dwarf planet Pluto, the Sun's oblateness, and relativistic effects were considered. The osculating elements of asteroid orbits from the AstDyS catalogue for the epoch MJD 60200 were taken as initial ones. The integration intervals were 5 and 2 Myr.

† http://adams.dm.unipi.it/orbfit/

Different variants of the evolution were considered: without considering the diurnal Yarkovsky effect and with this effect considered. The influence of the diurnal Yarkovsky effect was considered in the form of a constant drift rate da/dt of the semi-major axis of the asteroid's orbit a. The drift rate was estimated using known physical and thermophysical parameters or, when unavailable, using typical values for the asteroid's taxonomic class. We assumed that all asteroids, as well as the main asteroid of the (14627) Emilkowalski family, are of type D. Following Ribeiro et al. (2023), we used albedo $p_v = 0.05$, density $\rho = 1300$ kg m^{-3}, thermal conductivity $k = 0.5$ W m^{-1} K^{-1}, heat capacity $c = 500$ J kg^{-1} K^{-1}.

Estimation of the drift rate of the semi-major axis da/dt was performed using a modified version of the Orbit9 software included in the OrbFit‡ package (see Fenucci and Novakovic (2022)). The Orbit9 provides the possibility of taking into account the influence of the Yarkovsky effect and the YORP effect on the basis of setting the dynamic (period of axial rotation and inclination of the rotation axis), physical (density) and thermophysical (heat capacity, thermal conductivity, radiation and absorption coefficients of surface layers) parameters of the asteroid. In estimating the drift rate of the semi-major axes of orbits, we used the values of physical and thermophysical parameters characteristic ("average") for asteroids of taxonomic type D. For this reason, we assumed the drift rates to be constant over the entire interval under consideration.

The age of the pairs was estimated using a method based on the search for moments of convergence of the nodes and perihelion orbits (synchronous zeroing of the longitude differences of the ascending nodes $\Delta\Omega = 0$ and perihelion arguments $\Delta\omega = 0$) (see, e.g., Kuznetsov et al. (2020)). Estimates of the age of pairs T were obtained as the average of $T = (T_\Omega + T_\omega)/2$ moments of zeroing the differences between the longitudes of the ascending nodes T_Ω and the perihelion arguments T_ω. In the following, we used estimates for which the relative moment difference $\Delta T = |T_\Omega - T_\omega|/T < 0.1$. The difference ΔT provides the proximity of the zeroing moments of the differences between the longitudes of the ascending nodes and the perihelion arguments necessary for obtaining age estimates. When several evolution variants with different values of the drift rate of the semi-major axis were considered for an asteroid, the interval of possible ages of a pair was estimated. Obviously, the age estimates are given the chaotic nature of orbital evolution, and they sense to small changes in initial conditions. We will take this feature into account later when analyzing the results of probabilistic evolution. We now present simulation results for nominal orbits. Note that the uncertainty of the position of the rotation axis when the influence of the Yarkovsky effect is taken into account gives a scatter of results comparable to the uncertainty of the orbit elements (Safronova and Kuznetsov 2024). We have controlled the moments and time intervals estimates corresponding to the convergence of the nodes and perihelion of the orbits at different drift rates of the semi-major axes of the orbits. To do this, we calculated the differences metrics $\rho_2 - \rho_5$. A small value of the difference between the metrics $|\rho_2 - \rho_5|$ is a necessary condition for the youth of the pair as showed by Kuznetsov et al. (2020).

4. Scenario for the formation of the Emilkowalski asteroid family

The Emilkowalski family formation scenario was constructed based on a joint analysis of the moments of convergence of the nodes and perihelion orbits and the values of the Kholshevnikov metrics. The construction of the scenario was divided into three stages. At the first stage, the moments of simultaneous convergence of nodes and perihelion orbits were analyzed for cases of asteroid orbits convergence with the orbit of the main asteroid of the (14627) Emilkowalski family. As a result, groups of asteroids corresponding to

‡ `https://github.com/Fenu24/OrbFit`

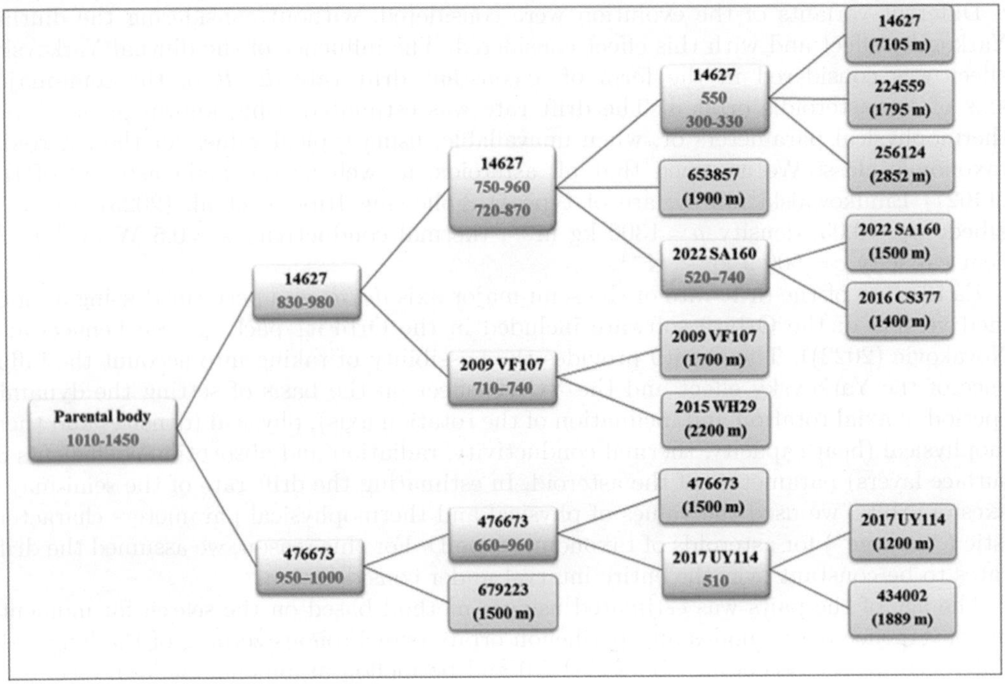

Figure 1. Emilkowalski family formation scenario.

the sequence of parent body destruction events were identified. At this stage, asteroids whose orbits do not converge with the orbit of (14627) Emilkowalski were excluded from consideration. At the second stage, the proximity of the selected pairs of orbits was analyzed based on estimates of the values of the Kholshevnikov metrics ρ_2 and ρ_5, as well as their difference $\rho_2 - \rho_5$. It was assumed that the values of metrics and their difference for the subsequent event were smaller than for the previous one. At the third stage, we analyzed the possibility of the formation of family members as a result of cascade fission of fragments of the parental body of the family. Figure 1 shows the scheme of the constructed scenario, where for each parent body the range of fragment ages in kyr (red font) is indicated in the order of their location on the scheme. For each family member, the number or name and size in meters are given.

Note that the constructed scenario of the Emilkowalski family formation is based on the analysis of the evolution of nominal asteroid orbits. When taking into account the influence of the Yarkovsky effect, the value of the drift rate of the semi-major axis was varied for asteroids for which the direction of the rotation axis is not known. To test the proposed scenario, the probabilistic evolution of all asteroids in the family will be studied. We are going to estimate the moments of low relative-velocity close encounters of asteroids and the attainment of minima of the Kholshevnikov metric ρ_2. Based on these data, it is planned to refine the age estimates of the pairs and, if necessary, to correct the developed scenario.

5. Conclusions

A scenario of the formation of the Emilkowalski family was constructed based on the analysis of the evolution of the nominal orbits of asteroids. We used the method of convergence of the nodes and perihelion orbits together with the Kholshevnikov metrics to estimate the age of the pairs. The constructed model shows that the family was formed as a result of cascading destruction of the parent body. It is possible that the family is significantly younger than previously thought ($3.020^{+1.232}_{-1.340}$ Myr (Pravec et al. 2018) or $3.447^{+0.524}_{-0.898}$ Myr (Fatka et al. 2020)), with an age of less than 1.5 Myr. We plan to test this conclusion through further studies on the probabilistic evolution of the family.

The study was funded by the Russian Science Foundation grant No. 24-22-00143, `https://rscf.ru/en/project/24-22-00143/`

References

Fatka, P., Pravec, P., Vokrouhlický, D. 2020, *Icarus*, 338, 113554

Fenucci, M., Novakovic, B. 2022, *Serbian Astronomical Journal*, 204, 51

Kholshevnikov, K.V., Kokhirova, G.I., Babadzhanov, P.B., Khamroev, U.H. 2016, *Monthly Notices of the Royal Astronomical Society*, 462, 2275

Knežević, Z., Milani, A. 2000, *Celestial Mechanics and Dynamical Astronomy*, 78, 17

Knežević, Z., Lemaître, A., Milani, A. 2002, Asteroids III / Eds.: W. Bottke, A. Cellino, P. Paolicchi, R.P. Binzel. Tucson: Univ. Arizona Press, 603

Knežević, Z., Milani, A. 2003, *A&A*, 403, 1165

Kuznetsov, E.D., Rosaev, A.E., Plavalova, E., *et al.* 2020, *Solar System Research*, 54, 236

Nesvorný, D., Vokrouhlický, D. 2006, *The Astronomical Journal*, 132, 1950

Perminov, A. Young families of asteroids. Mendeley Data, V1. 2024. doi: 10.17632/hs4rtk9np9.1

Pravec, P., Fatka, P., Vokrouhlický, D. *et al.* 2018, *Icarus*, 304, 110

Ribeiro, T.M., D'Ambrosio, A., Dominguez Calabuig, G.J., *et al.* 2023, *Acta Astronautica*, 212, 213

Safronova, V.S., Kuznetsov, E.D. 2024, *Solar System Research*, 58, 732

Planetary Science and Exoplanets in the Era of JWST
Proceedings IAU Symposium No. 393, 2026
Z. Benkhaldoun, Y. Moulane & A. Szentgyorgyi, eds.
doi:10.1017/S1743921325000572

Advances in Coronagraphy for Future Space Missions: Interferometric Apodization through Homothety (IAH) and its Application in APLC Coronagraphs

J. Chafi[1], Y. El Azhari[1,2], O. Azagrouze[1,2], A. Jabiri[1], Z. Benkhaldoun[3], A. Boskri[1] and A. Qazbour[1]

[1]Oukaimeden Observatory, High Energy Physics and Astrophysics Laboratory, Cadi Ayyad University, Marrakesh, Morocco
email: j.chafi.ced@uca.ac.ma

[2]CRMEF-Marrakech Av. Mouzdalifa, BP 797 Marrakech, Morocco

[3]Sharjah University, United Arab Emirates and University Cadi Ayyad, Morocco

Abstract. Our study addresses the challenges of direct exoplanet observation by introducing an innovative technique based on numerical simulations. We developed and evaluated a method combining Lyot coronagraphy with an Interferometric Apodization by Homothety (IAH) in segmented apertures, aimed at next-generation large telescopes. Simulations conducted using a wavelength range centered around 0.5 μm with a 20% bandwidth show significant improvements in contrast and reduced angular separations. The technique demonstrates low chromaticity, maintaining almost constant contrast across the wavelength range, an advantage over traditional methods. These findings represent a breakthrough in enhancing exoplanet imaging capabilities, enabling more precise detection even at low angular separations, and emphasize the critical role of numerical simulations in designing high dynamic range space imaging techniques.

Keywords. Homothety, Apodization, Apodized Pupil Lyot Coronagraphs (APLC), Habitable Worlds Observatory (HWO)

---

1. Introduction

The discovery and study of exoplanets are pivotal areas in modern astronomy, with direct observation playing a crucial role in identifying potentially habitable worlds. NASA's Habitable Worlds Observatory (HWO), recommended by the Astro2020 decadal survey (National Academies of Sciences, Engineering, and Medicine 2023), aims to directly image around 25 Earth-like planets and search for signs of life. To achieve this goal, specialized coronagraphs tailored to HWO's telescopes are essential (Stark et al. 2024b). The Coronagraph Design Survey (CDS) (Belikov et al. 2023, 2024) was initiated to assess global coronagraph architectures, inform HWO of their technological readiness, and identify promising emerging technologies for NASA's future investments.

Among the techniques evaluated, the Apodized Pupil Lyot Coronagraph (APLC) is noteworthy for its advanced light suppression capabilities. Widely used in ground-based systems like VLT/SPHERE (Guerri et al. 2011) and Gemini/GPI (Sivaramakrishnan et al. 2010), APLC has also been selected for future space missions, including the Roman Space Telescope (Zimmerman et al. 2016a) and LUVOIR-A (LUVOIR Team 2019; Zimmerman et al. 2016b). Upcoming projects such as LUVOIR (Laginja et al. 2019; LUVOIR Team 2019; Juanola-Parramon et al. 2022) and the

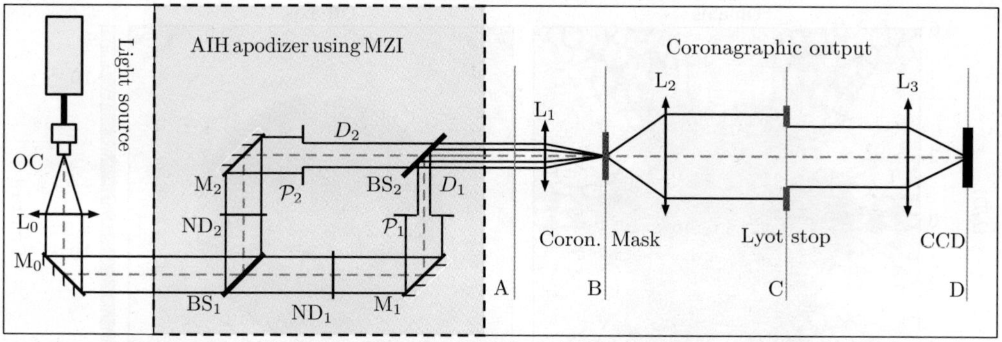

Figure 1. Schematic of an optical setup with IAH apodization in a coronagraph system. A broad-spectrum source (centered at 0.5 μm, 20% bandwidth) simulates stellar light using a microscopic objective OC and lens L_0. Mirror M_0 optimizes the optical path. The MZI apodizer includes apertures $\mathscr{P}_1$ and $\mathscr{P}_2$ with neutral density filters ND_1 and ND_2 to control intensity. The output is directed to the coronagraph with a mask in the focal plane of lens L_1. The four key planes are: A (pupil), B (focal mask), C (Lyot Stop), and D (CCD camera).

Habitable Exoplanet Observatory (HabEx) (Gaudi et al. 2020) underscore the critical importance of apodized coronagraphs in achieving extremely high contrast levels (10^{-10}) required for advanced astronomical exploration.

This study leverages the IAH technique to enhance the performance of the APLC in starlight suppression. Originally introduced by Habib et al. (2010), the IAH technique was further refined by Chafi et al. (2023a) and validated experimentally in the laboratory by Chafi et al. (2023b). This innovative integration aims to improve contrast levels and adaptability across various telescope apertures, positioning it as a promising approach for future exoplanet imaging missions.

2. Principles and formalism

In this section, we explore the fundamental principles and formalisms underlying our study, focusing on the Interferometric Apodization by Homothety (IAH) technique and its innovative design strategies. We discuss the mathematical and theoretical frameworks that support our analysis, highlighting the synergy between theory and practical application. Additionally, we emphasize the optimization processes crucial for enhancing the efficiency and effectiveness of our designs.

2.1. *Design of the IAH Coronagraph*

The IAH technique enhances a coronagraph's ability to reject starlight, improving contrast for direct imaging of exoplanets. Initially proposed by Habib et al. (2010), it uses the coherent superposition of two similar point spread functions (PSFs) to minimize diffraction wings, aligning their lobes through amplitude adjustment. As detailed by Chafi et al. (2023a), the process involves splitting a parallel light beam from the telescope using an interferometer, typically a Mach-Zehnder, and applying apodization to concentrate light energy in the central diffraction lobe. Key parameters, γ (amplitude ratio) and η (homothety ratio), are crucial for optimizing performance. The IAH technique is versatile, adapting to various aperture shapes, and is highly efficient in starlight suppression.

2.2. *Formalism*

In this section, we present the formalism of coronagraphy with apodization, inspired by Aime et al. (2002) and Soummer et al. (2003a) and tailored to our study. The

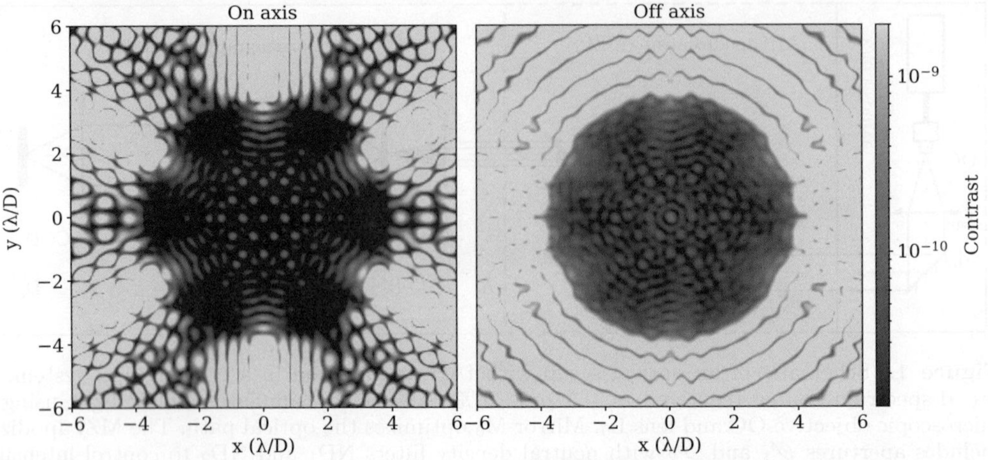

Figure 2. Simulation results in the final focal plane (plane D) of an APLC coronagraph apodized with IAH. The left panel shows the on-axis configuration for an HWO aperture, and the right panel shows the off-axis configuration. The color scale indicates the contrast levels, with values ranging from 1.32×10^{-11} to 1.9×10^{-9}.

coronagraphic system is organized into four planes: A (apodized pupil), B (focal plane with an occulting mask), C (Lyot stop), and D (final image plane). The electric field progression through these planes is described by sequential Fourier transforms.

The field amplitudes at each plane are:

$$\Psi_A(\vec{r}, \gamma, \eta) = P(\vec{r}, \gamma, \eta), \tag{1a}$$

$$\Psi_B(\vec{r}, \gamma, \eta) = \widehat{\Psi}_A(\vec{r}, \gamma, \eta)\,\mathrm{M(u)}, \tag{1b}$$

$$\Psi_C(\vec{r}, \gamma, \eta) = \widehat{\Psi}_B(\vec{r}, \gamma, \eta)\,\mathrm{S(u)}, \tag{1c}$$

$$\Psi_D(\vec{r}, \gamma, \eta) = \widehat{\Psi}_C(\vec{r}, \gamma, \eta), \tag{1d}$$

where $\vec{r}$ is the position vector, and u is the spatial frequency. The APLC mask $\mathrm{M_L(u)}$ has a transmission function defined as 1 for $u < u_m$ and 0 otherwise, with circular symmetry. The Lyot stop S(u) allows transmission within a frequency range $u \leq u_s$, blocking light beyond this limit.

3. Results

In this section, we present the simulation results of the APLC using the IAH technique applied to HWO telescope apertures. The results show that the IAH technique meets the starlight suppression levels required for imaging potentially habitable planets and enhances coronagraphic flexibility for various missions. These findings validate the IAH technique for the HWO mission, demonstrating its effectiveness in advancing exoplanet imaging capabilities.

3.1. *PSF in the final focal plane (Plane D)*

Figure 2 presents the simulation results at the final focal plane (Plane D) of an Apodized Pupil Lyot Coronagraph (APLC) using the IAH technique. The left panel shows the "on-axis" configuration, while the right panel displays the "off-axis" configuration of the HWO aperture. The axes are in units of λ/D, with contrast levels ranging from 1.32×10^{-11} to 1.9×10^{-9}, indicated by a color bar.

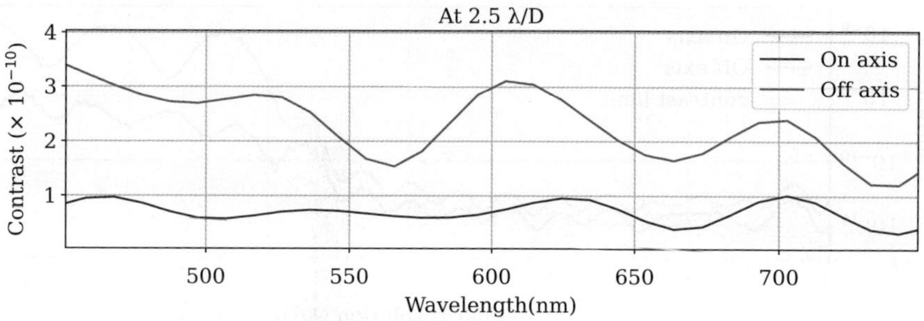

Figure 3. Contrast as a function of wavelength for the 'On axis' configuration shown in blue and the 'Off axis' configuration in red, evaluated at an angular separation of 2.5 λ/D. The results show that the contrast remains relatively stable across the wavelength range from 450 nm to 750 nm, with minimal chromatic variation observed.

This study aims to compare APLC results using the IAH technique with previous studies involving different apodizers, such as binary apodizers (Juanola-Parramon et al. 2022). Our approach utilizes optimized parameters γ and η tailored to the aperture geometry (Chafi et al. 2023a). Future work will further refine these parameters through advanced simulations for various segmented configurations. Notably, prior studies, including N'Diaye et al. (2018) and Nickson et al. (2022), achieved contrasts of 10^{-10} for various telescope architectures, highlighting the potential of APLC with optimized designs.

3.2. *Contrast and Wavelength*

Figure 3 shows the contrast as a function of wavelength for "on-axis" (blue curve) and "off-axis" (red curve) configurations at an angular separation of 2.5 λ/D. The contrast remains stable across the wavelength range of 450 nm to 750 nm, demonstrating a low chromatic effect, which is crucial for multi-spectral exoplanet observations. This indicates that the IAH-apodized APLC coronagraph effectively mitigates chromatic aberrations, maintaining high contrast necessary for detecting faint planetary signals.

The "off-axis" configuration generally achieves higher contrast than the "on-axis" due to its improved light suppression and reduced diffraction effects. These results highlight the advantages of the "off-axis" setup for missions targeting close angular separations. The minimal chromatic variation observed underscores the robustness of our approach, making it ideal for future space missions where consistent contrast across wavelengths is vital. The performance differences emphasize the need to optimize coronagraph design and orientation to enhance imaging capabilities in various observational contexts.

3.3. *Contrast and Angular Separation*

Figure 4 displays the azimuthal average contrast for the HWO aperture on a logarithmic scale, with red indicating the "off-axis" and blue the "on-axis" configurations. The dotted line marks the contrast threshold of 10^{-10}. The "off-axis" configuration achieves contrast below 10^{-10} at 5 λ/D, while the "on-axis" reaches this at 4 λ/D, demonstrating effective stray light suppression and improved exoplanet detection capabilities.

Comparatively, Stark et al. (2024a) focuses on bandwidth optimization in the ExoEarth study of HWO, emphasizing the need for precise detection and characterization wavelengths. This work underscores the importance of optimizing detection parameters to enhance the efficiency and accuracy of exoplanet missions, aligning with our goal of advancing direct exoplanet detection through advanced coronagraph technologies.

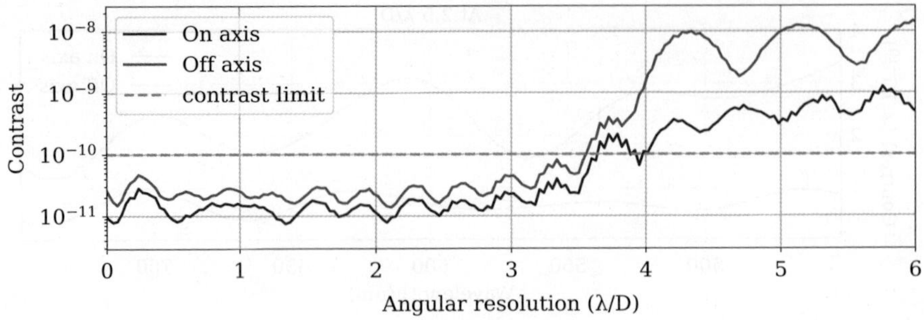

Figure 4. Azimuthal average for the HWO aperture configuration, shown on a logarithmic scale with blue for the On axis case and red for the Off axis case. The dotted line represents the contrast threshold of 10^{-10}. A contrast below 10^{-10} is observed for angular separations of 5 λ/D in the Off axis case, while for the On axis case, it is 4 λ/D. This highlights the effectiveness of the configuration in suppressing stray light and improving the detection of exoplanets.

4. Conclusion

This study showcases the advancement of APLC coronagraphy with the IAH technique in both on-axis and off-axis configurations, achieving contrast levels of at least 10^{-10} using specific apodization parameters ($\gamma = 1.928$, $\eta = 1.657$) adapted for the HWO mission. This combination improves starlight suppression and diffraction management, essential for detecting potentially habitable exoplanets near bright stars.

The IAH-apodized APLC minimizes chromatic and optical aberrations, enabling more precise exoplanet observations. While the results are based on simulations, upcoming experiments will validate performance under real conditions, reinforcing its potential for future missions.

However, the technique faces challenges, including signal loss from amplitude control filters and complexities with segmented apertures. Further optimization of parameters like γ and η is necessary to enhance overall coronagraph performance.

References

Aime, C., et al. A&A, 2002. doi:10.1051/0004-6361:20020419

Belikov, R., et al. In: Techniques and Instrumentation for Detection of Exoplanets XI, SPIE, vol. 12680, p. 126802G, 2023. doi:10.1117/12.2677732

Belikov et al. Space Telescopes and Instrumentation 2024: Optical, Infrared, and Millimeter Wave, SPIE, vol. 13092, 2024, doi:10.1117/12.3020614

Chafi, J., et al. MNRAS, vol. 523, no. 4, pp. 5442–5451, 2023

Chafi, J., et al. MNRAS, vol. 527, no. 3, pp. 7036–7046, 2023

Gaudi, B. S., et al. arXiv e-prints, 2020, arXiv:2001.06683

Guerri, G., et al. Experimental Astronomy, vol. 30, no. 1, pp. 59–81, 2011

Habib, A., et al. MNRAS, vol. 406, no. 4, pp. 2743–2748, 2010. doi:10.1111/j.1365-2966.2010.16873.x

Juanola-Parramon, R., et al. Journal of Astronomical Telescopes, Instruments, and Systems, vol. 8, p. 034001, 2022. doi:10.1117/1.JATIS.8.3.034001

Laginja, I., et al. SPIE Conference Series, vol. 11117, 2019, doi:10.1117/12.2530300

The LUVOIR Team. arXiv e-prints, 2019, arXiv:1912.06219

National Academies of Sciences, Engineering, and Medicine. The National Academies Press, Washington, DC, 2023, doi:10.17226/26141

Nickson, B. F., et al. In: Space Telescopes and Instrumentation 2022, SPIE, vol. 12180, p. 121805K, 2022. doi:10.1117/12.2630686

N'Diaye, M., et al. In: Space Telescopes and Instrumentation 2018, SPIE, vol. 10698, p. 106986A, 2018. doi:10.1117/12.2313225

Sivaramakrishnan, A., et al. Ground-based and Airborne Instrumentation for Astronomy III, SPIE, vol. 7735, 2010, doi:10.1117/12.858623

Soummer, R., et al. A&A, vol. 397, pp. 1161–1172, 2003. doi:10.1051/0004-6361:20021573

Stark, C. C., et al. Journal of Astronomical Telescopes, Instruments, and Systems, vol. 10, no. 1, 2024. doi:10.1117/1.jatis.10.1.014005

Stark, C. C., et al. Cornell University, 2024. doi:10.48550/arxiv.2405.19418

Zimmerman, N. T., et al. Journal of Astronomical Telescopes, Instruments, and Systems, vol. 2, 2016, doi:10.1117/1.JATIS.2.1.011012

Zimmerman, N. T., et al. Space Telescopes and Instrumentation 2016, SPIE, vol. 9904, 2016, doi:10.1117/12.2233205

Planetary Science and Exoplanets in the Era of JWST
Proceedings IAU Symposium No. 393, 2026
Z. Benkhaldoun, Y. Moulane & A. Szentgyorgyi, eds.
doi:10.1017/S1743921324001753

Spectral Characterization of a Complete Equatorial Sample of 615 K Dwarfs

Hodari-Sadiki Hubbard-James[1,2]

[1]Dept. of Physics and Astronomy, Agnes Scott College,
Decatur, Georgia, 30030, USA
email: hjames@agnesscott.edu

[2]RECONS Institute, Chambersburg, PA 17201, USA

Abstract. Our study analyzes nearby (within 33 pc) K dwarf stars to determine their stellar properties, activity levels, kinematics, and estimated ages, as part of the RKSTAR (RECONS K Star) Survey encompassing about 5000 K dwarf primaries within 50 pc. Utilizing high-resolution spectra from the CHIRON echelle spectrometer, we established a benchmark set of 35 K dwarfs aged 20 Myr to 5 Gyr, focusing on the Hα (6563 Å) and Li I (6707.9 Å) lines. We extended our analysis to 615 K dwarfs within 33 pc and found that about 8% exhibit spectroscopic features characteristic of young or active stars. As expected, Galactic *UVW* space motions indicate that most of the stars fall into the thin (80%) and thick (20%) disk populations, with a single outlier, HD 134439, which is a known halo star. Additionally, 4% of the stars were identified as metal-poor ([Fe/H] ≤ -0.5 dex). This work identifies 500 inactive and mature K dwarfs as prime targets for detecting terrestrial planets and serves as a vital resource for assessing host star suitability for exoplanet habitability.

Keywords. Late-type stars (909), Spectroscopy (1558), Stellar activity (1580)

---

1. Introduction

K dwarfs, characterized by surface temperatures between 3930 and 5270 K and masses of 0.59–0.88 solar masses (Henry & McCarthy (1993); Gray & Corbally (2009); Pecaut & Mamajek (2013)), sit between G-type and M-type stars on the main sequence. Their *Gaia* EDR3 absolute *BP* magnitudes range from M_{BP} = 5.9 to 9.6 and their colors span BP–K = 2.0 to 4.0. Despite their cooler temperatures and slightly lower masses compared to the Sun, K dwarfs are compelling candidates for exoplanet studies due to their longer lifespans and relative abundance. They account for around 12% of stars in the solar neighborhood, which is more than double the 5% represented by G stars (Henry et al. (2006, 2018)).

Despite this potential, K dwarfs have not been the primary focus of past exoplanet surveys, which have often targeted brighter F and G-type stars or smaller M dwarfs (Arney (2019); Richey-Yowell et al. (2019)). Observational biases have favored G stars due to their higher photon output, which leads to improved signal-to-noise ratios. M dwarfs have been selected because their smaller radii enhance the detectable signals from orbiting planets. This leaves mid and late-type K dwarfs, particularly those with temperatures below 4800 K, underexplored in the search for exoplanets. Figure 1 illustrates this observational gap, showing that within 25 pc, fewer K dwarfs are confirmed planet hosts than G or M dwarfs, not due to a lack of planets but rather limited survey focus.

The stellar age is a key factor in evaluating the habitability of exoplanets. Young, active K dwarfs, with their high ultraviolet luminosity and chromospheric flares, may

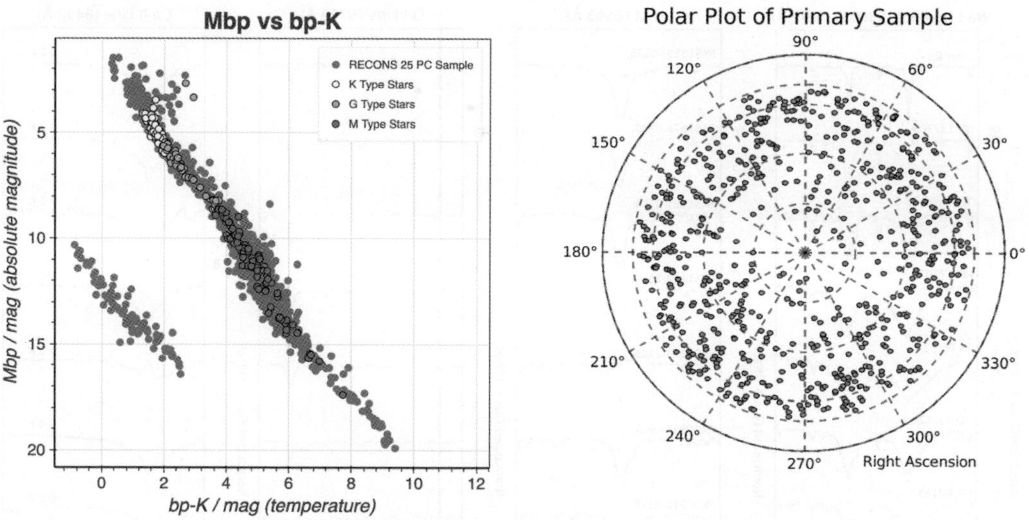

Figure 1. Left: HR diagram (Illustrated by: Daniela Garcia-Lara) displaying the RECONS sample of stars within 25 pc (grey dots), with confirmed exoplanet hosts as of June 30, 2024 from the NASA Exoplanet Archive highlighted. Right: Polar Plot illustrating right ascension (R.A.) (theta direction) and distance (radial direction) for the primary sample of 615 K dwarfs.

present environments hostile to life (Segura et al. (2010); Davenport et al. (2019)). Conversely, older, quieter K dwarfs provide more stable conditions conducive to habitability (Luger et al. (2015); Arney (2019)). Understanding the stellar properties, activity levels, and kinematics of these stars is essential for determining their ages and, consequently, their suitability as hosts for habitable worlds.

In summary, this paper investigates the spectral and kinematic properties of nearby K dwarfs to explore their potential as prime targets for habitable exoplanet searches.

2. The RECONS K Stars (RKSTAR) Sample

The RKSTAR Project aims to survey the ∼5,000 nearest K dwarfs within 50 parsecs of the Sun, based on *Gaia* DR2 parallax measurements and updates in subsequent *Gaia* data releases. The RECONS team is conducting four systematic surveys to study this sample comprehensively. The **Wide Field Survey** searches for stellar companions with separations greater than 1" using data from *Gaia* and the Washington Double Star (WDS) Catalog (Mason et al. (2001); Shea et al. (2023)). The **Speckle Survey** employs optical speckle cameras on 4 m to 8 m telescopes to detect companions within 0.5 to 100 AU (Horch et al. (2009, 2021)). The **Radial Velocity Survey** uses the CHIRON spectrograph on the SMARTS 1.5 m telescope to identify stellar, brown dwarf, and jovian companions orbiting within 3 AU (Paredes et al. (2021)).

The **Characterization Survey**, the primary focus of this paper, aims to evaluate the activity levels, ages, stellar properties, and kinematic motions of K dwarfs within 50 pc using high-resolution CHIRON spectra. This study examines two distinct samples: a primary set of 615 field K dwarfs located within 33 pc shown in Figure 1, and a benchmark sample of 35 stars with established age estimates for comparison (Hubbard-James et al. (2022)). Together, these surveys contribute to a comprehensive understanding of nearby K dwarfs as potential hosts for habitable planets.

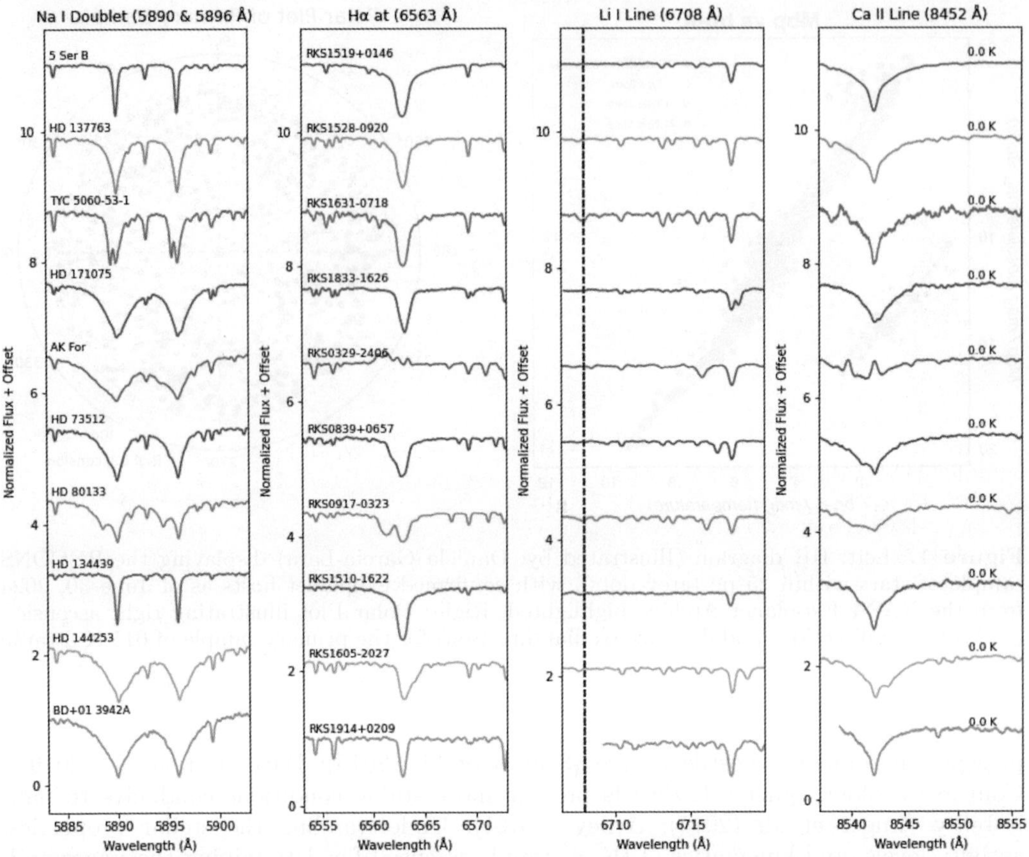

Figure 2. Example spectra from the spectral library showcasing key features (Na I doublet, Hα, Li I, and Ca II IRT lines) that reveal variations in surface gravity, magnetic activity, and age among K dwarfs.

3. The Spectral Library of K Dwarfs

Figure 2 showcases ten unique spectra from our growing spectral library, which currently includes high-resolution CHIRON spectra (R = 80,000) for 600 K dwarfs. The selected spectra highlight four significant spectral features: the Na I doublet (5891.58 and 5897.56 Å), the Hα line (6562.8 Å), the Li I line (6707.8 Å), and the Ca II infrared triplet (IRT) line (8542 Å). These features provide insight into key stellar properties, such as surface gravity, magnetic activity, and age, across a diverse range of effective temperatures ($T_{\rm eff}$) and metallicities ([Fe/H]). The library enables visual identification of active stars, spectroscopic binaries, and metal-poor K dwarfs. This resource is publicly accessible and will continue to expand with additional spectra from the RKSTAR sample. The spectral library can be found at: `https://hodarijames.github.io/spectral_library/page1.html`.

4. Spectroscopically Identifying Young and Active K Dwarfs

To identify young or active K dwarfs, we analyzed the equivalent widths (EWs) of key spectral lines, focusing on Hα and Li I. Figure 3's left plot, shows EW[Hα] versus EW[Li I] for 590 stars from our primary sample, with benchmark stars overlaid in various age groups for comparison. Stars within the dashed rectangular "old locus" box correspond to K dwarfs older than 750 Myr, such as those in the Hyades and field populations

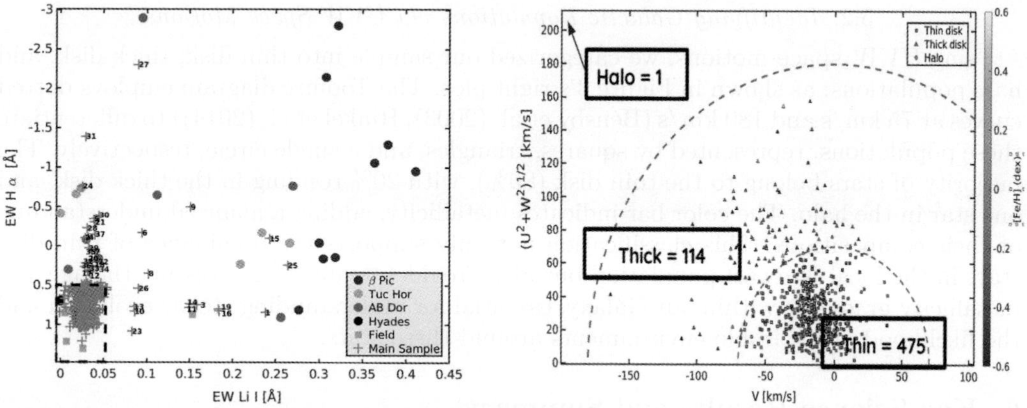

Figure 3. Left: The plot presents EW[Hα] vs. EW[LiI] for 590 K dwarfs from our main sample, including a set of 35 benchmark stars with estimated ages. Highlighted in the plot are 37 stars recognized as young or active. **Right:** A Toomre diagram of our 590-star sample helps distinguish Galactic regions. The dotted curves at 75 km/s and 180 km/s roughly delineate the thin disk, thick disk, and halo populations. The overlaid color bar indicates metallicity ([Fe/H]) for each star. The thin disk (squares), thick disk (triangles), and halo (circle) contain 475, 114, and 1 stars respectively.

(Hubbard-James et al. (2022)). Stars lying outside this region are flagged as either young or active, based on Li I EWs greater than 0.05 Å or Hα EWs below 0.5 Å.

Our analysis identified 37 stars from the primary sample as young, active, or both, with 16 showing signs of youth and 27 classified as active. Six of these stars display both characteristics. The stars identified as young or active include HD 17925, AK For, and LQ Hya, among others. These stars lie outside the old locus, indicating higher levels of activity or youthfulness. The remaining stars in our sample exhibit quiescent values for EW[Hα] and EW[Li I], suggesting older ages, lower chromospheric activity, and no significant close stellar companions that could induce activity.

5. Kinematics of the Primary Sample

Kinematic analysis plays a pivotal role in understanding the age, origin, and potential habitability of nearby K dwarfs. We analyze *UVW* space motions for our sample of 590 K dwarfs to identify young stars and classify them within the Galactic structure.

5.1. *Using Galactic Velocities to Identify Young K Dwarfs*

The *UVW* space motions of our sample were derived using values from Riedel et al. (2014) and the BANYAN Σ code (Gagné et al. (2018)), which aids in associating stars with known moving groups. Among the key kinematic groups analyzed, AB Dor and Hyades exhibit distinct velocities, $UVW = [-7, -27, -13]$ km/s and $UVW = [-41, -19, -1]$ km/s, respectively; Riedel et al. (2014). Our analysis identified four stars as members of the AB Dor moving group and seven as members of the Hyades. Interestingly, nine of these stars were not initially flagged as young or active through spectroscopic features alone, underscoring the value of kinematic data in identifying youthful characteristics. This expands the count of young or active stars in our primary sample to 46 (approximately 8%), highlighting the crucial role that space motions play in revealing stellar youth and activity beyond what is evident from spectral analysis.

5.2. *Identifying Galactic Populations via UVW Space Motions*

Using UVW space motions, we categorized our sample into thin disk, thick disk, and halo populations, as shown in Figure 3's right plot. The Toomre diagram employs dotted curves at 75 km/s and 180 km/s (Bensby et al. (2003); Hinkel et al. (2014)) to differentiate these populations, represented by squares, triangles, and a single circle, respectively. The majority of stars belong to the thin disk (80%), with 20% residing in the thick disk, and one star in the halo. The color bar indicates metallicity, adding a nuanced understanding of their compositions. This classification not only supports the dominance of thin disk stars in the local K dwarf population but also provides context for assessing the age and metallicity gradients within our Galaxy, essential for understanding stellar evolution and the likelihood of habitable environments around these stars.

6. Key Science Results and Summary

Spectral Library: We have developed an extensive spectral library consisting of high-resolution CHIRON spectra for 600 nearby K dwarfs. This library offers a valuable resource for the astronomical community, providing data for identifying spectroscopic binaries, evaluating activity levels, and determining stellar properties. The spectral library facilitated the discovery of two new double-lined spectroscopic binaries (SB2s) and serves as a foundation for broader kinematic and spectral analyses.

Youth and Activity: By applying both spectroscopic and kinematic techniques, we identified approximately 8% of the stars in our sample as young or chromospherically active. These identifications were made based on the equivalent widths of key spectral lines, such as Hα and Li I, alongside their space motions within the Milky Way. This subset provides insights into the prevalence of youth and activity within the K dwarf population, offering clues about their evolutionary status and potential impact on planetary systems.

Galactic Populations: Utilizing the Galactic UVW space velocities, we classified our sample of 590 stars into three distinct populations. The vast majority of stars belong to the thin disk (475 stars, 80%), with a significant fraction in the thick disk (114 stars, 20%), and a single star attributed to the Galactic halo.

Metal-Poor Population: Among our 590 K dwarfs, we identified a subset of 21 stars with metallicity values ([Fe/H]) of -0.5 dex or lower, making up about 4% of our sample. This metal-poor population provides a unique opportunity to study the chemical properties of K dwarfs in the local neighborhood and their role in the broader context of Galactic chemical evolution.

The Calm K Dwarfs: A key outcome of this work is the identification of a set of 500 Calm K dwarfs. This group was determined by excluding 46 stars flagged as young or active and 81 close-binary systems from our sample of 615 K dwarfs. These calm stars, which show low levels of activity and do not possess close stellar companions, are optimal targets for future exoplanet surveys aiming to find habitable terrestrial planets. This subset stands as a significant contribution to the field, highlighting K dwarfs as prime candidates for the search for potentially life-supporting exoplanets.

References

Arney, G. N. 2019, ApJL, 873, L7. doi:10.3847/2041-8213/ab0651

Bensby, T., Feltzing, S., & Lundström, I. 2003, A&A, 410, 527. doi:10.1051/0004-6361:20031213

Cuntz, M., & Guinan, E. F. 2016, ApJ, 827, 79. doi:10.3847/0004-637X/827/1/79
Davenport, J. R. A., Covey, K. R., Clarke, R. W., et al. 2019, ApJ, 871, 241. doi:10.3847/1538-4357/aafb76
Gagné, J., Mamajek, E. E., Malo, L., et al. 2018, ApJ, 856, 23. doi:10.3847/1538-4357/aaae09
Gaidos, E., Fischer, D. A., Mann, A. W., et al. 2013, ApJ, 771, 18. doi:10.1088/0004-637X/771/1/18
Gray, R. O., & Corbally, C. 2009, Stellar Spectral Classification, Princeton University Press, ISBN: 978-0-691-12511-4
Henry, T. J., & McCarthy, D. W. 1993, AJ, 106, 773. doi:10.1086/116685
Henry, T. J., Jao, W.-C., Subasavage, J. P., et al. 2006, AJ, 132, 2360. doi:10.1086/508233
Henry, T. J., Jao, W.-C., Winters, J. G., et al. 2018, AJ, 155, 265. doi:10.3847/1538-3881/aac262
Hinkel, N. R., Timmes, F. X., Young, P. A., et al. 2014, AJ, 148, 54. doi:10.1088/0004-6256/148/3/54
Horch, E. P., Veillette, D. R., Baena Gallé, R., et al. 2009, AJ, 137, 5057. doi:10.1088/0004-6256/137/6/5057
Horch, E. P., Broderick, K. G., Casetti-Dinescu, D. I., et al. 2021, AJ, 161, 295. doi:10.3847/1538-3881/abf9a8
Hubbard-James, H.-S., Lesley, D. X., Henry, T. J., et al. 2022, AJ, 164, 174. doi:10.3847/1538-3881/ac8d6a
Kunimoto, M., & Matthews, J. M. 2020, AJ, 159, 248. doi:10.3847/1538-3881/ab88b0
Lillo-Box, J., Santos, N. C., Santerne, A., et al. 2022, A&A, 667, A102. doi:10.1051/0004-6361/202243898
Luger, R., Barnes, R., Lopez, E., et al. 2015, Astrobiology, 15, 57. doi:10.1089/ast.2014.1215
Mason, B. D., Wycoff, G. L., Hartkopf, W. I., et al. 2001, AJ, 122, 3466. doi:10.1086/323920
Paredes, L. A., Henry, T. J., Quinn, S. N., et al. 2021, AJ, 162, 176. doi:10.3847/1538-3881/ac082a
Pecaut, M. J., & Mamajek, E. E. 2013, ApJS, 208, 9. doi:10.1088/0067-0049/208/1/9
Richey-Yowell, T., Shkolnik, E. L., Schneider, A. C., et al. 2019, ApJ, 872, 17. doi:10.3847/1538-4357/aafa74
Riedel, A. R., Finch, C. T., Henry, T. J., et al. 2014, AJ, 147, 85. doi:10.1088/0004-6256/147/4/85
Segura, A., Walkowicz, L. M., Meadows, V., et al. 2010, Astrobiology, 10, 751. doi:10.1089/ast.2009.0376
Shea, M., Hansen, C., Horch, E., et al. 2023, American Astronomical Society

Planetary Science and Exoplanets in the Era of JWST
Proceedings IAU Symposium No. 393, 2026
Z. Benkhaldoun, Y. Moulane & A. Szentgyorgyi, eds.
doi:10.1017/S1743921324001741

Detecting and characterising exoplanets with HARPS-N

K. Rice[1], A. Mortier[2], L. Malavolta[3,4], F. Pepe[5], A. Cameron[6], A. Ghedina[4], D. Latham[7], S. Udry[4], L. Affer[4], S. Aigrain[8], A. S. Bonomo[4], V. Bourrier[5], L. A. Buchhave[9], H. Cegla[10], P. Cortes-Zuleta[6], R. Cosentino[4], M. Cretignier[8], M. Damasso[4], X. Dumusque[5], D. Ehrenreich[5], A. A. John[6], B. Klein[8], A. Leleu[5], M. Lopez-Morales[7], C. Lovis[5], J. Maldonado[4], A. Martínez Fiorenzano[4], G. Micela[4], E. Molinari[4], B. Nicholson[11], N. O'Sullivan[8], M. Pinamonti[4], G. Piotto[3,4], D. Sasselov[7], A. M. Silva[12], A. Sozzetti[4], M. Stalport[13], A. Vanderburg[14], S. Vissapragada[7], C. Watson[15] and T. Wilson[10]

[1]University of Edinburgh, [2]University of Birmingham, [3]Università degli Studi di Padova, [4]INAF, [5]University of Geneva, [6]University of St Andrews, [7]Center for Astrophysics | Harvard & Smithsonian, [8]University of Oxford, [9]Technical University of Denmark, [10]University of Warwick, [11]University of Southern Queensland, [12]IA-Porto/CAUP, [13]University of Liège, [14]MIT, [15]Queen's University Belfast
email: wkmr@roe.ac.uk

Abstract. Exoplanet follow-up with JWST requires precise masses and radii. HARPS-N is a high-resolution spectrograph on the Telescopio Nazionale Galileo (TNG), predominantly used to detect and characterize exoplanets using the radial velocity (RV) method. The HARPS-N Collaboration has been characterising exoplanets with HARPS-N for over a decade. In this short paper we highlight the contributions that the HARPS-N Collaboration has made to the characterisation of small exoplanets.

Keywords. Exoplanets, Radial velocities, Transits, Exoplanet characterisation

1. Introduction

The High Accuracy Radial Velocity Planet Searcher for the Northern hemisphere (HARPS-N) is a high-precision radial-velocity spectrometer located on the Telescopio Nazionale Galileo (TNG) on La Palma. It has a spectral resolution of $R = 115000$, covers the wavelength range from 378 nm to 691 nm (Cosentino et al. 2012) and is a close copy of the HARPS instrument on the 3.6-m ESO telescope at La Silla, Chile (e.g., Mayor et al. 2003).

The HARPS-N Project was initially a collaboration between the Department of Astrophysics of the University of Geneva, Center for Astrophysics – Harvard & Smithsonian, University of St-Andrews, University of Edinburgh, Queen's University Belfast, and the Fundación Galileo Galilei - Instituto Nazionale di Astrofisica, but is now also open to researchers from other institutions. This collaboration, often referred to as the GTO Collaboration, still contributes to maintenance of the instrument and development of the data reduction pipeline.

The HARPS-N Collaboration has three broad scientific goals:†

- to characterize "super-Earths" in various orbits with enough precision to distinguish between volatile-rich and predominantly iron-silicate planets;
- to measure accurate masses to a precision of 10%, or better, for planets near the transition between super-Earths and Ice Giants (i.e., ∼10 $M_\oplus$);
- to confirm an Earth-twin planet in the habitable zone of a G5V star or later, with a precision of 30% in mass.

The HARPS-N Collaboration also has a secondary scientific goal of detecting Earth-like planets around nearby stars. Referred to as the Rocky Planet Search (RPS), this involves observing a sample of the brightest, least active K-dwarfs in the Northern hemisphere.

HARPS-N started making observations in late 2012. In this short paper, we highlight some of the successes of just over a decade of work by the HARPS-N Collaboration and also the impact that HARPS-N has had on the detection and characterisation of exoplanets, in particular presenting precise masses and radii of exoplanets that will be key targets for JWST. We focus mostly on the work done by the Collaboration, but should stress that HARPS-N has also been very successfully used by other groups (e.g., Covino et al. 2013; Luque et al. 2023).

2. Synergy with *Kepler*

One of the motivations for HARPS-N was to follow-up targets first detected by NASA's *Kepler* mission (Borucki et al. 2010). The *Kepler* satellite launched in March 2009 and initially monitored a single field in Cygnus/Lyra, identifying more than 2300 transiting exoplanets.

The HARPS-N Collaboration published mass estimates for a number of these *Kepler* targets, the first being Kepler-78 b, an Earth-sized planet with an Earth-like density (Pepe et al. 2013). Even for this relatively faint K-star ($m_V = 11.75$) the mass was determined with a precision of 20% ($m_p = 1.86^{+0.38}_{-0.28}$ $M_\oplus$), helped by the planet having an orbital period of only 8.5 hours. This was followed by intensive observations of Kepler-10, which hosts Kepler-10 b, the first rocky planet detected by *Kepler*, and Kepler-10 c (Batalha et al. 2011). The discovery paper presented a 3σ mass estimate for Kepler-10 b, and an upper limit for Kepler-10 c. Our HARPS-N campaign allowed us recover a mass precision of 15% for Kepler-10 b (3.33 ± 0.49 $M_\oplus$), and determine a mass for Kepler-10 c with a precision of 11% (17.2 ± 1.9 $M_\oplus$) (Dumusque et al. 2014). Given the interest in this system, the HARPS-N Collaboration is continuing to observe this system (Bonomo et al. 2024 in prep.).

Other Kepler systems characterised by the HARPS-N Collaboration include Kepler-101 (Bonomo et al. 2014), Kepler-93 (Dressing et al. 2015), Kepler-454 (Gettel et al. 2016), Kepler-20 (Buchhave et al. 2016), Kepler-21 (López-Morales et al. 2016), and Kepler-19 (Malavolta et al. 2017). In many of these systems, we were able to determine the masses of the small exoplanets with precisions of 20%, or better. We were also able to confirm the low densities of the planets in the Kepler-9 system (Borsato et al. 2019).

A particularly interesting system is Kepler-107, which hosts 4 sub-Neptune sized planets (Rowe et al. 2014) the inner two of which have very similar radii. Using HARPS-N, we were able to show that although these two planets have almost the same radii, they have very different masses, with the outer (Kepler-107 c) being almost twice as dense as the inner (Kepler-107 b) (Bonomo et al. 2019). The most likely scenario is that the outer part of Kepler-107 c's silicate mantle was stripped off by a giant impact.

† https://plone.unige.ch/HARPS-N/science-with-harps-n

2.1. *K2 mission*

When the second of four reaction wheels failed in May 2013, the primary *Kepler* mission came to an end. However, by pointing along the ecliptic and balancing the spacecraft against radiation pressure, it was possible to extend the mission. This mission extension was referred to as K2. The HARPS-N Collaboration then focussed on following up K2 targets, which were typically brighter than *Kepler* targets.

Some early examples were HIP 116454 (Vanderburg et al. 2015), which hosted the first planet discovered by the K2 mission, and K2-141 (Malavolta et al. 2018). K2-141 b still has the most precise mass of all small planets and, until recently, was the shortest period ultra short period (USP) planet known. Other examples are K2-3 (Damasso et al. 2018), a system that likely hosts 3 sub-Neptunes, HD 80653/K2-312 (Frustagli et al. 2020), a system with a short-period rocky Earth, and K2-111 (Mortier et al. 2020), the first time that detailed stellar, planetary and orbital characteristics were used to infer the internal structure of a transiting planet around a star with a chemical composition different to that of the Sun. This latter analysis confirmed a link between stellar composition and planet composition.

The HARPS-N Collaboration also used HARPS-N RVs to further constrain the properties of the GJ 9827 system (Rice et al. 2019), a 3 planet systems with planets on either side of the radius gap (Fulton et al. 2017) that separates rocky super-Earths from those that probably still retain substantial volatile atmospheres. Recent observations have indicated the presence of water in the atmosphere of GJ 9827 d (Roy et al. 2023).

2.2. *Combining almost 10 years of HARPS-N observations*

Given that the HARPS-N Collaboration has been observing since 2012, we eventually had a very large sample of radial velocities (RVs) for numerous *Kepler* and K2 targets. This allowed us to carry out a homogeneous analysis of 38 *Kepler* and K2 small planet systems (Bonomo et al. 2023). This resulted in improved physical and orbital parameters for 64 *Kepler* and K2 small planets, but also allowed us to search for the presence of cold Jupiters, exoplanets with semimajor axes of $a \sim 1 - 10$ AU and masses of $m_p = 0.5 - 13\ \mathrm{M_{Jup}}$.

We found 5 cold Jupiters in 3 of the systems, including a highly eccentric one in HD 80653/K2-312, and linear trends compatible with an outer planet in 2 others. Overall, we inferred a cold Jupiter occurrence rate of $9.3^{+7.7}_{-2.9}\%$. This is lower than some earlier estimates and does not confirm previous claims of an excess of cold Jupiters in systems with small inner planets (e.g., Bryan et al. 2019).

3. Synergy with TESS and CHEOPS

With the launch of NASA's Transiting Exoplanet Survey Satellite (TESS - Ricker et al. 2015), the HARPS-N Collaboration started to contribute to the follow-up of TESS targets. Of particular interest was the origin of the radius valley (e.g., Fulton et al. 2017), especially around M-dwarfs (Cloutier and Menou 2020). The HARPS-N Collaboration published a number of analyses of *keystone* "super-Earths". These are planets that lie in a region of radius-orbital period space where their properties can help to constrain the different formation pathways, such as radius evolution through thermally-driven mass loss, direct formation in a gas-poor environment, and formation in a gas-depleted environment. Examples are TOI-1235 b (Cloutier et al. 2020) and TOI-1634 b (Cloutier et al. 2021). Although thermally-driven mass loss is a favoured mechanism around stars with $T_{\rm eff} > 4700$ K, the results in these papers suggest that around cooler M-dwarfs, close-in "super-Earths" might preferentially form in a gas depleted environment.

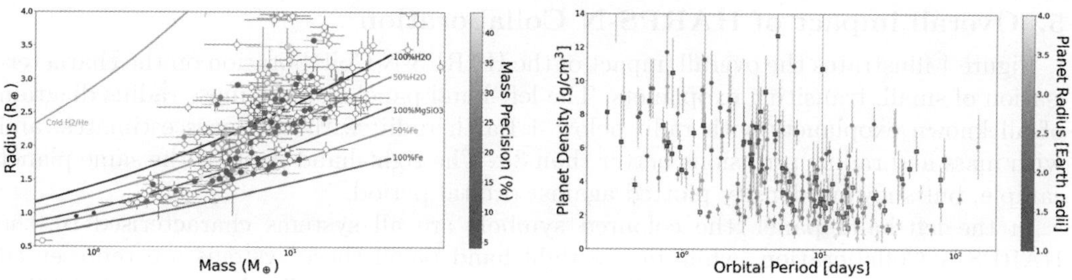

Figure 1. Mass-radius (left-hand panel) and density-orbital period (right-hand panel) plots for all planets with sizes below 4 Earth radii and with mass and radius precisions better than 3σ, taken from the Encyclopedia of Exoplanetary Systems. In the left-hand panel, the coloured symbols are those characterised by the HARPS-N Collaboration, while these planets are indicated by squares in the right-hand panel. The colours also indicate the mass precision.

The HARPS-N Collaboration also started taking advantage of the CHaracterising ExOPlanet Satellite (CHEOPS - Benz et al. 2021) to further constrain the properties of the systems being studied. CHEOPS is an ESA mission with one of its main scientific goals being to refine the radii of known exoplanets. This was especially useful when trying to determine the properties of the TOI-561 system. It was initially presented as a system with 3 planets, the outer of which transited twice. This interpretation was supported by the analysis in Weiss et al. (2021).

The HARPS-N Collaboration's analysis, however, concluded that TOI-561 actually had 4 planets, with the outer 2 only transiting once each (Lacedelli et al. 2021). CHEOPS played a crucial role in resolving this discrepancy. CHEOPS observed numerous transits of the inner two planets and one transit of an outer planet at a time consistent with this being a 4-planet system, but not a 3-planet system (Lacedelli et al. 2022). This analysis also indicated that the innermost planet was the lowest density ultra-short period planet detected to date. The low metallicity of the host star makes this consistent with the general bulk density-stellar metallicity trend initially highlighted in our analysis of K2-111 (Mortier et al. 2020).

The HARPS-N Collaboration also successfully combined HARPS-N RVs with TESS and CHEOPS photometry in the analysis of HD 77946 (TOI-1778), which hosts a planetary-mass object that lies in a region of mass-radius space where there is compositional confusion. This analysis suggests that HD 77946 b is probably a sub-Neptune with a low-mass H/He atmosphere that makes up a significant fraction of its radius, but can't entirely rule out a water world (Palethorpe et al. 2024).

4. Rocky Planet Search

As mentioned earlier, the HARPS-N Collaboration have also been observing 52 of the brightest, least active K-dwarfs in the Northern hemisphere to detect an Earth-like planet around a nearby star. The first result from this survey was the detection of a 4-planet system around HD 219134, the inner planet of which is a transiting rocky planet (Motalebi et al. 2015). The second innermost planet has since also been shown to transit and is also rocky (Gillon et al. 2017). This is still the closest known transiting system.

This dataset has since been used to review the planetary systems around HD 99492, HD 147379 and HD 190007 (Stalport et al. 2023), to show that it is possible to detect RV signals with amplitudes below 1 m/s (John et al. 2023), and to present a trio of super-Earths around HD 48948, one of which is in its star's habitable zone (Dalal et al. 2024).

5. Overall impact of HARPS-N Collaboration

Figure 1 illustrates the overall impact of the HARPS-N Collaboration on the characterisation of small, transiting exoplanets. The left-hand panel shows a mass-radius diagram of all known exoplanets with radii below 4 Earth radii, with RV mass estimates, and with mass and radius precisions better than 3σ. The right-hand panel is the same planet sample, but showing density plotted against orbital period.

In the left-hand panel, the coloured symbols are all systems characterised by the HARPS-N Collaboration, while in the right-hand panel these systems are represented by squares. The colours also indicate the mass precision. To-date, the HARPS-N Collaboration has characterised 25% of $R_p < 4$ $R_\oplus$ planets with mass precisions better than 30% and 25% of those with mass precisions better than 10%. For planets with $R_p < 2$ $R_\oplus$, the HARPS-N Collaboration has characterised one-third of those with mass precisions better than 30% and 40% of those with mass precisions better than 10%.

6. Summary

Exoplanet follow-up observations with JWST require precise mass and radius estimates, with detailed atmospheric characterisation requiring mass precision of 20%, or better (Batalha et al. 2019). The HARPS-N Collaboration has been collecting RVs with HARPS-N since late 2012. The Collaboration has made substantial contributions to our understanding of the composition of small exoplanets, the origin of the radius gap, the frequency of cold Jupiters, and the presence of rocky planets around nearby bright stars. It has also characterised a significant fraction of the small exoplanets with mass and radius estimates that are precise enough for them to be key targets for JWST.

Acknowledgements

The author list is the current HARPS-N Collaboration science team. We would like to also acknowledge the past science team members and collaborators who have also contributed to the success of the HARPS-N Collaboration.

References

Batalha, N. E., Lewis, T., Fortney, J. J., *et al.* 2019, *ApJL*, 885(1), L25
Batalha, N. M., Borucki, W. J., Bryson, S. T., *et al.* 2011, *ApJ*, 729(1), 27
Benz, W., Broeg, C., Fortier, A., *et al.* 2021, *Experimental Astronomy*, 51(1), 109–151
Bonomo, A. S., Dumusque, X., Massa, A., *et al.* 2023, *A&A*, 677, A33
Bonomo, A. S., Sozzetti, A., Lovis, C., *et al.* 2014, *A&A*, 572, A2
Bonomo, A. S., Zeng, L., Damasso, M., *et al.* 2019, *Nature Astronomy*, 3, 416–423
Borsato, L., Malavolta, L., Piotto, G., *et al.* 2019, *MNRAS*, 484(3), 3233–3243
Borucki, W. J., Koch, D., Basri, G., *et al.* 2010, *Science*, 327(5968), 977
Bryan, M. L., Knutson, H. A., Lee, E. J., *et al.* 2019, *AJ*, 157(2), 52
Buchhave, L. A., Dressing, C. D., Dumusque, X., *et al.* 2016, *AJ*, 152(6), 160
Cloutier, R., Charbonneau, D., Stassun, K. G., *et al.* 2021, *AJ*, 162(2), 79
Cloutier, R. & Menou, K. 2020, *AJ*, 159(5), 211
Cloutier, R., Rodriguez, J. E., Irwin, J., *et al.* 2020, *AJ*, 160(1), 22
Cosentino, R., Lovis, C., Pepe, F., *et al.* In McLean, I. S., Ramsay, S. K., & Takami, H., editors, *Ground-based and Airborne Instrumentation for Astronomy IV* 2012, volume 8446 of *Society of Photo-Optical Instrumentation Engineers (SPIE) Conference Series*, 84461V
Covino, E., Esposito, M., Barbieri, M., *et al.* 2013, *A&A*, 554, A28
Dalal, S., Rescigno, F., Cretignier, M., *et al.* 2024, *MNRAS*, 531(4), 4464–4481
Damasso, M., Bonomo, A. S., Astudillo-Defru, N., *et al.* 2018, *A&A*, 615, A69
Dressing, C. D., Charbonneau, D., Dumusque, X., *et al.* 2015, *ApJ*, 800(2), 135
Dumusque, X., Bonomo, A. S., Haywood, R. D., *et al.* 2014, *ApJ*, 789(2), 154

Frustagli, G., Poretti, E., Milbourne, T., *et al.* 2020, *A&A*, 633, A133
Fulton, B. J., Petigura, E. A., Howard, A. W., *et al.* 2017, *AJ*, 154(3), 109
Gettel, S., Charbonneau, D., Dressing, C. D., *et al.* 2016, *ApJ*, 816(2), 95
Gillon, M., Demory, B.-O., Van Grootel, V., *et al.* 2017, *Nature Astronomy*, 1, 0056
John, A. A., Collier Cameron, A., Faria, J. P., *et al.* 2023, *MNRAS*, 525(2), 1687–1704
Lacedelli, G., Malavolta, L., Borsato, L., *et al.* 2021, *MNRAS*, 501(3), 4148–4166
Lacedelli, G., Wilson, T. G., Malavolta, L., *et al.* 2022, *MNRAS*, 511(3), 4551–4571
López-Morales, M., Haywood, R. D., Coughlin, J. L., *et al.* 2016, *AJ*, 152(6), 204
Luque, R., Osborn, H. P., Leleu, A., *et al.* 2023, *Nature*, 623(7989), 932–937
Malavolta, L., Borsato, L., Granata, V., *et al.* 2017, *AJ*, 153(5), 224
Malavolta, L., Mayo, A. W., Louden, T., *et al.* 2018, *AJ*, 155(3), 107
Mayor, M., Pepe, F., Queloz, D., *et al.* 2003, *The Messenger*, 114, 20–24
Mortier, A., Zapatero Osorio, M. R., Malavolta, L., *et al.* 2020, *MNRAS*, 499(4), 5004–5021
Motalebi, F., Udry, S., Gillon, M., *et al.* 2015, *A&A*, 584, A72
Palethorpe, L., John, A. A., Mortier, A., *et al.* 2024, *MNRAS*, 529(4), 3323–3341
Pepe, F., Cameron, A. C., Latham, D. W., *et al.* 2013, *Nature*, 503(7476), 377–380
Rice, K., Malavolta, L., Mayo, A., *et al.* 2019, *MNRAS*, 484(3), 3731–3745
Ricker, G. R., Winn, J. N., Vanderspek, R., *et al.* 2015, *Journal of Astronomical Telescopes, Instruments, and Systems*, 1, 014003
Rowe, J. F., Bryson, S. T., Marcy, G. W., *et al.* 2014, *ApJ*, 784(1), 45
Roy, P.-A., Benneke, B., Piaulet, C., *et al.* 2023, *ApJL*, 954(2), L52
Stalport, M., Cretignier, M., Udry, S., *et al.* 2023, *A&A*, 678, A90
Vanderburg, A., Montet, B. T., Johnson, J. A., *et al.* 2015, *ApJ*, 800(1), 59
Weiss, L. M., Dai, F., Huber, D., *et al.* 2021, *AJ*, 161(2), 56

Planetary Science and Exoplanets in the Era of JWST
Proceedings IAU Symposium No. 393, 2026
Z. Benkhaldoun, Y. Moulane & A. Szentgyorgyi, eds.
doi:10.1017/S1743921324002692

Anticipating additional exoplanets in multi-planetary systems

M. Mousavi-Sadr[1,2], G. Gozaliasl[3] and D. M. Jassur[2]

[1]School of Astronomy, Institute for Research in Fundamental Sciences (IPM), Tehran, Iran
email: mahdiyar@ipm.ir

[2]Department of Theoretical Physics and Astrophysics, Faculty of Physics, University of Tabriz, Tabriz, Iran

[3]Department of Computer Science, Aalto University, P. O. Box 15400, Espoo, FI-00 076, Finland

Abstract. We present a search for the presence of additional exoplanets in previously identified multi-planetary systems, using the Titius-Bode (TB) relation, a logarithmic distance between planets in our solar system. We analyze 229 multi-planetary systems, each containing a minimum of three confirmed planets, by modeling the TB relation employing the Markov Chain Monte Carlo (MCMC) technique. Our findings indicate that ~53% of these systems exhibit a logarithmic spacing relation more pronounced than that observed in the solar system. We predict the existence of 426 supplementary exoplanets within these 229 systems, identifying 197 candidates through interpolation and 229 through extrapolation. Among the predicted planets, 47 are located in the habitable zone of their respective host stars. Within this subset, five planets have a maximum radius below $1.25R_{\oplus}$ and an estimated mass between 0.1 and $2M_{\oplus}$. We minimize inaccuracies in the orbital period estimation for the anticipated planets compared to the earlier studies.

Keywords. planets and satellites: detection, planets and satellites: dynamical evolution and stability, planets and satellites: formation, planets and satellites: general

1. Introduction

The Titius-Bode (TB) law, which has been known for more than 200 years, describes a roughly simple logarithmic spacing between planets in our solar system. Its classical relation is as follows:

$$a_n = 0.4 + 0.3 \times 2^n, \tag{1}$$

where a_n represents the semi-major axis of the nth planet in AU. The planet Mercury corresponds to $n = -\infty$, Venus to $n = 0$, Earth to $n = 1$ and so on (Nieto & Davis 1974). The undetected objects in our solar system were accurately anticipated using the TB relation. This relation is thought to be useful for making similar predictions in exoplanet systems. A pioneering application of the TB relation to multiple-exoplanet systems was carried out by Poveda & Lara (2008) on the five-planet 55 Cnc system. They found that the semi-major axes of the five observed planets fit well with a simple exponential TB connection, and they further predicted two additional planets at distances of 2 and 15 AU from the host star. A study by Bovaird & Lineweaver (2013) (hereafter, Bovaird & Lineweaver (2013)) examined a sample of 68 multiple-exoplanet systems, each with at least four planets, including confirmed and candidate systems. They discovered that most

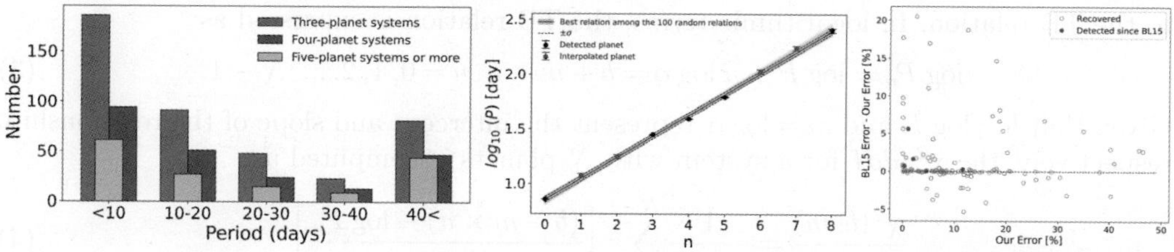

Figure 1. Left panel: distribution of the orbital period of exoplanets in 229 systems hosting different numbers of planets. Middle panel: the TB relation applied to the data from the GJ 667 C system. For this system, the fourth step is the best combination of detected and predicted planets with a γ value of 2.6. The two black dashed lines show $\pm 1\sigma$ uncertainties around the best scaling relation (solid black line). Right panel: Comparison of the uncertainty of the orbital period between Bovaird, Lineweaver, & Jacobsen (2015) and this study. The blue points indicate various planet removal scenarios in the systems, while the five red points represent detected planets after Bovaird, Lineweaver, & Jacobsen (2015) predictions. We achieve improved precision in uncertainties of the orbital period.

of these exoplanetary systems adhered to the TB relation more closely than the solar system. Using the Bovaird & Lineweaver (2013) method, Bovaird, Lineweaver, & Jacobsen (2015) (hereafter, Bovaird, Lineweaver, & Jacobsen (2015)) predicted the existence of 228 additional planets within 151 Kepler multiple-exoplanet systems, finding an average of 2 ± 1 planets in the HZ of each star. Due to intrinsic noise from the host stars, the planets' radii, their highly inclined orbits, and the detection methods, they established a completeness factor, which led them to conclude that not all predicted planets would be detected. The goal of this research is to test the adherence of 229 multiple-exoplanet systems to the TB relation and to predict whether any more undiscovered planets will exist in these systems. Moreover, we highlight the predicted exoplanets located within the HZ of their host stars. Finally, we compare our predictions with those of Bovaird, Lineweaver, & Jacobsen (2015).

2. Data and method

We incorporate physical characteristics of exoplanets, including their orbital periods, radii, and masses, alongside the masses and radii of their respective stars, sourced from the NASA Exoplanet Archive and the Extrasolar Planets Encyclopedia. Furthermore, we apply conservative and optimistic boundaries of the HZ of accessible stars, as provided by the Habitable Zone Gallery (Kane & Gelino 2012). There are 230 known multi-planetary systems, each consisting of at least three confirmed planets, totaling 818 planets. Of these, 142 systems have three planets, 59 systems have four, 20 systems host five planets, and seven systems contain six planets. The Kepler-132 system has four planets, where two of them, Kepler-132 b and Kepler-132 c, possess nearly identical orbital periods of 6.1782 days and 6.4149 days, respectively. However, it has been determined that Kepler-132 b and Kepler-132 c cannot orbit the same star (Lissauer et al. 2014). Therefore, we exclude Kepler-132 from our analysis, reducing our sample to 229 systems. The distribution of the orbital period of 813 member exoplanets in 229 multiple-exoplanet systems is shown in the left panel of Figure 1.

We use the TB relation to predict additional planets in multiple-exoplanet systems. The TB relation expressed through the orbital periods as follow:

$$P_n = P\alpha^n, n = 0, 1, 2, ..., N-1. \quad (2)$$

Let P_n denote the orbital period of the nth planet, with P and α as the fitting parameters. The solar system serves as a benchmark to assess the adherence of exoplanetary systems to the TB relation. In logarithmic terms, the TB relation is expressed as:

$$\log P_n = \log P + n \log \alpha = b + m \times n, n = 0, 1, 2, ..., N-1. \quad (3)$$

Given that $b = \log P$ and $m = \log \alpha$ represent the intercept and slope of the relationship, respectively, the χ^2/dof for a system with N planets is computed as:

$$\frac{\chi^2(b,m)}{N-2} = \frac{1}{N-2} \sum_{n=0}^{N-1} \left[\frac{(b + m \times n) - \log P_n}{\sigma} \right]^2, \quad (4)$$

N-2 represents the degrees of freedom (N planets, 2 fitted parameters, b and m), while σ measures the system's density. The density σ is given by $\sigma = 0.273\ S_p$ (see Figure 5 in Mousavi-Sadr, Gozaliasl, & Jassur (2021) (hereafter Sadr2021)), where S_p is defined as the average logarithmic period spacing between planets:

$$S_p = \frac{\log P_{N-1} - \log P_0}{N}, \quad (5)$$

Consider N as the number of planets within the system, where P_{N-1} and P_0 denote the largest and smallest orbital periods, respectively. By selecting an appropriate σ, $\chi^2/\mathrm{dof} = 1$ is calibrated for the solar system. An extrapolated planet beyond the outermost observed planet is predicted only if the system's planets adhere to the TB relation more accurately than those in the solar system. For cases where $\chi^2/\mathrm{dof} > 1$, indicating poorer adherence, interpolation begins. Initially, we incorporate a new planet chosen from 5,000 hypothetical candidates with random periods between the innermost and outermost detected planets, examining all potential intervals between adjacent planets to recalculate χ^2/dof. The specific planet minimizing χ^2/dof is selected. Up to 10 new specific planets may be inserted iteratively, considering all possible intervals and combinations. To determine the optimal combination of observed and predicted planets, we use the highest value of the parameter γ, which measures the enhancement in χ^2/dof per added planet, defined as:

$$\gamma = \frac{\left(\frac{\chi_i^2 - \chi_j^2}{\chi_j^2}\right)}{n_{ins}}, \quad (6)$$

where χ_i^2 and χ_j^2 are the χ^2 values before and after inserting of n_{ins} planets. The analysis of the system GJ 667 C is presented in Figure 1 (middle panel), where we have inserted four additional planets into the system. We independently analyze the data for each system, determining the uncertainty of the optimal parameters via the MCMC method. The geometric transit probability (P_{tr}) for predicted planets is approximated by $d_s/2a_p$, where d_s represents the stellar diameter and a_p denotes the planet's orbital radius (Koch & Borucki 1996). The upper limits on the mass or radius of the predicted planets are derived by selecting the lowest signal-to-noise ratio (SNR) from the observed planets within that system and applying it to the predicted planet's orbital period. For transiting planets, the maximum radius is calculated by:

$$R_{max} = R_{minSNR} \left(\frac{P_{predicted}}{P_{minSNR}} \right)^{0.25}, \quad (7)$$

and for the planets detected by radial velocity, the maximum mass is calculated by:

$$M_{max} = M_{minSNR} \left(\frac{P_{predicted}}{P_{minSNR}} \right)^{\frac{7}{6}}. \quad (8)$$

Here, R_{minSNR}, M_{minSNR}, and P_{minSNR} denote the radius, mass, and orbital period of the planet with the minimum SNR, respectively. We employ the mass-radius relationship from Bashi et al. (2017) to interchange radius and mass values.

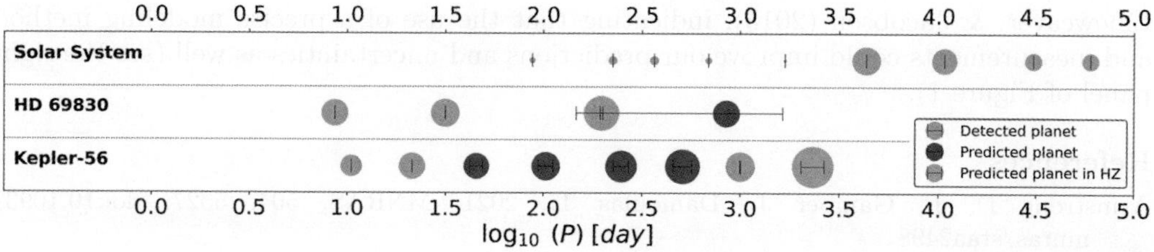

Figure 2. Orbital periods and scaled radii of planets in the solar system and two exoplanet systems with predicted planets. A list of all systems is provided in Figures 7 and 8 of Sadr2021.

Table 1. Exoplanets predicted to lie in the HZ of host stars, with estimated maximum mass and radius similar to Earth's.

Panel	Host name	Dis.[1]	Period (days)	a(AU)	ON[2]	R_{Max} ($R_\oplus$)	M_{Max} ($M_\oplus$)	P_{tr} (%)	HZ_{Cons} (AU)	HZ_{Opt} (AU)
1	GJ 3138	RV	$41.0^{+11.4}_{-9.4}$	$0.20^{+0.01}_{-0.03}$	2	1.0	1.2	1.12	0.22-0.42	0.17-0.44
2	Kepler-186	Tr	$41.2^{+1.7}_{-1.7}$	$0.19^{+0.01}_{-0.01}$	4	0.9	1.1	1.26	0.24-0.46	0.19-0.49
3	Kepler-186	Tr	$73.6^{+3.5}_{-3.5}$	$0.28^{+0.01}_{-0.01}$	5	1.0	1.4	0.85	0.24-0.46	0.19-0.49
4	Wolf 1061	RV	$61.9^{+2.5}_{-2.5}$	$0.20^{+0.01}_{-0.01}$	2	1.2	1.8	0.70	0.10-0.21	0.08-0.22
5	YZ Cet	RV	$7.2^{+0.4}_{-0.3}$	$0.04^{+0.01}_{-0.01}$	3 E	1.2	1.9	1.96	0.05-0.10	0.04-0.10

Notes:
[1]Discovery method of the system: 'Tr' and 'RV' represent transit and radial velocity, respectively.
[2]Orbital numbers (ON) followed by 'E' indicate the extrapolated planets.

3. Results

Our analysis reveals that 122 systems align with the TB relation more closely than the solar system without the need for interpolation. For the 107 systems that align less well with the TB relation compared to the solar system, we inserted up to 10 additional planets. Post-interpolation, these systems follow the TB relation as well or better than the solar system. Specifically, within these 107 systems, 50 needed one additional planet, 33 needed two, and the remaining 24 required more than two additional planets. Figure 2 shows the orbital periods and scaled radii of exoplanets in HD 69830 and Kepler-56 along with the solar system. For all systems studied, we also predict an extrapolated planet beyond the outermost observed one. In total, we foresee 426 potential additional exoplanets, 197 predicted through interpolations. Detailed information on all predicted exoplanets is provided in Sadr2021 and Mousavi-Sadr (2024). Notably, six planets predicted in Kepler-1388, Kepler-1542, Kepler-164, Kepler-374, Kepler-402, and Kepler-403 are labeled "Planetary Candidate" by the NASA Exoplanet Archive. Of the predicted exoplanets, 47 lie within the HZ of their host stars (see Table 2 in Sadr2021). In the Kepler-167 system, we predict three interpolated exoplanets, two within the conservative HZ with orbital periods of 157.0 and 373.5 days. Kepler-186 is projected to host two additional planets with periods of 41.2 and 73.6 days, residing within the optimistic and conservative HZ, respectively. Among these 47 HZ exoplanets, five have estimated maximum mass and radius within Earth's range as defined by Borucki et al. (2011) and Stevens & Gaudi (2013): the fourth and fifth planets of Kepler-186, the second planets of GJ 3138 and Wolf 1061, and the YZ Cet extrapolated planet (see Table 1). There are new planet detection for seven exoplanetary systems (Kepler-1388, Kepler-150, Kepler-1542, Kepler-20, Kepler-80, Kepler-82, and KOI-351) after the predictions made by Bovaird, Lineweaver, & Jacobsen (2015). We find that for five of them, the detected and predicted orbital periods agree very well within the errors.

We remove planets from systems with different combinations and apply our method and Bovaird, Lineweaver, & Jacobsen (2015)'s to recover the orbital periods of removed planets. As a result, our predictions agree roughly better than those made by Bovaird,

Lineweaver, & Jacobsen (2015), indicating that the use of a precise modeling method and measurements could improve our predictions and uncertainties as well (see the right panel of Figure 1).

References

Armstrong D. J., Gamper J., Damoulas T., 2021, MNRAS, 504, 5327. doi:10.1093/mnras/staa2498

Bashi D., Helled R., Zucker S., Mordasini C., 2017, A&A, 604, A83. doi:10.1051/0004-6361/201629922

Borucki W. J., Koch D. G., Basri G., Batalha N., Brown T. M., Bryson S. T., Caldwell D., et al., 2011, ApJ, 736, 19. doi:10.1088/0004-637X/736/1/19

Bovaird T., Lineweaver C. H., 2013, MNRAS, 435, 1126. doi:10.1093/mnras/stt1357

Bovaird T., Lineweaver C. H., Jacobsen S. K., 2015, MNRAS, 448, 3608. doi:10.1093/mnras/stv221

Chang H.-Y., 2008, JASS, 25, 239. doi:10.5140/JASS.2008.25.3.239

Chang H.-Y., 2010, JASS, 27, 1. doi:10.5140/JASS.2010.27.1.001

Deeg H. J., Alonso R., 2018, haex.book, 117. doi:10.1007/978-3-319-55333-7_117

Foreman-Mackey D., Farr W., Sinha M., Archibald A., Hogg D., Sanders J., Zuntz J., et al., 2019, JOSS, 4, 1864. doi:10.21105/joss.01864

Goodman J., Weare J., 2010, CAMCS, 5, 65. doi:10.2140/camcos.2010.5.65

Kane S. R., Gelino D. M., 2012, PASP, 124, 323. doi:10.1086/665271

Koch D., Borucki W., 1996, chz..conf, 229

Lissauer J. J., Marcy G. W., Bryson S. T., Rowe J. F., Jontof-Hutter D., Agol E., Borucki W. J., et al., 2014, ApJ, 784, 44. doi:10.1088/0004-637X/784/1/44

Mousavi-Sadr M., Gozaliasl G., Jassur D. M., 2021, PASA, 38, e015. doi:10.1017/pasa.2021.9

Mousavi-Sadr M., 2024, PhDT. doi:10.48550/arXiv.2402.17898

Nieto M. M., Davis M. S., 1974, PhT, 27, 54. doi:10.1063/1.3128596

Poveda A., Lara P., 2008, RMxAA, 44, 243. doi:10.48550/arXiv.0803.2240

Stevens D. J., Gaudi B. S., 2013, PASP, 125, 933. doi:10.1086/672572

Discussion

We present a list of 229 analyzed exoplanetary systems (123 of them have not been analyzed by Bovaird & Lineweaver (2013) and Bovaird, Lineweaver, & Jacobsen (2015) before) containing their unique TB relations and a total of 426 additional predicted exoplanets, of which 47 are located within HZ of their parent stars. Additionally, we predict that the maximum mass and radius of five of the predicted planets in HZ fall within the mass and radius range of Earth. As an important result, the planets of ~53% of our sample follow a logarithmic spacing relation like the TB relation. Thus, to identify the possible dynamical or gravitational aspects of the TB relation, we must await exoplanet surveys such as TESS to examine more comprehensive data on multi-planetary systems.

Planetary Science and Exoplanets in the Era of JWST
Proceedings IAU Symposium No. 393, 2026
Z. Benkhaldoun, Y. Moulane & A. Szentgyorgyi, eds.
doi:10.1017/S1743921324003168

Refining the exoplanet mass-radius relation using machine learning

M. Mousavi-Sadr[1,2], G. Gozaliasl[3] and D. M. Jassur[2]

[1]School of Astronomy, Institute for Research in Fundamental Sciences (IPM), Tehran, Iran
email: mahdiyar@ipm.ir

[2]Department of Theoretical Physics and Astrophysics, Faculty of Physics, University of Tabriz, Tabriz, Iran

[3]Department of Computer Science, Aalto University, P. O. Box 15400, Espoo, FI-00 076, Finland

Abstract. The rising number of exoplanet discoveries and advances in machine learning (ML) techniques present new possibilities for exploring and understanding the characteristics of worlds beyond our solar system. This research examines the exoplanet dataset by applying ML techniques to categorize these systems, uncover relationships among their physical features, and predict the exoplanet radius. We group the data into two primary categories: 'small' and 'giant' planets, with thresholds at $R_p = 8.13 R_\oplus$ and $M_p = 52.48 M_\oplus$. Our study indicates that the planetary mass, orbital period, and stellar mass play critical roles in predicting the exoplanet radius. A notable finding of our research is that small planets exhibit a positive linear mass-radius relationship, consistent with other studies. Conversely, for giant planets, we observe a strong correlation between planetary radius and the mass of their host stars, potentially providing significant insights into the relationship between giant planet formation and stellar properties.

Keywords. planets and satellites: fundamental parameters, planets and satellites: composition, planets and satellites: formation, planets and satellites: dynamical evolution and stability, planets and satellites: general, software: data analysis

1. Introduction

The exoplanet data generated by the state-of-the-art telescopes is accessible to all. Researchers skilled in data analysis and data science analyze these data to anticipate, discover, and characterize exoplanets, and to improve theories of planet formation and evolution (Armstrong, Gamper, & Damoulas 2021; Mousavi-Sadr, Gozaliasl, & Jassur 2021; Mousavi-Sadr 2024). A polytropic relationship can be used to describe the relationship that exists between the mass and radius of a planet (Chabrier & Baraffe 2000). Many studies have examined the correlation between planetary parameters, specifically mass and radius, in order to infer the composition and structure of exoplanets (Otegi, Bouchy, & Helled 2020). Weiss et al. (2013) separated the planets into two categories based on mass, those with masses higher than and below $150 M_\oplus$, then provided a power law relation for the mass-radius distribution of each group. The radius of a giant planet is related to other parameters such as the planetary equilibrium temperature (T_{equ}), the semi-major orbital axis, the tidal heating rate, and stellar metallicity and irradiation (Enoch, Collier Cameron, & Horne 2012). Bhatti et al. (2016) assessed the impact of several physical parameters on the radii of the planet using a random forest regression model. They found that the parameters that most influence the radius of a hot Saturn ($0.1 < M_p < 0.5M_J$) are the planet's mass and equilibrium temperature. Moreover, Ulmer-Moll et al. (2019) used the Random Forest model to predict the exoplanet radii based on several stellar and planetary parameters. The important parameters were, as in earlier findings, the mass and equilibrium temperature of an exoplanet. In this work, we apply different machine learning methods to identify plausible correlations between the physical properties of exoplanet systems. Furthermore, the exoplanets are grouped and their features are studied using various machine learning clustering algorithms.

2. Data and method

We use the NASA Exoplanet Archive and the Extrasolar Planets Encyclopedia to extract the data of exoplanets with reported physical parameters including the orbital period (P) and eccentricity (e), the planetary mass (M_p), and the radius (R_p), and the stellar mass (M_s), the radius (R_s), the metallicity (Fe/H), and the effective temperature ($T_{\rm eff}$). We have eight solar system planets and 762 confirmed exoplanets in our data set. The target variable is the planetary radius (R_p), and the features are the orbital period (P), eccentricity (e), planetary mass (M_p), stellar mass (M_s), stellar radius (R_s), metallicity (Fe/H), and effective temperature ($T_{\rm eff}$). It is generally accepted that unnecessary and ineffective features can negatively impact the performance of an ML model and produce unreliable results (Amr Abdallah & de La Iglesia 2015). Therefore, we use five Feature Selection (FS) approaches to identify and exclude the least significant planetary, stellar, and orbital features: Spearman's rank correlation test is a filter method, where features are selected independently of any induction algorithm, relying on specific performance evaluation metrics that are computed directly from the data, the Forward Selection and Backward Elimination as two wrapper methods, where the selection process relies on the effectiveness of a particular ML algorithm functioning with a specific set of features, and the CART and XGBoost as two Embedded methods. Embedded methods integrate the advantages of both filter and wrapper techniques. They conduct the FS during the training process and are typically specific to given learning algorithms (Belanche & González 2011). Similar to exclusive clustering, where each data point belongs exclusively to one cluster (Jain & Dubes 1988), we want to group exoplanets into distinct, non-overlapping clusters. Data clustering algorithms can be categorized into two main types: hierarchical and partitional. Hierarchical algorithms identify clusters by utilizing previously formed clusters, whereas partitional algorithms determine all clusters simultaneously. We apply the following ML clustering algorithms to identify potential exoplanet groups and explore their properties: Affinity Propagation, BIRCH, DBSCAN, Gaussian Mixture Model, Hierarchical Clustering, K-Means, Mean Shift, Mini Batch K-Means, OPTICS, and Spectral Clustering (Kaufman & Rousseeuw 1990). We also apply ML models to predict planet radii when other parameters are given. The algorithms used to perform regression tasks are Decision Tree, K-Nearest Neighbors, Linear Regression, Multilayer Perceptron, M5P, and Support Vector Regression (SVR). Bootstrap Aggregation and Random Forest serve as ensemble algorithms that integrate the predictions from various models (Breiman 2001). The hyperparameters are tuned to achieve optimal algorithm performance. A 10-fold cross-validation procedure is also used to evaluate the performance of the models. The mean absolute error (MAE), the root mean square error (RMSE), and the coefficient of determination (ρ^2) are calculated as validation metrics. Linear regression and M5P are two algorithms that can extract parametric equations. To quantify the uncertainties of the best-fitting parameters of linear regression and M5P, we use the Markov Chain Monte Carlo (MCMC) (Goodman & Weare 2010; Foreman-Mackey et al. 2019).

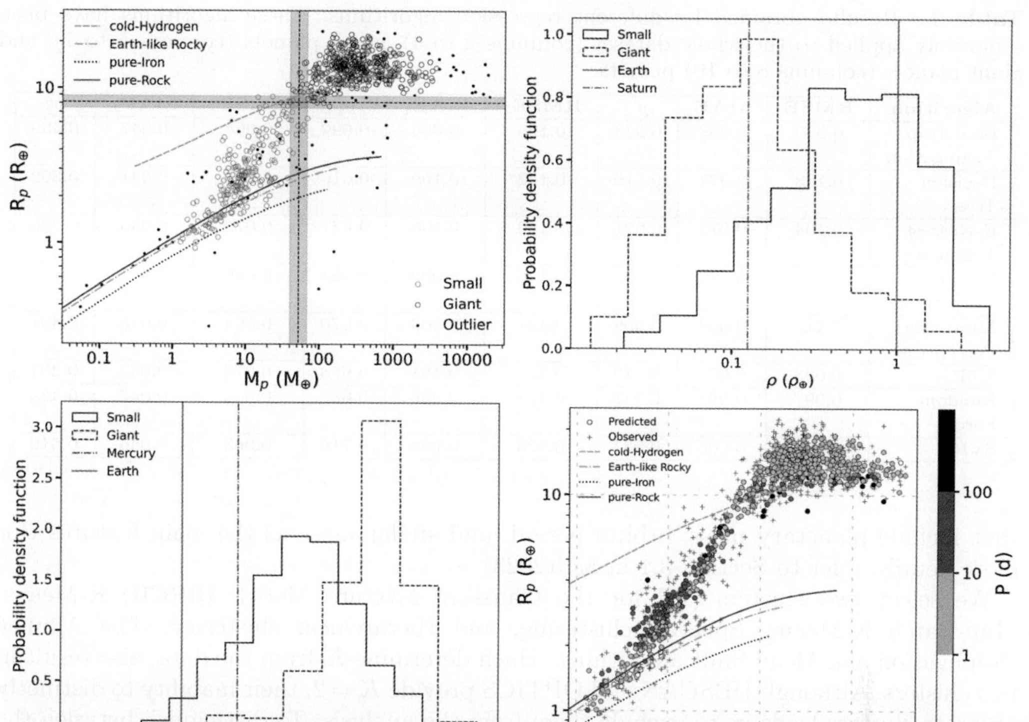

Figure 1. The mass-radius distribution for clustered data is shown in the upper left panel. The dataset is divided into two categories by $R_p = 8.13 R_\oplus$ (horizontal dashed line) and $M_p = 52.48 M_\oplus$ (vertical dashed line). Mean errors in mass and radius are represented by gray regions. The upper right and lower left panels show histograms of the average density (ρ) and equilibrium temperature ($T_{\rm equ}$), respectively, for small (solid-border bars) and giant (dashed-border bars) planets. A comparison of the average densities and equilibrium temperatures of these planets reveals that giant planets tend to be hotter and less dense than their smaller counterparts. The lower right panel represents the predicted and observed radius as a function of mass and orbital period obtained by the SVR model for all planets in the sample. Iso-density curves are plotted based on Marcus et al. (2010) and Becker et al. (2014).

3. Results

Before applying the clustering and regression models, we use the Local Outlier Factor (LOF) method to identify outliers, which marks 76 data points as outliers in our sample. The LOF employs two hyperparameters: the neighborhood size, which defines the neighborhood for local density calculation, and contamination, which indicates the percentage of outliers in the dataset. Then, it identifies and filters out data points that deviate significantly from the overall dataset. To enhance the efficacy of predictive models, we remove these outliers and transform the data into a logarithmic space. The impact of data-cleaning and data re-scaling on prediction accuracy are discussed in Appendixes A and B of Mousavi-Sadr, Jassur, & Gozaliasl (2023) (hereafter, Sadr2023). FS methods highlight that planetary mass plays an important role in predicting the radius of exoplanets, whereas metallicity and eccentricity demonstrate relatively less significance. So, we exclude metallicity and eccentricity from the feature list. Moreover, all stellar parameters (M_s, R_s, and $T_{\rm eff}$) are of similar importance. Eventually, we choose a set of parameters

Table 1. Results obtained by different regression algorithms. These algorithms have been separately applied to the whole dataset (columns 2 to 4), small planets (columns 5 to 7), and giant planets (columns 8 to 10) planets.

Algorithm	RMSE	MAE	ρ^2	RMSE$_1$	MAE$_1$	ρ_1^2	RMSE$_2$	MAE$_2$	ρ_2^2
Bootstrap Aggregation	0.096	0.068	0.933	0.126	0.096	0.692	0.064	0.047	0.489
Decision Tree	0.108	0.077	0.916	0.142	0.109	0.616	0.071	0.051	0.392
K-Nearest Neighbors	0.104	0.075	0.921	0.140	0.106	0.627	0.072	0.053	0.390
Linear Regression	0.155	0.127	0.822	0.128	0.099	0.683	0.069	0.051	0.402
Multilayer Perceptron	0.112	0.084	0.909	0.131	0.103	0.670	0.064	0.048	0.491
M5P	0.098	0.071	0.930	0.130	0.101	0.688	0.073	0.053	0.391
Random Forest	0.097	0.068	0.932	0.128	0.096	0.682	0.064	0.047	0.485
SVR	0.093	0.065	0.937	0.123	0.088	0.710	0.063	0.046	0.510

that include planetary mass, orbital period, and stellar mass as the main features (for more details, refer to Section 3.1 in Sadr2023).

We select two clusters (K) for the Gaussian Mixture Model, BIRCH, K-Means, Mini-Batch K-Means, Spectral clustering, and Hierarchical clustering. The Affinity Propagation and Mean Shift algorithms, which determine K from the data, also result in two clusters. Although DBSCAN and OPTICS provide $K = 2$, their inability to distinctly separate clusters leads us to exclude them from the analysis. To distinguish between the two groups in the mass-radius space of the planet, we construct a Gaussian kernel density estimate for each cluster and identify the intersection point. The clustering results are largely consistent across the algorithms except for DBSCAN and OPTICS (refer to Table 1 in Sadr2023). We used an average value of $\log M_p = 1.72$ ($M_p = 52.48 M_\oplus$) for the mass break point and $\log R_p = 0.91$ ($R_p = 8.13 R_\oplus$) for the radius breakpoint to categorize the planets into two groups. Planets with $R_p > 8.13 R_\oplus$ and $M_p > 52.48 M_\oplus$ are classified as giant planets, while those with $R_p \leq 8.13 R_\oplus$ and $M_p \leq 52.48 M_\oplus$ are termed small planets. There are 254 small planets and 440 giant planets. The mass-radius distribution of small and giant planets is illustrated in the upper left panel of Figure 1. Our findings indicate significant differences between small and giant planets. Giant planets exhibit larger masses, larger radii, and lower densities (see the upper right panel of Figure 1), suggesting a higher frequency of volatile-rich exoplanets in this category. Additionally, giant planets typically orbit closer to their parent stars and possess higher equilibrium temperatures (see the lower left panel of Figure 1). In contrast, small planets consist mainly of elements heavier than hydrogen and helium and have lower equilibrium temperatures.

We apply predictive models separately to entire, small, and giant planets. The SVR model delivers the best performance among the ML regression models, achieving an RMSE of 0.093 (see Table 1). Using Linear Regression and M5P, which can derive parametric equations, we find a positive linear mass-radius relationship for small planets and a positive linear correlation between the radius of giant planets and the mass of their host stars (see Table 4 in Sadr2023 for more details). The left panel of Figure 2 displays the correlation between the radius and mass of small planets derived from the MCMC. The best scaling relation, represented by the red line, is illustrated using Equation 3.1. The correlation between the radius of giant planets and the stellar mass is displayed in the right panel of Figure 2, where the best scaling relation is illustrated using Equation 3.2.

$$\log \frac{R_p}{R_\oplus} = 0.497^{+0.023}_{-0.022} \times \log \frac{M_p}{M_\oplus} - 0.050^{+0.024}_{-0.024}. \tag{3.1}$$

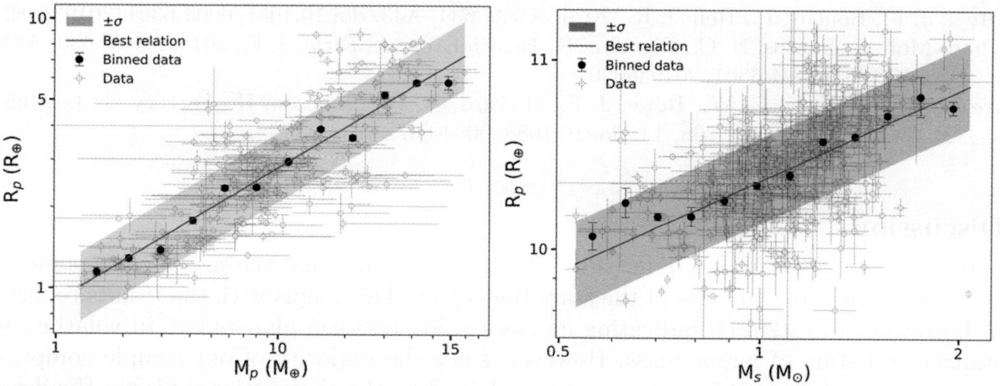

Figure 2. The mass-radius relation of small planets (left panel) and the relation between the radius of giant planets and the host star's mass (right panel) obtained by the MCMC method. Equation 3.1 and Equation 3.2 represent the best scaling relations (red lines). The highlighted area is $\pm 1\sigma$ uncertainties around the best relation. For better illustration, the data points have been binned.

$$\log \frac{R_p}{R_\oplus} = 0.480^{+0.036}_{-0.037} \times \log \frac{M_s}{M_\odot} + 1.109^{+0.004}_{-0.004}. \tag{3.2}$$

Supplementary material

To view supplementary material for this article, please visit https://doi.org/10.1017/S1743921324003168

References

Amr Abdallah, T., de La Iglesia, B., 2015, arXiv, arXiv:1510.02892. doi:10.48550/arXiv.1510.02892

Armstrong, D. J., Gamper, J., Damoulas, T., 2021, MNRAS, 504, 5327. doi:10.1093/mnras/staa2498

Becker, A., Lorenzen, W., Fortney, J. J., Nettelmann, N., Schöttler, M., Redmer, R., 2014, ApJS, 215, 21. doi:10.1088/0067-0049/215/2/21

Belanche, L. A., González, F. F., 2011, arXiv, arXiv:1101.2320. doi:10.48550/arXiv.1101.2320

Bhatti, W., Bakos, G. Á., Hartman, J. D., Zhou, G., Penev, K., Bayliss, D., Jordán, A., et al., 2016, arXiv, arXiv:1607.00322. doi:10.48550/arXiv.1607.00322

Breiman, L., 2001, MachL, 45, 5. doi:10.1023/A:1010933404324

Chabrier, G., Baraffe, I., 2000, ARA&A, 38, 337. doi:10.1146/annurev.astro.38.1.337

Enoch, B., Collier Cameron, A., Horne, K., 2012, A&A, 540, A99. doi:10.1051/0004-6361/201117317

Foreman-Mackey, D., Farr, W., Sinha, M., Archibald, A., Hogg, D., Sanders, J., Zuntz, J., et al., 2019, JOSS, 4, 1864. doi:10.21105/joss.01864

Goodman, J., Weare, J., 2010, CAMCS, 5, 65. doi:10.2140/camcos.2010.5.65

Jain, A. K., Dubes, R. C., 1988, acd..book

Kaufman, L., Rousseeuw, P. J., 1990, fgda.book

Lozovsky, M., Helled, R., Pascucci, I., Dorn, C., Venturini, J., Feldmann, R., 2021, A&A, 652, A110. doi:10.1051/0004-6361/202140563

Marcus, R. A., Sasselov, D., Hernquist, L., Stewart, S. T., 2010, ApJL, 712, L73. doi:10.1088/2041-8205/712/1/L73

Mousavi-Sadr, M., Gozaliasl, G., Jassur, D. M., 2021, PASA, 38, e015. doi:10.1017/pasa.2021.9

Mousavi-Sadr, M., Jassur, D. M., Gozaliasl, G., 2023, MNRAS, 525, 3469. doi:10.1093/mnras/stad2506

Mousavi-Sadr, M., 2024, PhDT. doi:10.48550/arXiv.2402.17898

Otegi, J. F., Bouchy, F., Helled, R., 2020, A&A, 634, A43. doi:10.1051/0004-6361/201936482
Ulmer-Moll, S., Santos, N. C., Figueira, P., Brinchmann, J., Faria, J. P., 2019, A&A, 630, A135. doi:10.1051/0004-6361/201936049
Weiss, L. M., Marcy, G. W., Rowe, J. F., Howard, A. W., Isaacson, H., Fortney, J. J., Miller, N., et al., 2013, ApJ, 768, 14. doi:10.1088/0004-637X/768/1/14

Discussion

In our study of a sample comprising 770 planets, we find that the size of giant planets is positively linked to the mass of the stars they orbit. This aligns with the results reported in Lozovsky et al. (2021), indicating an association between planets rich in volatile substances and stars of larger mass. However, since the majority of our sample comprises transiting exoplanets, this result can be explained by the observational bias in the detection method, in addition to naturally correlated parameters. To understand exoplanet composition and structure and test theories about how planets form and evolve, further follow-up programs like the Atmospheric Remote-sensing Infrared Exoplanet Large-survey (ARIEL), the Planetary Transits and Oscillations of Stars (PLATO) mission, and the Extremely Large Telescope (ELT) are needed.

Planetary Science and Exoplanets in the Era of JWST
Proceedings IAU Symposium No. 393, 2026
Z. Benkhaldoun, Y. Moulane & A. Szentgyorgyi, eds.
doi:10.1017/S1743921324002709

Dynamic evolution of the compact planetary system K2-72

E. D. Kuznetsov and A. S. Perminov

Ural Federal University, Yekaterinburg, Russia
email: eduard.kuznetsov@urfu.ru

Abstract. We consider the dynamic evolution of the compact four-planetary system K2-72. We considered a few scenarios for the evolution of the K2-72 system over 100 Myr using the Posidonius software, which considers tidal interactions. We showed that the compact planetary system K2-72 likely evolves beyond low-order resonances. A significant change in the large semi-major axes of the orbits of the K2-72 b and K2-72 d planets leads to the moving of the adjacent planets b-d and d-c out of the 7/5 and 8/5 resonance regions, respectively. The adjacent planets K2-72 d and K2-72 c are located far from the 2/1 resonance, which excludes the possibility of forming chains of mean motion resonances and, hence, 3-planet mean motion resonances. If the orbital eccentricities do not exceed 0.03, the evolution of the compact planetary system K2-72 over 100 Myr remains stable even in the presence of tidal perturbations.

Keywords. Compact planetary system, mean-motion resonances, tides, dynamic evolution, stability.

1. Introduction

At present, several dozen compact planetary systems containing more than two planets with masses of the order of the Earth's mass are known. It is shown that the stable evolution of compact planetary systems requires the presence of resonances that prevent close encounters of planets moving in neighboring orbits (see, for example, the five-planet systems Kepler-80 (MacDonald et al. 2016), K2-138 MacDonald et al. (2022) and the seven-planet system TRAPPIST-1 (Luger et al. 2017)). In this case, resonances between pairs of planets can form chains. The longest resonance chain known to date is realized in the TRAPPIST-1 system: 8:5 – 5:3 – 3:2 – 3:2 – 4:3 – 3:2 (Murphy and Armitage 2022). In the K2-138 system, five planets form the longest chain consisting of identical 3:2 resonances (MacDonald et al. 2022). On the other hand, simulation results show that in wide systems with massive planets, chains of high-order resonances can lead to the destruction of planetary systems (Murphy and Armitage 2022).

We consider the dynamic evolution of the compact four-planetary system K2-72. Star K2-72 is an M-type dwarf. The system contains three Earth-like planets and one super-Earth. We searched for low-order resonances within the uncertainty of determining the periods of the planets. We considered a few scenarios for the evolution of the K2-72 system over 100 Myr using the Posidonius software (Blanco-Cuaresma and Bolmont 2017), which considers tidal interactions.

2. Resonant arguments

When analyzing the resonance properties of planetary systems, it is necessary to study the behavior of the resonance arguments. For two planets i and $i+1$, which are in resonance of mean motions $k_i/(k_i - l_i)$, the resonance argument can be given as follows

Table 1. Parameters of the planetary system K2-72 (Dressing et al. 2017)

Parameter	K2-72 b	K2-72 c	K2-72 d	K2-72 e
R $[R_\oplus]$	1.08 ± 0.11	1.16 ± 0.13	1.01 ± 0.12	$1.29^{+0.14}_{-0.13}$
m $[M_\oplus]$	$1.39^{+0.58}_{-0.44}$	$1.80^{+0.85}_{-0.63}$	$1.09^{+0.55}_{-0.39}$	$2.65^{+1.18}_{-0.85}$
P [day]	5.577212 ± 0.00042	15.189034 ± 0.0031	$7.760178^{+0.001496}_{-0.001496}$	$24.158868^{+0.003726}_{-0.003850}$
e	$0.11^{+0.02}_{-0.09}$	$0.11^{+0.02}_{-0.09}$	$0.11^{+0.02}_{-0.09}$	$0.11^{+0.02}_{-0.09}$
i [deg]	$89.15^{+0.59}_{-0.86}$	$89.54^{+0.32}_{-0.44}$	$89.26^{+0.5}_{-0.7}$	$89.68^{+0.22}_{-0.32}$
g [deg]	7.49^{+120}_{-134}	16.83^{+113}_{-138}	14.28^{+114}_{-137}	11.39^{+117}_{-136}

(Huang and Ormel 2022):

$$\varphi_{i,\,i+1,\,i+s} = (k_i - l_i)\lambda_i - k_i\lambda_{i+1} + l_i\varpi_{i+s} \tag{1}$$

where l_i is the order of resonance, λ_i, λ_{i+1} are the mean longitudes of planets i and $i+1$, respectively, ϖ_{i+s} is the longitude of the pericenter of the orbit of planet i ($s=0$) or $i+1$ ($s=1$). For the next pair of planets $i+1$ and $i+2$, which are in resonance $k_{i+1}/(k_{i+1}-l_{i+1})$, the resonance argument is defined as follows:

$$\varphi_{i+1,\,i+2,\,i+1+s} = (k_{i+1} - l_{i+1})\lambda_{i+1} - k_{i+1}\lambda_{i+2} + l_{i+1}\varpi_{i+1+s} \tag{2}$$

Instead of two resonances between two adjacent pairs of planets, one can consider three-body resonances. In this case the three-body resonance is a chain of two two-body resonances $k_i/(k_i - l_i)$ and $k_{i+1}/(k_{i+1} - l_{i+1})$ with the resonance argument which does not depend on the pericenter longitude (Huang and Ormel 2022),

$$\Phi^{p,\,p+q,\,q}_{i,\,i+1,\,i+2} = p\lambda_i - (p+q)\lambda_{i+1} + q\lambda_{i+2} \tag{3}$$

where $p = l_{i+1}(k_i - l_i)$, $q = l_i k_{i+1}$. For three-body resonances we will use the following notation: $(p,\ -(p+q),\ q)$.

3. Compact planetary system K2-72

Host star K2-72 is an M2 dwarf with a mass of $0.27^{+0.08}_{-0.09}$ solar masses $M_\odot$ and a radius of $0.33^{+0.03}_{-0.03}$ radii of the Sun $R_\odot$. Table 1 gives the parameters of the K2-72 planetary system. The radii R of the planets are expressed in Earth radii $R_\oplus$. Planetary masses m are expressed in Earth masses $M_\oplus$ and calculated from the radii of the planets R, assuming that the density of a planet ρ depends on its radius R as Weiss and Marcy (2014):

$$\rho = 2.43 + 3.39\frac{R}{R_\oplus}\ \text{g cm}^{-3} \text{ for } R < 1.5R_\oplus. \tag{4}$$

The orbital elements P, e, i, g are the period, eccentricity, inclination, periapsis argument, respectively. The moment T_{conj} corresponds to the conjunction of the planet with the star.

4. Numerical simulations

For compact systems, an important factor affecting evolution is tidal interaction; therefore, we considered a few scenarios for the evolution of the K2-72 system over 100 Myr using the Posidonius software (Blanco-Cuaresma and Bolmont 2017), which takes into account tidal interactions. At nominal values of the periods and eccentricities, the simulation ended with the decay of the system 3 Myr after the start of integration. This is because at eccentricities of 0.11, the distance at the apocenter of the orbit of the planet K2-72 b becomes greater than the distance at the pericenter of the orbit of the planet K2-72 d. If the initial values of eccentricities of orbits of all planets are set equal to the

Table 2. Initial values of orbital eccentricities e_0, at which the planetary system does not disintegrate

Initial	K2-72 b	K2-72 c	K2-72 d	K2-72 e
e_0	0.04	0.04	0.04	0.03
e_0	0.04	0.04	0.03	0.04
e_0	≤ 0.03	≤ 0.03	≤ 0.03	≤ 0.03

Table 3. Mean-motion resonances and resonance offsets for the K2-72 system

Adjacent planets		MMR	Δ
K2-72 b – K2-72 d	(1–2)	7/5	0.0086
K2-72 d – K2-72 c	(2–3)	2/1	0.0427
K2-72 c – K2-72 e	(3–4)	8/5	0.0095

minimum value of 0.02, the system remains stable over the entire considered interval of 100 Myr for all values of star masses from 0.18 to 0.35 $M_\odot$.

The results of simulations including tidal interactions show that the compact four-planet system K2-72 can have a stable dynamical evolution in the absence of low-order resonances. Based on the results of numerical simulations with the Posidonius software, we obtained limits on the maximum values of eccentricities of the orbits of the planets, at which the system remains stable over 100 Myr (see Table 2). If the initial eccentricity values are greater than those given in Table 2, the system decays. If the orbital eccentricities do not exceed 0.03, the evolution of the compact planetary system K2-72 over 100 Myr remains stable even in the presence of tidal perturbations. In case the initial eccentricities of the orbits of the three planets are equal to 0.04, the eccentricities of the orbits of one of the planets K2-72 d or K2-72 e should not exceed 0.03 to ensure the stability of the system.

5. Resonant properties of K2-72

In (Dressing et al. 2017) it was noted that the orbital periods of the planets are close to commensurability, and it was suggested that mean-motion resonances (MMR) may be realised in the K2-72 system. We performed a search for possible resonances in the system for the case when the numerator and denominator in the rational approximation $k_i/(k_i - l_i)$ of the ratio of the periods P_j/P_i of adjacent planets i and j do not exceed 10. For each resonance, we estimated of the resonance offset (Charalambous et al. 2023)

$$\Delta_{ij} = \frac{P_j}{P_i} - \frac{k_i}{k_i - l_i} \tag{5}$$

Table 3 gives potential mean-motion resonances and resonance offsets Δ (5) for adjacent planets of the K2-72 system. The arrangement of planets in the system in order of distance from the star: 1 – K2-72 b, 2 – K2-72 d, 3 – K2-72 c, 4 – K2-72 e.

We showed that the compact planetary system K2-72 likely evolves beyond low-order resonances.

A significant change in the semi-major axes of the orbits of the K2-72 b and K2-72 d planets leads to the moving of the adjacent planets b-d and d-c out of the 7/5 and 8/5 resonance regions, respectively. The adjacent planets K2-72 d and K2-72 c are located far from the 2/1 resonance, which excludes the possibility of forming chains of mean motion resonances and, hence, three-planet mean motion resonances (see Fig. 1).

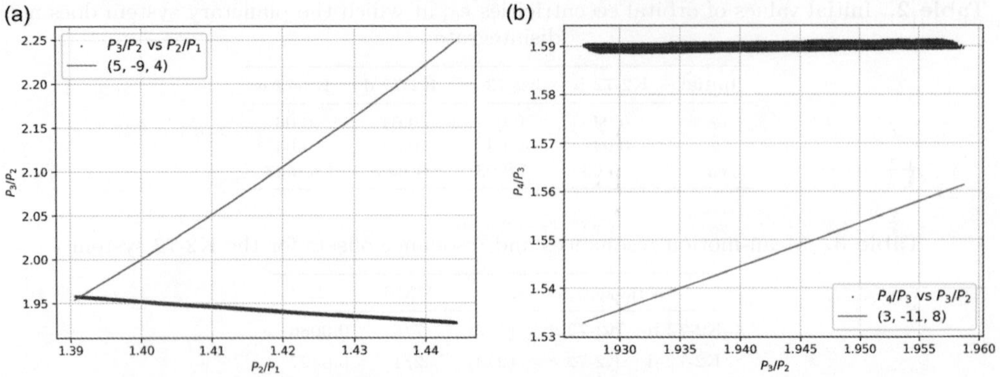

Figure 1. Evolution of the period ratios of the adjacent planets of the K2-72 system over 100 Myr: a) P_3/P_2 vs P_2/P_1, b) P_4/P_3 vs P_3/P_2. Blue dots correspond to the ratio of periods of adjacent planets, red lines show the positions of the three-body resonances. The system evolves from left to right.

6. Conclusions

The study shows that the compact planetary system K2-72 is not resonant. Under initial conditions corresponding to the masses and orbital elements of the planets determined from observations with their errors taken into account, the evolution of the K2-72 system is stable and regular over of 100 Myr for initial orbital eccentricities not exceeding 0.3.

The study was supported by the Russian Ministry of Science and Higher Education via the State Assignment Project FEUZ-2020-0038.

References

Blanco-Cuaresma, S., Bolmont, E. 2017, EWASS Spec. Session 4 (2017): Star-planet interactions (EWASS-SS4-2017), doi:10.5281/zenodo.1095095.
Charalambous, C., Teyssandier, J., Libert, A.S. 2023, *Astronomy and Astrophysics*, 677, A160.
Dressing, C.D., Vanderburg, A., Schlieder, J.E., *et al.* 2017, *The Astronomical Journal*, 154, 207.
Huang, S., Ormel, C.W. 2022, *Monthly Notices of the Royal Astronomical Society*, 511, 3814
Luger, R., Sestovic, M., Kruse, E., *et al.* 2017, *Nature Astronomy*, 1, 0129
MacDonald, M.G., Ragozzine, D., Fabrycky, D.C., *et al.* 2016, *The Astronomical Journal*, 152, 105
MacDonald, M.G., Feil, L., Quinn, T., Rice, D. 2022, *The Astronomical Journal*, 163, 162
Murphy, M.M., Armitage, P.J. 2022, *Monthly Notices of the Royal Astronomical Society*, 512, 2750
Weiss, L.M., Marcy, G.W. 2014, *The Astrophysical Journal Letters*, 783, L6.

Planetary Science and Exoplanets in the Era of JWST
Proceedings IAU Symposium No. 393, 2026
Z. Benkhaldoun, Y. Moulane & A. Szentgyorgyi, eds.
doi:10.1017/S1743921324002722

Advancing Apodization Techniques: Optimizing Interferometric Apodization by Homothety for Exoplanet Detection

A. Qazbour[1,3], Y. El Azhari[1,2,3], J. Chafi[1,3], Z. Benkhaldoun[4], O. Azagrouze[1,2,3] and A. Boskri[1,3]

[1]High Energy Physics and Astrophysics Laboratory, FSSM Cadi Ayyad University, Marrakech, Morocco
email: azizqazbourr@gmail.com

[2]CRMEF-Marrakech Av. Mouzdalifa, BP 797 Marrakech, Morocco

[3]Oukaimeden Observatory, Cadi Ayyad University, Marrakech, Morocco

[4]Sharjah University, United Arab Emirates and University Cadi Ayyad, Morocco

Abstract. New apodization techniques are emerging rapidly to enhance the coronagraphs' rejection capabilities and refine the optics for directly detecting exoplanets. One such technique, Interferometric Apodization by Homothety (IAH), involves splitting the incident Point Spread Function (PSF) into two using a 50 : 50 beamsplitter. One of the resulting PSFs has its amplitude reduced by a factor γ and its transverse dimension expanded by a factor η. By combining these two PSFs, an apodized PSF is generated. In this study, we will use the standard values of γ and η for both rectangular and circular apertures. We implement this approach in the laboratory using a Mach-Zehnder Interferometer with Cube Beamsplitters, chosen for their advantages over Plate Beamsplitters, including easy integration at a $0°$ angle of incidence and equal optical path lengths for reflected and transmitted light. This technique shows significant promise, achieving a contrast of approximately 5.10^{-3} at small angular separations around 2.8 λ/D.

Keywords. High angular resolution, Apodization, Interferometric Apodization by Homothety "IAH", Exoplanets, Interferometry, Mach-Zehnder interferometer

1. Introduction

Direct detection of exoplanets, especially Earth-like ones, remains a major challenge in modern astronomy due to the high brightness contrast between a star and its orbiting planets (Guyon et al. 2005). The light from a star can be 10^6 more intense than that reflected by its exoplanet in the infrared spectrum and 10^{10} in the visible (Ridgway 2004), which complicate direct imaging. Even advanced instruments struggle to isolate exoplanet light from that of its host star (Traub et al. 2010).

Coronagraphy is a promising solution, it consist to blocks starlight to reveal faint objects like nearby exoplanets. However, managing diffracted light remains a challenge, highlighting the need for apodization (Aime et al. 2001).

Apodization modulates light intensity at the edges of an optical aperture, reducing diffraction and enhancing contrast necessary for exoplanet detection. Integrating an apodizer into a coronagraph significantly improves its abilities to suppress starlight and detect faint planets (Aime et al. 2002).

Improving apodization techniques is central to the scientific community to enhance the telescopes capabilities for future projects. the Habitable Worlds Observatory (HWO)

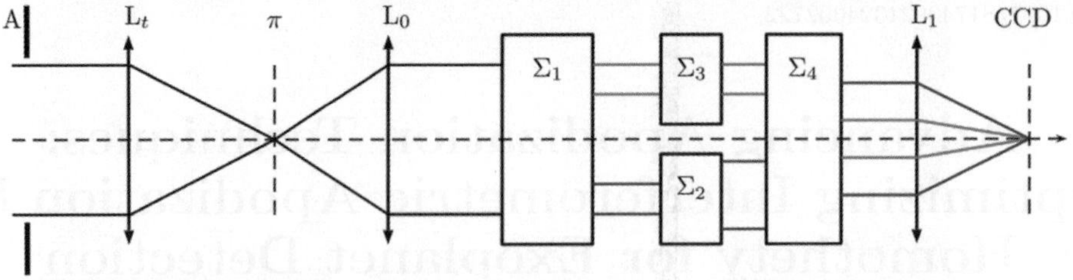

Figure 1. The optical layout of the IAH Principle; A and L_t representing the telescope aperture, π its focal plane. Σ_1 where the amplitude of the incident PSF has splited into two equal PSFs, Σ_2 controle the transverse dimension and the amplitude of one of the two PSFs, Σ_3 componsate for the phase shift introduced by Σ_2, Σ_4 recombine the two PSFs, and finally, the CCD camera where the apodized PSF is formed (Chafi et al. 2023a).

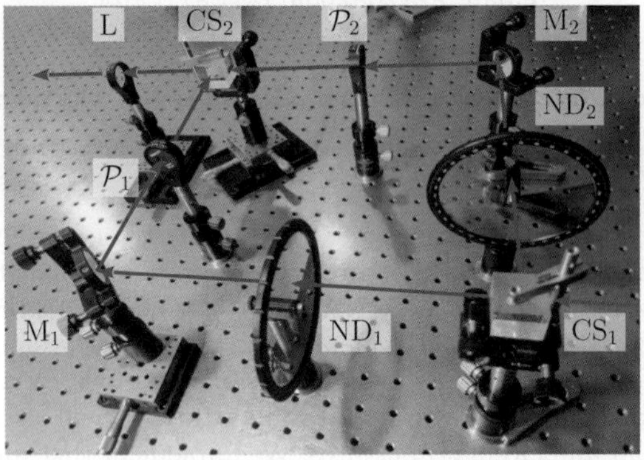

Figure 2. Illustration of the experimental interferometric part of the IAH: the laser light enters into the apodizer from the left (CS_1) and exits from the right (CS_2). The apertures P_1 & P_2 are accurately placed between the first cube beamsplitter CS_1 and the two MZI mirrors M_1 and M_2 respectively, while the two neutral density filters (ND_1 & ND_2) are placed between the last cube beamsplitter (CS_2) and the two MZI mirrors M_1 and M_2 respectively.

project, as recommended in the Astro2020 report (National Academies of Sciences, Engineering, and Medicine 2023). This project aims to directly detect and characterize habitable exoplanets in their star's circumstellar zone (Stark et al. 2024). By combining next-generation coronagraphs with advanced apodization techniques, HWO seeks to overcome technological barriers and usher in a new era of Earth-like exoplanet detection (Belikov et al. 2023).

In our study, we reproduced experimental results reported in Chafi et al. (2023b), using cube beam splitters instead of plate beam splitters, which offer advantages such as easy integration at a 0° angle of incidence, making them a more suitable alternative for direct exoplanet detection experiments.

2. Principle of the technique

IAH consist of splitting the incident PSF into two equal PSFs using amplitude division, The amplitude of one of the resulting PSFs has reduced by a factor γ and its transverse dimension expanded by a factor η, to match its negative lobes with the positive lobes of the second PSF and . By combining these two PSFs, an apodized PSF is generated.

3. Experimental setup

In order to generate the interferometric apodization we use a Mach-Zehnder interferometer with cube beamsplitters instead of plate beamsplitters used by Chafi et al.

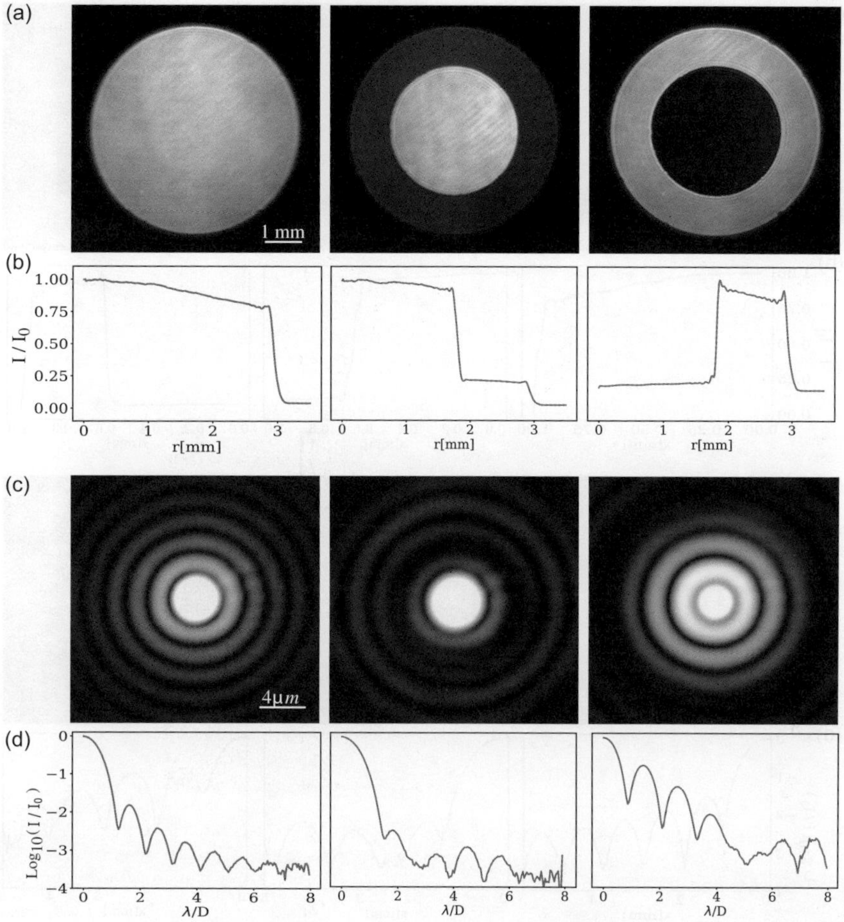

Figure 3. Experimental results for circular aperture; images in the pupil plane (a) and them radial profile (b) PSFs in the focal plane (c) and them radial profile (d). Unapodized (First column), apodized (Second column), and antiapodized (Third column)

(2023b) due to them advantages. The incident beam is splitted into two equal beams, the transverse dimension of one of the PSFs is stretched by the factor η and its amplitude is reduced by the factor γ. The superposition of those two PSFs at the finale focal plane gives an apodized PSF. The Homothety parameters that optimaze the apodization are calculated by Chafi et al. (2023a) for both rectangular and circular apertures.

Figure 2 shows the experimental setup used in this study. As a beam, we use a monochromatic He-Ne laser with a wavelength of 543.5 nm, and a high-resolution camera (SBIG) equipped with a 3326 × 2504 pixel array KAF-8300 CCD sensor with a surface area of approximately 17.96 × 13.52 mm^2 and 65,000 levels of grayscale.

3.1. *Cube Beamsplitters Advantages*

For this work we choose to use cube beam splitters 'CBS' instead of plate beam splitters 'PBS' due to them advantages. They have an easy integration at a 0° angle of incidence, which permit to avoid the prismatic effect caused by multi reflection inside the PBS. In addition no beam shift, equal optical path lengths, and a shortened optical path.

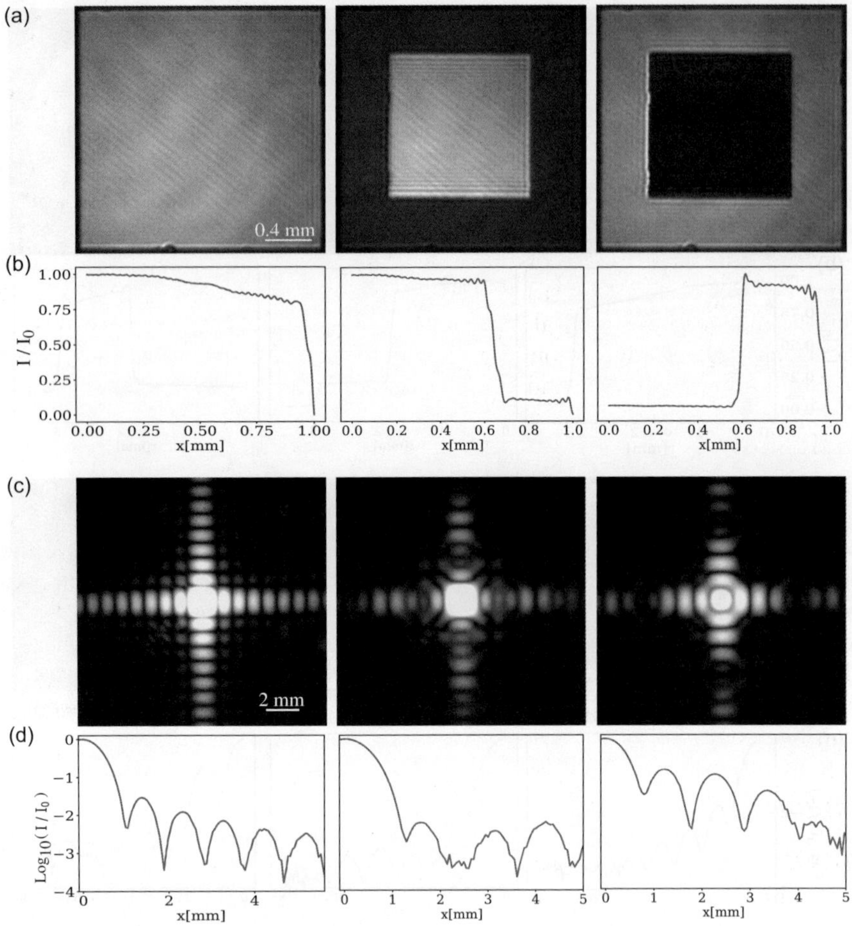

Figure 4. Experimental results for rectangular aperture; images in the pupil plane (a) and them transverse cross-sections along the horizontal axis (b) PSFs in the focal plane (c) and them transverse cross-sections along the horizontal axis (d). Unapodized (First column), apodized (Second column), and antiapodized (Third column)

4. Experimental results

At first, we place the CCD camera at the pupil plane, then we move it to the focal plane to capture the PSFs.

For this experimental study, we use the standard values calculated by Chafi et al. (2023a) for the parameters γ and η for both rectangular and circular apertures. All the images are taken with 0.9s exposure time.

4.1. *Circular aperture*

The objectif of this paragraph is to see the final results in the focal and the pupil planes, to compare them with Chafi et al. (2023b) results.

The CCD camera is initially placed at the pupil plane. Line (a) shows the pupil images, and line (b) presents their corresponding radial profiles. The CCD camera is then moved to the focal plane; line (c) shows the PSFs (Point Spread Functions), and line (d) presents their radial profiles. The first column corresponds to the unapodized pupil, the second column to the apodized pupil (destructive interferometry), and the third column to the anti-apodized pupil (constructive interferometry) (see figure 3).

As shown in the figure, this technique can attend a contrast of approximately 5.10^{-3} at small angular separations around 2.8 λ/D.

4.2. *Rectangular aperture*

We will do the same as in the case of circular aperture; we replace the circular apertures by the rectangular apertures and we change the values of the filters to correspond to those used by Chafi et al. (2023a).

By repeating the same process as in the case of circular aperture the CCD camera is placed in the pupil plane. Line (a) presents the pupil images, line (b) them transverse cross-sections along the horizontal axis, while Line (c) presents the PSFs, and line (d) them transverse cross-sections along the horizontal axis. First column unapodized pupil, second column apodized pupil, and third column antiapodized pupil (see figure 4).

5. Conclusion

Our study demonstrates the effectiveness of interferometric apodization by homothety in enhancing the capabilities of exoplanet detection. By employing a Mach-Zehnder interferometer with cube beamsplitters, we tried to optimize the concentration of light in the central diffraction lobe, and achieve notable efficiency for both circular and rectangular apertures. The experimental results show a contrast of approximately 5.10^{-3} around 2.8 λ/D for circular aperture, which can improve the sensitivity and rejection capabilities of coronagraphs. The use of CBS instead of PBS, this technique proves to be promising for future integration with advanced telescopes. By increasing the contrast and detection capabilities at small angular separations, our method holds significant potential for improving upcoming exoplanet observation missions. In the coming months, we will focus on evaluating the performance of this approach using a white light source.

References

Aime, C., Soummer, R., & Ferrari, A. 2001, AAP, 379, 697. doi:10.1051/0004-6361:20011293

Aime, C., et al. A&A, 2002. doi:10.1051/0004-6361:20020419

Belikov, R., et al. In: Techniques and Instrumentation for Detection of Exoplanets XI, SPIE, vol. 12680, p. 126802G, 2023. doi:10.1117/12.2677732

Chafi, J., et al. MNRAS, vol. 523, no. 4, pp. 5442–5451, 2023

Chafi, J., et al. MNRAS, vol. 527, no. 3, pp. 7036–7046, 2023

Gaudi, B. S., et al. arXiv e-prints, 2020, arXiv:2001.06683

Guerri, G., et al. Experimental Astronomy, vol. 30, no. 1, pp. 59–81, 2011

Guyon, O., et al. (2005). "Exoplanet Imaging with a Phase-induced Amplitude Apodization Coronagraph. I. Principle." The Astrophysical Journal, 622(1), 744–758

Habib, A., et al. MNRAS, vol. 406, no. 4, pp. 2743–2748, 2010. doi:10.1111/j.1365-2966.2010.16873.x

Juanola-Parramon, R., et al. Journal of Astronomical Telescopes, Instruments, and Systems, vol. 8, p. 034001, 2022. doi:10.1117/1.JATIS.8.3.034001

National Academies of Sciences, Engineering, and Medicine. The National Academies Press, Washington, DC, 2023, doi:10.17226/26141

Ridgway, S. T. 2004, EAS Publications Series, 12, 49. doi:10.1051/eas:2004023

Stark, C. C., Mennesson, B., Bryson, S., et al. 2024, arXiv:2405.19418. doi:10.48550/arXiv.2405.19418

Traub, W. A., Oppenheimer, B. R. (2010). "Direct Imaging of Exoplanets." In Exoplanets, ed. S. Seager (Tucson: University of Arizona Press), 111–156

Planetary Science and Exoplanets in the Era of JWST
Proceedings IAU Symposium No. 393, 2026
Z. Benkhaldoun, Y. Moulane & A. Szentgyorgyi, eds.
doi:10.1017/S1743921325001590

Interferometric Apodization using Deformable Mirror in a Michelson Interferometer to Improve Exoplanet Detection – I. Theoretical Foundations

A. Qazbour[1,3], **Y. El Azhari**[1,2,3], **J. Chafi**[1,3], **Z. Benkhaldoun**[4], **O. Azagrouze**[1,2,3] **and A. Boskri**[1,3]

[1]High Energy Physics and Astrophysics Laboratory, FSSM Cadi Ayyad University, Marrakech, Morocco
email: azizqazbourr@gmail.com
[2]CRMEF-Marrakech Av. Mouzdalifa, BP 797 Marrakech, Morocco
[3]Oukaimeden Observatory, Cadi Ayyad University, Marrakech, Morocco
[4]Sharjah University, United Arab Emirates and University Cadi Ayyad, Morocco

Abstract. To capture images of Earth-like planets orbiting distant stars, advanced instruments with exceptional contrast ratios are imperative. While coronagraphs play a crucial role, they often lack the capability to achieve the requisite contrast levels independently. Hence, supplementary apodization techniques are indispensable for augmenting their rejection capabilities. In this context, we introduce an innovative apodization method that harnesses interferometry, seamlessly integrating a deformable mirror into the Michelson interferometer setup. This sophisticated approach entails splitting the incident Point Spread Function (PSF) into two components, introducing an additional inhomgenious phase $\phi(x, y)$ to one of them via a deformable mirror, and subsequently recombining them to yield an apodized PSF. We illustrate, in particular, the influence of several parameters of the deformable mirror on the optimization of the additional phase profile.

Keywords. High angular resolution, Apodization, Interferometric Apodization using Deformable Mirror, Exoplanets, Interferometry, Michelson interferometer

---

1. Introduction

In order to detect exoplanets and gather more information about them, it is more suitable to use direct detection techniques. These allow for the direct study of the light coming from exoplanets, enabling various analyses such as spectroscopy. Despite these advantages it faces a significant challenge due to the considerable difference in brightness between the planet and its host-star, the ratio can up to 10^6 in the infrared spectrum and 10^{10} in the visible (Ridgway 2004), which renders the direct observation of the planet theoretically impossible (Fogarty et al. 2014). To overcome this difficulty, scientific community propose several approaches. One such approach is the use of coronagraphy, which blocks the star light and allows the exoplanet's light to pass. scientists tried to elevate the direct imaging systems contrast, and use them in the upcoming telescopes, including LUVOIR (The LUVOIR Team 2019), HabEx (Gaudi et al. 2020)...

The first coronagraph designed by Lyot (1931) was dedicated to produce an artificial eclipse, its aim to study the closest objects to the sun. Researchers tried to apply coronagraphy on extra-solar systems to study objects orbiting around stars such as exoplanets (Traub et al. 2010). By the time its appear that coronagraphs alone are not sufficient attend the needed contrast, due to the diffraction generated by the entrance pupil of the telescope which often hide the faint light from the exoplanets. So the need for Apodization to enhance the coronagraph's rejection (Aime et al. 2001).

Apodization is a primordial technique used to overcome this difficulty by concentrating the diffracted energy in the central diffraction lope. Traditionally, a Michelson interferometer is employed to produce a cosine transmission profile, as demonstrated by El Azhari et al. (2005) in their study on interferometric apodization of rectangular apertures.

Pueyo et al. (2011) has produced apodization by replacing the mirror of Phase-induced Amplitude Apodization Coronagraph 'PIAA' (Guyon et al. 2005) by a deformable mirror 'DM', which shows good results.

We propose replacing one of the flat mirrors in the Michelson interferometer used by El Azhari et al. (2005) with a DM, to produce an interferometric apodization similar to Interferometric Apodization by Homothety 'IAH' proposed by Habib et al. (2010), and reproduced by Chafi et al. (2023a) who verified it experimentally in Chafi et al. (2023b). This innovation allows for a dynamic control of the profile transmitted, enabling the creation of a customized apodization tailored to specific observational needs. By adjusting the DM shape, we can achieve a transmission profile that enhances the contrast and detection capabilities for exoplanets.

This approach aims to offer a more flexible and effective solution for direct exoplanet detection, surpassing the limitations of the fixed cosine profile produced by traditional Michelson interferometer.

Our proceeding presents the conceptual framework and the proposed setup, demonstrating the potential of this method in advancing exoplanets detection.

2. Experimental setup

Figure 1 shows the proposed setup used in this study, We use a Michelson interferometer with a flat mirror in the first arm and a DM in the second arm. The aperture P is illuminated by a monochromatic and perfectly collimated beam, the diffracted beam intensity is splitted by a 50:50 beamsplitter S_p into two beams, the first one is reflected by the plate mirror M_1 while an additional inhomgenious phase $\phi(x, y)$ is introduced to the second one. Combining these two beams generate an apodized beam.

3. Theory

The set-up is illuminated by a monochromatic punctual source of amplitude Ψ_0 placed in the object focal plane of the lens L_e to get a perfectly collimated beam. our work consist to find the expression of the amplitude at each component of the set-up using the Fresnel propagation.

3.1. *Equivalent setup*

The setup can be reduced to two equivalent configurations, each representing the light propagation in the interferometer arms, which simplifies the calculations. Figure 2 shows the light propagation in the first arm of the interferometer with a plate mirror, while figure 3 shows the light propagation in the arm with deformable mirror.

The PSF in the first arm with a plate mirror is given by:

$$\Psi_{\pi_{f,1}}(x,y) = C_1 \cdot \exp\left[\frac{i\pi}{\lambda f_s}\left(x^2+y^2\right)\right] \cdot TF_{\frac{x}{\lambda f_s},\frac{y}{\lambda f_s}}\left\{t_p\left(x',y'\right)\right\} \cdot \exp\left[\frac{i\pi}{\lambda d}\left(\left(\frac{x}{\lambda f_s}\right)^2+\left(\frac{y}{\lambda f_s}\right)^2\right)\right] \tag{1}$$

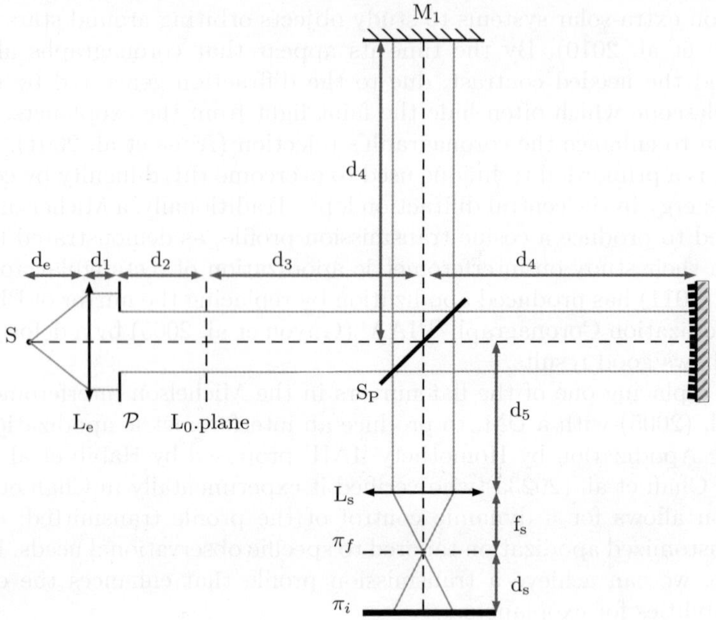

Figure 1. Illustration of the apodization principle using a deformable mirror in a Michelson interferometer. The punctual source S illuminate the lens L_e of focal distance $f = d_e$ to produce a perfectly collimated beam, the aperture P simulate the entrance pupil of the telescope, the light diffracted by P is splitted by the beamsplitter S_P into two equal PSFs, the first PSF is then reflected by a flat mirror M_1 toward the lens L_s, while the other PSF is reflected by the deformable mirror M_2 toward the beamsplitter where its reflected toward the lens L_s which in its focal plane π_f we can observe the apodization effect, and π_i is the image plane. L_0 plane is where we will place later a lens to form the image of P in the mirrors plane.

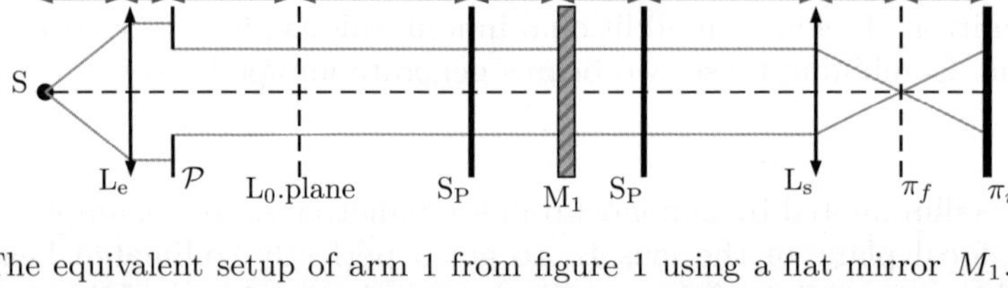

Figure 2. The equivalent setup of arm 1 from figure 1 using a flat mirror M_1. This diagram represents the main components and the distances traversed by the light from the source S until the finale focal plane.

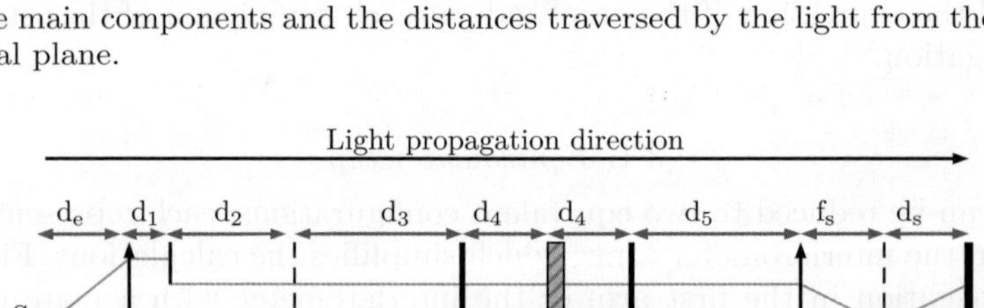

Figure 3. The equivalent setup of the second arm with the deformable mirror (M_2) from the figure 1.

with :

$$C_1 = \frac{\Psi_0 \exp[ik(d_e+d_1+d+f_s)]}{2(i\lambda)^4 f_e d_1 d f_s} \qquad \text{and the distance} \qquad d = d_2 + d_3 + 2d_4 + d_5$$

The second PSF in the arm with a DM is given by :

$$\begin{aligned}\Psi_{\pi_{f,2}}(x,y) =& C_2 \cdot \exp\left[\frac{i\pi}{\lambda(d_4+d_5)}\left(\left(\frac{x}{\lambda f_s}\right)^2 + \left(\frac{y}{\lambda f_s}\right)^2\right)\right] \cdot \exp\left[\frac{i\pi\,(x^2+y^2)}{\lambda f_s}\right] \\ &\cdot \left[TF_{\frac{x}{\lambda f_s},\frac{y}{\lambda f_s}}\left\{\left[t_p(x',y')\right]\right\} \cdot \exp\left[\frac{i\pi}{\lambda(d_2+d_3+d_4)}\left(\left(\frac{x}{\lambda f_s}\right)^2 + \left(\frac{y}{\lambda f_s}\right)^2\right)\right]\right] \\ &* TF_{\frac{x}{\lambda f_s},\frac{y}{\lambda f_s}}\left\{t_{M_2}(x',y')\right\}\end{aligned} \tag{2}$$

With :

$$C_2 = \frac{\Psi_0 \exp\left[ik\left(d_e+d_1+d+f_s\right)\right]}{(i\lambda)^5 d_e d_1\left(d_2+d_3+d_4\right)\left(d_4+d_5\right) f_s}$$

The amplitude at the final focal plane is the sum of the two amplitudes, $\Psi_{\pi_{f,1}}(x,y)$ and $\Psi_{\pi_{f,2}}(x,y)$. It is given by:

$$\Psi_{\pi_f}(x,y) = C_1 B \cdot TF_1 \cdot D_1 + C_2 B \cdot D_2\left[TF_1 \cdot D_3\right] * TF_2 \tag{3}$$

where :

$$\begin{aligned} B &= \exp\left[\frac{i\pi}{\lambda f_s}\left(x^2+y^2\right)\right] \\ D_1 &= \exp\left[\frac{i\pi}{\lambda d}\left(\left(\frac{x}{\lambda f_s}\right)^2 + \left(\frac{y}{\lambda f_s}\right)^2\right)\right] \\ D_2 &= \exp\left[\frac{i\pi}{\lambda(d_4+d_5)}\left(\left(\frac{x}{\lambda f_s}\right)^2 + \left(\frac{y}{\lambda f_s}\right)^2\right)\right] \\ D_3 &= \exp\left[\frac{i\pi}{\lambda(d_2+d_3+d_4)}\left(\left(\frac{x}{\lambda f_s}\right)^2 + \left(\frac{y}{\lambda f_s}\right)^2\right)\right] \end{aligned} \tag{4}$$

$$\begin{aligned} TF_1 &= TF_{\frac{x}{\lambda f_s},\frac{y}{\lambda f_s}}\left\{t_p(x',y')\right\} \\ TF_2 &= TF_{\frac{x}{\lambda f_s},\frac{y}{\lambda f_s}}\left\{t_{M_2}(x',y')\right\} \end{aligned} \tag{5}$$

The equation 3 describes how the amplitude function $\Psi_{\pi_f}(x,y)$ at the final plane π_f is obtained from the parameters of the interferometric system with a deformable mirror. Where the variables and terms used can be explained as fellow :

* $\Psi_{\pi_f}(x,y)$: The light amplitude at the final plane, depending on the coordinates (x,y).
* C_1, C_2: Constants that depend on the optical parameters of the system (the focal distances (f_e, f_s), the distances between different components $(d_1, d_2, d_3, d_4, d_5)$, the wavelength λ, and the incident amplitude Ψ_0.
* B: Complex phase term due to the last lens L_s, it's depending on the coordinates (x,y) and the wavelength λ.
* D_1, D_2, D_3: Phase terms due to propagation between the different parts of the interferometric system.
* TF_1, TF_2: Fourier transforms of the transmission functions $t_p(x',y')$ of the aperture P, and $t_{M_2}(x',y')$, the transmission of the deformable mirror (DM), respectively.

The deformable mirror shape is adjustable, which allow for a wide shapes possibilities. It has been demonstrated that linear prolate functions are the most suitable apodizers for rectangular apertures in stellar coronagraphy. A total extinction of starlight can be produced by prolate functions using Roddier & Roddier phase mask (Aime et al. 2002).

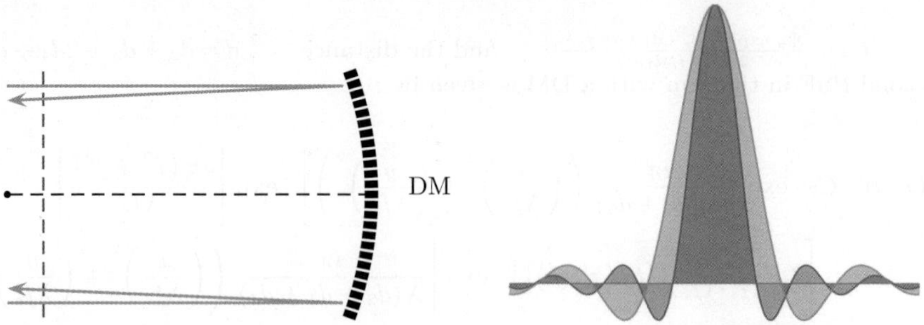

Figure 4. The deformable mirror taking the shape of a concave spherical mirror (left), and a radial cut of the superposition of the two PSFs (right)

4. Case of spherical mirror shape

Our future work will be focused on giving the deformable mirror the shape of a concave spherical mirror, this shape permit to concentrate one of the two PSF inside the other, in the manner of which the positive secondary diffraction wings of the first annihilate with the second to carry out the apodization. Figure 4 shows the deformable mirror shape (left) and a radial cut of the superposition of the two PSFs (right).

5. Conclusion

In this proceeding, we have presented a novel method for enhancing the detection of exoplanets through the use of interferometric apodization with a deformable mirror within a Michelson interferometer setup. Our approach introduces dynamic control over the transmission profile, allowing for a customized apodization that can significantly improve contrast and detection capabilities.

The theoretical foundations laid out demonstrate how adjusting the deformable mirror's shape can optimize the phase profile, leading to more effective suppression of starlight and better isolation of the faint signals from exoplanets. This method shows great potential in surpassing the limitations of traditional fixed cosine profiles used in Michelson interferometers.

Future work will involve numerical modeling to simulate the performance of this approach, followed by experimental validation to confirm the theoretical predictions. By advancing this technology, we aim to contribute to the broader goal of directly imaging Earth-like exoplanets and expanding our understanding of distant planetary systems.

References

Aime, C., Soummer, R., & Ferrari, A. 2001, *AAP*, 379, 697. doi:10.1051/0004-6361:20011293

Aime, C., et al. *A&A*, 2002. doi:10.1051/0004-6361:20020419

Chafi, J., et al. *MNRAS*, vol. 523, no. 4, pp. 5442–5451, 2023

Chafi, J., et al. *MNRAS*, vol. 527, no. 3, pp. 7036–7046, 2023

El Azhari, Y., Azagrouze, O., Martin, F., Soummer, R., & Aime, C. (2005). "Interferometric Apodization of Rectangular Aperture - Laboratory Experiments"

Fogarty, K., Pueyo, L., & Mawet, D. (2014). "Optimal Apodizations for On-Axis Vector Vortex Coronagraphs." *SPIE*, vol. 9143, p. 914326. doi:10.1117/12.2056760

Gaudi, B. S., Seager, S., Mennesson, B., et al. 2020, arXiv:2001.06683. doi:10.48550/arXiv.2001.06683

Guyon, O., et al. (2005). "Exoplanet Imaging with a Phase-induced Amplitude Apodization Coronagraph. I. Principle." *The Astrophysical Journal*, 622(1), 744–758

Habib, A., et al. *MNRAS*, vol. 406, no. 4, pp. 2743-2748, 2010. doi:10.1111/j.1365-2966.2010.16873.x

Lyot, B. (1931). "Étude de la couronne hors éclipse." L'Astronomie, 45, juin

Pueyo, L., Kasdin, N. J., Carlotti, A., & Vanderbei, R. (2011). "Design of Phase Induced Amplitude Apodization Coronagraphs over Square Apertures." *The Astrophysical Journal Supplement Series*, 195(2), 25. https://doi.org/10.1088/0067-0049/195/2/25

Ridgway, S. T. 2004, EAS Publications Series, 12, 49. doi:10.1051/eas:2004023
Traub, W. A., Oppenheimer, B. R. (2010). "Direct Imaging of Exoplanets." In Exoplanets, ed. S. Seager (Tucson: University of Arizona Press), 111–156
The LUVOIR Team 2019, arXiv:1912.06219. doi:10.48550/arXiv.1912.06219

Planetary Science and Exoplanets in the Era of JWST
Proceedings IAU Symposium No. 393, 2026
Z. Benkhaldoun, Y. Moulane & A. Szentgyorgyi, eds.
doi:10.1017/S1743921324001935

Apatite Formation in Shergottite NWA 7397: Implications for the Mapping of Martian Meteorites

Amber Bonsall-Pisano[1], Geoffrey H. Howarth[1], Chad Peel[1] and Gelu Costin[2]

[1]Department of Geological Sciences, University of Cape Town, South Africa
email: Awbonsall@gmail.com

[2]Department of Earth, Environmental and Planetary Sciences, Rice University, Houston, TX, USA

Abstract. Martian meteorites are currently our only existing samples from Mars. They are divided into two primary types, the shergottites and the nakhlite-chassignite types. The shergottites are by far the most abundant of the Martian meteorites. Apatite in particular is the only volatile bearing phase in these and thus is crucial for understanding volatile cycles on Mars. The primary goal of the study is to understand the effects of shock metamorphism on the volatile content of apatite. In particular, looking at intergrown apatite-merrillite grains to observe the Cl variation within apatite and to determine if the merrillite is in fact merrillite or if its tuite. Here we used various chemical analyses to accurately map the mineralogy of shergottite NWA 7397 to learn more about volatile content and shock effects to constrain its petrogenesis.

For NWA 7397, what we focused on was the phosphates. Mainly on the calcium phosphates, apatite ($Ca_5[PO_4]_3[F, CL, OH]$) and merrillite ($Ca_9Na(Fe, Mg)(PO_4)_7$). We can identify apatite based on its volatile components (F and Cl) as part of its crystal structure which merrillite lacks. Apatite is, however, the only volatile bearing phase in shergottites, making it our only was to constrain water and Cl content of martian magmas and by extension martian mantle reservoirs. However, the volatile contents can be effected by shock ejection when the rock was blasted off the martian surface. This means we need to understand the effects of shock before using apatite to estimate magma volatile content. From our research so far we have been able to identify apatite and merrillite, but no tuite. This will require further analysis to identify tuite, if any is present.

Keywords. Mars, meteorite, apatite

1. Introduction

Mars has been the one of the most studied planets in our Solar system. There have been countless optical observations done, but also about 50 missions to the red planet. These vary from flybys to actual landers such as *InSight* and *Perseverance.*

Despite the numerous missions to Mars, we have no direct samples from the Martian surface. Many probes have been sent to Mars to analyze the geology (See Figure 1), but it is not the same as having a human analyze the sample. The technology on Earth is also exponentially more advanced than anything we could send. The tests that the rovers and probes can do is limited to what technology was able to be placed on the machine. If something new or exciting is discovered from these samples, there is no extra analysis that can be done and no followup can be completed.

To date, approximately 280 Martian meteorites have been discovered (Meteoritical Bulletin Database, May 2024). In general, the samples are relatively young (mid to late

Figure 1. A topographic map of Mars from the Mars Orbiter Laser Altimeter (MOLA) an instrument on the Mars Global Surveyor (MGS). This map shows the main regions of the Martian surface and all the current successful landing sites and dates Udry et al. (2020).

Amazonian) (Udry et al. 2020). These samples are believed to have been ejected from the surface of Mars at an escape velocity 5 km/s Fritz et al. (2005). Martian meteorites can be classified into one of three classes: shergottites, nakhlites, and chassignites (SNC). Most SNCs are basaltic in composition, usually containing olivine, pyroxene and plagioclase. Shergottites make up the largest group of Martian meteorites and are the most geochemically diverse. They can be classified into three sub-groups based on mineralogy and petrology: basaltic, olivine-phyric, and poikilitic, as well as new grouping known as gabbroic and olivine-gabbro (Yoshida et al. 2016). Poikilitic shergottites are mineralogically the most complex sub-group among shergottites comprising multiple generations of both olivine and pyroxene and characterized by the presence of coarse-grained (~1cm) compositionally zoned pyroxene containing abundant olivine and chromite. Shergottites can be further classified geochemically using incompatible trace elements (ITE) and radiogenic isotopes into light rare earth elements (LREE) enriched, intermediate or depleted.

The aim of this project is to focus on the phosphates in NWA 7397 and particularly the Cl zoning in apatite. This will also tell us whether that is related to phase change during high-P to tuite. Tuite is believed to be formed from the decomposition of apatite or the phase transformation of merrillite Darling et al. (2021). Apatite is the only volatile bearing phase in shergottites. However, this can be tricky since we need to know if the volatile contents were modified by the ejection that removed them from the surface of Mars.

2. Techniques

The specific meteorite we worked with was Northwest Africa 7397 (henceforth referred to as NWA 7397). NWA 7397 is considered a poikilitic shergottite, with characteristics of being both poikilitic and non-poikilitic. Poikilitic shergottites contain olivine chadacrysts, which are encased by large pyroxene oikocrysts and later formed olivine, pyroxene, and maskelynite (Udry et al. 2020). Three different analytical techniques were used for this sample.

QEMSCAN: Quantitive Evaluation of Minerals by Scanning Electron Microscopy (QEMSCAN) A QEMSCAN is a form of X-ray electron microscopy designed specifically for the analysis and evaluation of minerals. The QEMSCAN 650F in the Chemical Engineering department of University of Cape Town (UCT) uses a field emission gun-scanning electron microscope (FEG-SEM) from FEI Company, combined with high-resolution Back scattered electron (BSE), Bruker energy Dispersive Spectrometers (EDS), and a Spectral Analysis Engine (SAE) to analyze phases. Minerals are classified by using SIP files (Species Identification Protocol). In the case of our samples, we started with a broader existing SIP list and condensed it to a list more appropriate for Martian meteorites. This required the removal and addition of certain minerals species.

Electron Microprobe Analysis: Electron microprobe analysis is used for determining the chemical compositions of minerals. The instrument is called an electron probe micro-analyzer (EMPA). Each element emits its own spectrum with various peaks and peak intensity in the X-ray. Each grain was mapped and then point analysis was done on 2-4 spots in each grain. These spots were chosen by using the QEMSCAN maps to determine what might be apatite and what might be merrillite, but clearly a phosphate. The probe itself is 15kV. The beam has a current of 50nA and a beam size of 5 microns.

Raman Spectroscopy: Raman spectroscopy is a non-destructive form of chemical compositional analysis. In this case, monochromatic radiation is direction at the sample, where it is either reflected, absorbed, or scattered. The scattered light is then used to analyze vibrational energy of the sample and produce spectra. Each spectral peak indicates something about the sample. The frequencies of the vibrations are dependent on atomic mass and bond strength. For example, heavy atoms and weak bonds create a low frequency shift and lighter atoms with strong bonds create higher frequency shifts. Raman shift is measured in cm-1. Spectra are then compared to existing or known spectra from the RRuff ^{TM} database. Each spot was analyzed for 30 seconds to produce a spectra. After only 30 seconds some spectra were hard to determine what they were but there were hints of possibilities of them being merrillite. They appeared to show the barest hints of merrillites trademark bimodal peaks, it is believed with longer analysis time this will become more prevalent and the bimodal peaks will become more defined. The spectra that turned out to be apatite, were very clear and very easily discerned.

3. Results

QEMSCAN Analysis: In the original QEMSCAN analysis, merrillite was not a mineral of choice to scan for. Post scan, we were able to add in the data for merrillite based on its comparisons with apatite. The QEM map shows relatively large phosphates grains that we were then able to analyze more directly and take a look for apatite and merrillite transitions (See Figure 2). In Figure 2 you are not able to easily see the difference between apatite and merrillite unless you zoom in on individual grains. In the enlarged grain 7 you are able to see the difference between the apatite and merrillite.

Microprobe Anaylsis: After acquiring a QEM and BSE map we were able to do microprobe mapping and elemental analysis on individual grains (See Figure 3). We select various grains that had also been analyzed in detail from the QEMSCAN and did a selected map analysis on them. With the microprobe we were able to map Ca, P, Cl, and F. We know both apatite and merrillite are both phosphates but there is a difference in the calcium content and merrillite lacks chlorine and fluorine. You can see in Figure 3 which parts were chlorine and fluorine-rich and which were not, making it quite easy to distinguish what was apatite and what was merrillite. This also matched with what we had on the grain map in the QEMSCAN.

Raman Analysis: Once we had confirmation of the relative locations of apatite and merrillite (See Figure 3) we performed Raman analysis. We used existing standards for

Figure 2. Full QEMSCAN map of NWA-7397 (left) and phosphate grain 7 (right). Pink and lime green indicate detected phosphates. Pink being apatite and lime green identifying possible merrillites.

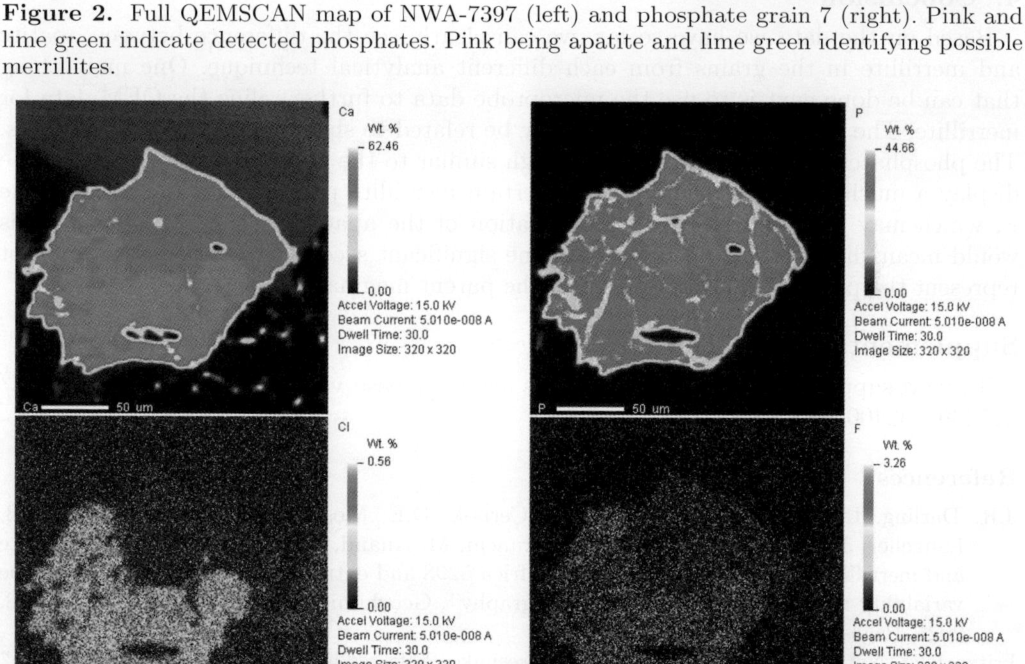

Figure 3. Microprobe analysis shows Ca, P, Cl, and F concentrations in the same grain used for QEMSCAN

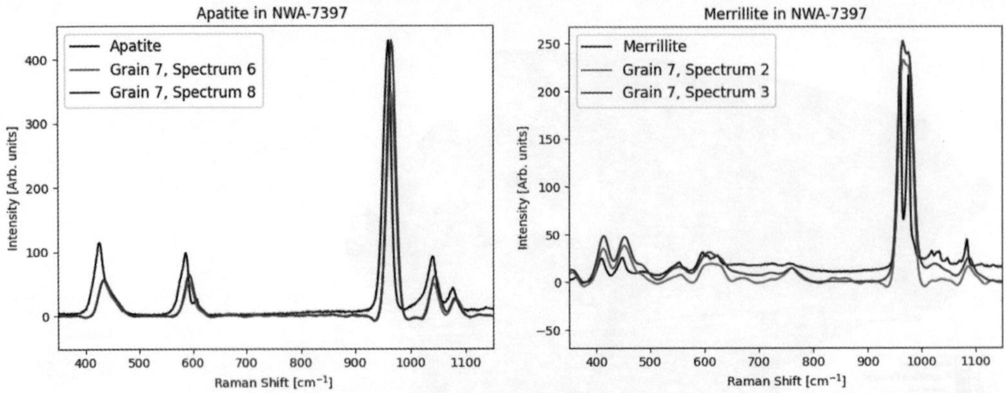

Figure 4. Raman data for the same grain. Apatite standard is black and the grain data for 2 points is red and blue in image on left. Merrillite image on the right showing standard in black and point data in red and blue.

both apatite and merrillite. In Figure 4 you are able to see the spectra for apatite are pretty much an identical match to the standard. When it came to merrillite which is most identifiable peak is the bimodal peak at around 950, the individual spectra do not show as defined bimodal peaks. It is believed this is due to the brevity of the anaylsis and if this experiment were to be repeated for longer the peaks would become more defined. We were not able to find any identifiable tuite in any Raman scans that were done; all scans were clearly apatite or merrillite.

4. Conclusion

Based on the data we have so far, we can clearly see the difference between apatite and merrillite in the grains from each different analytical technique. One main thing that can be done next is to use the microprobe data to further refine the QEM data for merrillite. The Cl-rich parts of apatite may be related to shock mobilisation of volatiles. The phosphates studied showed that though similar to the apatites here on earth, these display a much higher Cl content. Also certain merrillite phases seem to contain some F, which may represent a high-P modification of the apatite to tuite transition. This would mean that the apatite has undergone significant secondary processes and do not represent the primary volatile content of the parent magma.

Supplementary material

To view supplementary material for this article, please visit https://doi.org/10.1017/S1743921324001935

References

J.R. Darling, L.F. White, T. Kizovski, A. Černok, D.E. Moser, K.T. Tait, J. Dunlop, B. Langelier, J.O. Douglas, X. Zhao, I.A. Franchi, M. Anand, "The shocking state of apatite and merrillite in shergottite Northwest Africa 5298 and extreme nanoscale chlorine isotope variability revealed by atom probe tomography", Geochimica et Cosmochimica Acta, 293, 2021, Pages 422–437

Fritz, Jörg, Natalia Artemieva, & Ansgar Greshake 2005, "Ejection of Martian meteorites." Meteoritics & Planetary Science 40.9-10 (2005): 1393–1411

Udry, Arya, et al. "What Martian meteorites reveal about the interior and surface of Mars." (2020): e2020JE006523

Yoshida, Masashi, et al. "Petrology and mineralogy of Northwest Africa 7397 lherzolitic shegottite"

Planetary Science and Exoplanets in the Era of JWST
Proceedings IAU Symposium No. 393, 2026
Z. Benkhaldoun, Y. Moulane & A. Szentgyorgyi, eds.
doi:10.1017/S174392132400262X

Discovery of Three Sub-Neptune Exoplanets Orbiting M Dwarfs with TESS

A. Soubkiou[1], K. Barkaoui[2,3,4], T. Gan[5] and Z. Benkhaldoun[6]

[1]Oukaimeden Observatory, HEPA Lab, FSSM, Cadi Ayyad University, Marrakech, Morocco
email: a.soubkiou.ced@uca.ac.ma

[2]Astrobiology Research Unit, Université de Liège, Liège, Belgium

[3]Department of Earth, Atmospheric and Planetary Science, MIT, Cambridge, MA 02139, USA

[4]IAC, Calle Vía Láctea s/n,38200, La Laguna, Tenerife, Spain

[5]Department of Astronomy, Tsinghua University, Beijing 100084, People's Republic of China

[6]Sharjah University, United Arab Emirates and University Cadi Ayyad, Morocco

Abstract. We present the discovery of three sub-Neptune exoplanets orbiting M dwarf stars: TOI-2136 b (period 7.85 days, radius $R_p = 2.19 \pm 0.17 R_\oplus$, mass $M_P = 6.4 \pm 2.4 M_\oplus$), TOI-2084 b (period 6.08 days, radius $R_p = 2.47 \pm 0.13 R_\oplus$), and TOI-4184 b (period 4.9 days, radius $R_p = 2.43 \pm 0.21 R_\oplus$). All discoveries utilized TESS data combined with ground-based follow-up observations for confirmation. The location of these planets near the "radius valley" suggests diverse planetary formations around low-mass stars and holds promise for future atmospheric characterization with instruments like JWST.

1. Introduction

The discovery of thousands of transiting exoplanets by the Kepler mission (Borucki *et al.* 2010) opened a new era in exoplanet research. Among its most significant findings is the abundance of super-Earths and sub-Neptunes ($1 R_\oplus < R_p < 4 R_\oplus$) in close-in orbits around other stars, while such planets are absent in our solar system (Howard *et al.* 2012; Fressin *et al.* 2013; Petigura *et al.* 2013). Further demographic studies, such as the California-Kepler Survey (Petigura *et al.* 2017; Johnson *et al.* 2017), revealed a bimodal distribution in the radii of small planets, with a valley centered at around 1.8 $R_\oplus$ (Fulton *et al.* 2017, 2018). This radius gap is thought to result from atmospheric loss mechanisms like photoevaporation (Owen *et al.* 2013; Lopez *et al.* 2014; Jin *et al.* 2014; Chen *et al.* 2016; Owen *et al.* 2017), where high-energy radiation from the host star strips planets of their hydrogen/helium envelopes, as well as core-powered mass-loss (Ginzburg *et al.* 2018; Lopez *et al.* 2018; Gupta *et al.* 2019, 2020, 2021), which shrinks planetary radii over time.

While most studies have focused on F, G, and K-type stars (Van Eylen *et al.* 2018; Martinez *et al.* 2019), increasing attention is being paid to M dwarfs, which are the most common stars in our galaxy (Henry *et al.* 1994; Kirkpatrick *et al.* 1999). M dwarfs are particularly promising for detecting small, temperate exoplanets due to their small sizes, low masses, and luminosities, which make transit signals deeper and radial-velocity semi-amplitudes larger compared to Sun-like stars (Nutzman *et al.* 2008; Kaltenegger *et al.* 2009; Winters *et al.* 2015). Additionally, M dwarf systems are ideal for atmospheric characterization through transmission spectroscopy, including with JWST (Kempton *et al.* 2018), allowing for detailed studies of their planetary atmospheres.

Table 1. Ground-based observational parameters: date of observation, filter used, telescope, exposure time(s), photometric aperture size, and FWHM of the point-spread function. *a*: Burdanov et al. (2022); *b*: Barkaoui et al. (2019); *c*: Narita et al. (2020); *d*: Demory et al. (2020); *e*: Brown et al. (2013); *f*: Dominik et al. (2010); *g*: Bonfils et al. (2015).

Planet	Date (UT)	Filter	Telescope	Exptime [second]	Aperture size [arcsec]	FWHM [arcsec]
TOI-2084.01	Aug 13 2020	$I+z$	SPECULOOS-N-1.0m [*a*]	33	2.8	1.4
TOI-2084.01	Jan 30 2021	$I+z$	TRAPPIST-N-0.6m [*b*]	65	7.6	3.1
TOI-2084.01	May 20 2021	g', r', i', zs	MuSCAT3-2.0m [*c*]	337,80,53,70	4.0	2.0
TOI-2084.01	May 26 2021	g', r', i', zs	MuSCAT3-2.0m	337,80,53,70	4.0	2.0
TOI-2084.01	Jun 25 2021	$I+z$	SPECULOOS-1.0m	33	3.2	1.6
TOI-2084.01	Jun 25 2021	$I+z$	TRAPPIST-N-0.6m	65	5.6	3.7
TOI-2084.01	Jul 13 2021	r'	SAINT-EX-1.0m [*d*]	141	3.2	1.4
TOI-4184.01	Jul 28 2021	ip	LCO-CTIO-1.0m [*e*]	240	3.1	1.3
TOI-4184.01	Aug 02 2021	$I+z$	TRAPPIST-S-0.6m	150	3.5	2.4
TOI-4184.01	Sep 15 2021	I_C	Danish-1.54m [*f*]	60	2.8	1.3
TOI-4184.01	Sep 15 2021	1.21μm	ExTrA-0.6m [*g*]	60	4.0	1.5
TOI-4184.01	Sep 20 2021	I_C	Danish-1.54m	60	5.5	4.5
TOI-4184.01	Sep 20 2021	$I+z$	TRAPPIST-S-0.6m	150	6.2	2.7
TOI-4184.01	Sep 20 2021	i'	LCO-CTIO-1.0m	240	4.3	2.2
TOI-4184.01	Sep 25 2021	z'	SPECULOOS-S-1.0m	42	3.9	1.7
TOI-4184.01	Sep 25 2021	Rc	Danish-1.54m	100	4.0	2.8
TOI-4184.01	Oct 10 2021	g', i'	LCO-SAAO-1.0m	400,240	3.9	2.3
TOI-2136.01	May 12 2021	z'	TRAPPIST-N-0.6m	20	7.4	1.5
TOI-2136.01	Jun 28 2021	z'	TRAPPIST-N-0.6m	20	9.2	1.5
TOI-2136.01	Jul 06 2021	z'	TRAPPIST-N-0.6m	20	10.1	1.4
TOI-2136.01	Jun 21 2021	z_s	LCO-CTIO-1.0m	80	11.0	2.1
TOI-2136.01	Aug 22 2021	z_s	LCO-CTIO-1.0m	80	15.0	3.1
TOI-2136.01	Oct 24 2021	z'	SPECULOOS-N-1.0m	16	8.0	1.3

In this conference, we present the discovery and validation of three sub-Neptune exoplanets–TOI-2136 b (Gan *et al.* 2022), TOI-2084 b, and TOI-4184 b (Barkaoui *et al.* 2023) using data from TESS. We present follow-up observations including high-precision ground-based photometry, radial velocity measurements from SPIRou, and high-resolution imaging to confirm the planetary nature of these systems. These discoveries provide valuable insight into the radius valley and atmospheric evolution of small planets around M dwarfs, contributing to the broader understanding of planetary formation around these common stars.

2. Observations

2.1. *Photometry*

The discovery of TOI-2136 b, TOI-2084 b, and TOI-4184 b was initiated through TESS (Ricker *et al.* 2015) photometry. TOI-2136 (TIC 336128819) was observed in Sector 26 and re-observed in Sector 40. TOI-4184 (TIC 394357918) was observed in Sectors 1, 28, and 39, while TOI-2084 (TIC 441738827) was monitored in a 2-minute cadence across multiple sectors. The TESS light curves were processed using the SPOC pipeline, with the Presearch Data Conditioning (PDC) module effectively removing systematics to yield clean light curves suitable for transit analysis. Ground-based follow-up observations confirmed the transits and refined the orbital and planetary parameters, with detailed observational parameters provided in Table 1. The transit light curves for TOI-2136 b, TOI-2084 b and TOI-4184 b are presented in Figure 1.

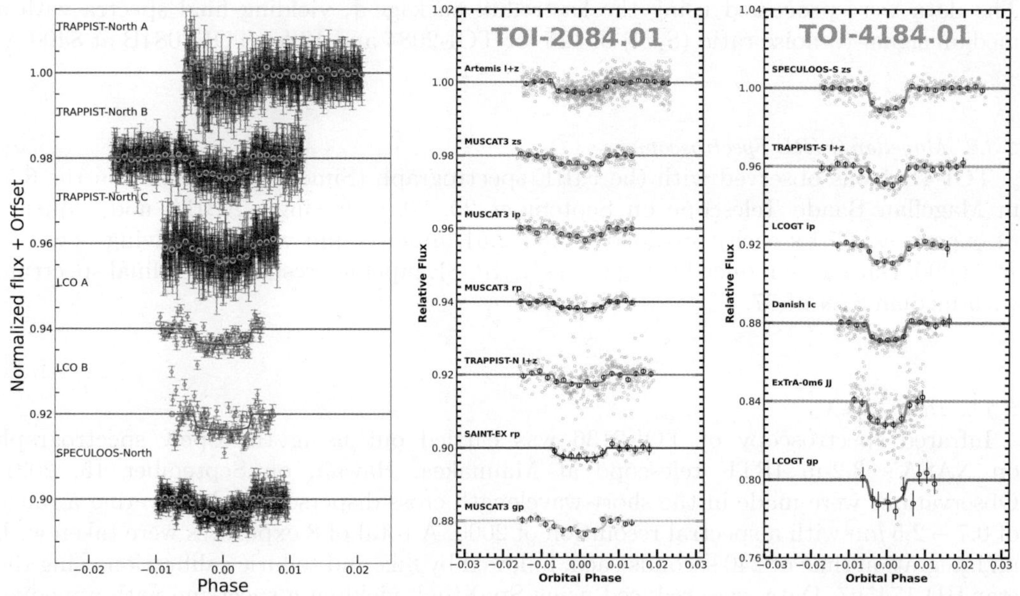

Figure 1. Ground-based photometric light curves of the three planets. *Left panel* shows the TOI-2136 b transit light curves. The black points with error bars and orange circles show the unbinned and binned data, respectively. The red solid lines are the best-fitting transit model. *middle* and *right panels* show the transit light curves of TOI-2084 b and TOI-4184 b. The gray points show the unbinned data and the black points are data binned at 10 minutes. The coloured lines are the best-fitting transit model. The light curves are shifted along the y-axis for visibility.

2.2. *High-angular-resolution imaging*

High-resolution imaging is crucial for identifying spatially close companions to exoplanet host stars, which can lead to false-positive transit signals and skew planetary radius estimates (Ciardi *et al.* 2015). Since Gaia cannot reliably detect binaries with separations smaller than 0.7" (Ziegler *et al.* 2020), we performed high-angular-resolution imaging for TOI-2136, TOI-2084, and TOI-4184.

For TOI-2136, Robo-AO (Lamman *et al.* 2020) observations confirmed no companions on July 29, 2016, with a Strehl ratio of 4.2% and a FWHM of approximately 0.12″. Further observations with the Shane 3-meter telescope (Kupke *et al.* 2012) on April 30, 2021, also found no nearby companions. For TOI-2084 and TOI-4184, high-resolution imaging was conducted using the Alopeke and Zorro speckle (Scott *et al.* 2021) instruments on the Gemini North and South 8-m telescopes, respectively. Both stars were confirmed to be single, with no companions brighter than 4-6 magnitudes below the targets, corresponding to spatial limits of 2.3 to 137 au for TOI-2084 and 1.4 to 83 au for TOI-4184.

2.3. *Spectroscopy*

2.3.1. *Shane/Kast Optical Spectroscopy*

Optical spectra of TOI-2084 and its co-moving companion were obtained on November 13, 2021, using the Kast double spectrograph (Miller & Stone 1994) on the 3-meter Shane Telescope at Lick Observatory. Six exposures of 600 seconds each were collected, covering the wavelength range of 6000–9000 Å at a spectral resolution of $\lambda/\Delta\lambda = 1900$.

The data were processed using the kastredux package †, yielding final spectra with a median signal-to-noise ratio (S/N) of 125 for TOI-2084 and 12 for TOI-2084B at 8400 Å.

2.3.2. Magellan/FIRE Spectroscopy

TOI-4184 was observed with the FIRE spectrograph (Simcoe et al. 2008) on the 6.5-m Magellan Baade Telescope on September 23, 2021. A single ABBA nod sequence (4 exposures, 95.1,s each) provided a 0.82–2.51μm spectrum with a resolving power of $R \sim 6000$. Data were reduced using the FIREHOSE pipeline, resulting in a final spectrum with median S/N of 77.

2.3.3. IRTF/SpeX

Infrared spectroscopy of TOI-2136 was carried out using the SpeX spectrograph on NASA's 3.2-m IRTF telescope at Maunakea, Hawaii, on September 15, 2021. Observations were made in the short-wavelength cross-dispersed mode, covering a range of $0.7 - 2.5\,\mu$m with a spectral resolution of 2000. A total of 8 exposures were taken with an integration time of 240 seconds each, followed by flux and telluric calibration using the star HD 174567. Data were reduced using SpeXtool, yielding a spectrum with a median SNR of 200, peaking at 250–300 in the JHK bands.

2.3.4. CFHT/SPIRou

TOI-2136 was observed with SPIRou at CFHT from April 24 to June 28, 2021. A total of 69 spectra were collected with a spectral resolution of 75,000, covering $0.98 - 2.5\,\mu$m. Observations spanned over 50 days with 2–4 exposures per night. The final spectra had a median SNR of 85 per pixel in the H band. Simultaneous drift calibration using the Fabry-Pérot mode was employed to achieve a radial velocity precision better than 10 m/s.

3. Results and discussion

We outline our analysis steps using available light curve and radial velocity (RV) data. For TOI-2136, we conduct a joint fit to determine the best-fit parameters for TOI-2136 b, utilizing the stellar parameters listed in Table 2. In contrast, the analyses for TOI-2084 and TOI-4184 follow a different approach, as detailed in their respective sections.

For a more detailed discussion on the analysis and results, refer to the original paper by Gan *et al.* (2022); Barkaoui *et al.* (2023).

3.1. *TOI-2136 b*

We perform a joint fit for TOI-2136 using the `juliet` package, simultaneously modeling the detrended *TESS* and ground-based light curves along with the SPIRou radial velocities (RVs). Gaussian priors for the linear limb darkening coefficients are applied, while dilution factors for *TESS* photometry are fitted to account for possible light contamination. The phase-folded light curves with the best-fit transit models are illustrated in Figure 1. The RV time series and best-fit model are presented in Figure 2. Our joint fit results indicate that TOI-2136 b has a radius of $2.20 \pm 0.17\,R_\oplus$ and a mass of $6.37 \pm 2.45\,M_\oplus$. A summary of the derived parameters is available in Table 3.

† `https://github.com/aburgasser/kastredux`

Table 2. Astrometry, photometry, and spectroscopy stellar properties of TOI-2084, TOI-4184, and TOI-2136. Parameters in bold are used as priors for our global analysis.

	Star Information		
Parameter	TOI-2084	TOI-4184	TOI-2136
Identifying information:			
TIC	441738827	394357918	336128819
GAIA DR2 ID	1652137995942479744	4620574887039870720	2096535783864546944
2MASS ID	2MASS J17170094+7244486	2MASS J02551841-7924554	2MASS J18444236+3633445
Parallax and distance:			
RA [J2000]	17:17:01.09	02:55:18.83	18:44:42.32
Dec [J2000]	+72:44:49.28	-79:24:52.93	36:33:47.27
Plx [*mas*]	8.750 ± 0.017	14.451 ± 0.027	29.976 ± 0.017
μ_{RA} [mas yr^{-1}]	47.73 ± 0.02	79.23 ± 0.03	-33.81 ± 0.02
μ_{Dec} [mas yr^{-1}]	36.76 ± 0.02	165.93 ± 0.03	177.05 ± 0.02
Distance [pc]	114.29 ± 0.22	69.20 ± 0.13	33.36 ± 0.02
Photometric properties:			
$\mathrm{TESS_{mag}}$	13.326 ± 0.007	14.261 ± 0.007	11.737 ± 0.007
V_{mag} [UCAC4]	15.115 ± 0.065	17.12 ± 0.20	-
B_{mag} [UCAC4]	16.668 ± 0.033	-	-
J_{mag} [2MASS]	11.961 ± 0.021	12.617 ± 0.023	10.184 ± 0.024
H_{mag} [2MASS]	11.356 ± 0.018	12.111 ± 0.026	9.604 ± 0.028
K_{mag} [2MASS]	11.148 ± 0.020	11.867 ± 0.025	9.343 ± 0.022
G_{mag} [Gaia DR3]	14.4096 ± 0.0005	15.5939 ± 0.0008	12.946 ± 0.011
$W1_{\mathrm{mag}}$ [WISE]	11.017 ± 0.023	11.685 ± 0.023	9.194 ± 0.022
$W2_{\mathrm{mag}}$ [WISE]	10.927 ± 0.020	11.472 ± 0.020	9.050 ± 0.021
$W3_{\mathrm{mag}}$ [WISE]	10.763 ± 0.061	11.371 ± 0.120	8.924 ± 0.027
Spectroscopic and derived parameters			
T_{eff} [K]	3550 ± 50	3225 ± 75	3342 ± 100
$\log g_\star$ [dex]	4.75 ± 0.05	5.01 ± 0.04	4.91 ± 0.03
$[Fe/H]$ [dex]	-0.13 ± 0.20	-0.27 ± 0.09	0.03 ± 0.07
$M_\star$ [$M_\odot$]	0.49 ± 0.03	0.240 ± 0.012	0.34 ± 0.02
$R_\star$ [$R_\odot$]	0.475 ± 0.016	0.242 ± 0.013	0.34 ± 0.02
$L_\star$ [$L_\odot$]	$0.0322^{+0.0028}_{-0.0025}$	$0.00544^{+0.00082}_{-0.00073}$	0.013 ± 0.003
$\rho_\star$ [$\rho_\odot$]	$4.21^{+0.17}_{-0.16}$	$16.32^{+1.25}_{-1.09}$	12.20 ± 2.53
Age [Gyr]	$7.5^{+4.4}_{-5.1}$	$6.7^{+4.9}_{-4.7}$	4.6 ± 1.0

R (1): Gaia EDR3 Gaia Collaboration et al. (2021); **(2)** *TESS* Input Catalog Stassun *et al.* (2018); **(3)** UCAC4 Zacharias *et al.* (2012); **(4)** 2MASS Skrutskie *et al.* (2006); **(5)** WISE Cutri *et al.* (2014).

3.2. *Joint-fit for TOI-2084 and TOI-4184*

We performed a joint fit of the light curves for TOI-2084 and TOI-4184 using the Metropolis-Hastings algorithm in an updated MCMC code (Gillon et al. 2012). The transit light curves were modeled with a quadratic limb-darkening model and a baseline model to correct for systematic variations, selected based on minimizing the Bayesian information criterion (BIC). The phase-folded light curves with the best-fit transit models are illustrated in Figures 1. We also used the relationship from Chen, J. & Kipping (2017) to predict the plausible mass of TOI-4184 b and TOI-2084 b. The derived parameters for both systems are summarized in Table 3.

3.3. *TOI-2136 b, TOI-2084, and TOI-4184 b and radius valley*

The bimodal distribution of small planets around FGK stars, which distinguishes super-Earths from sub-Neptunes, marks a transition between planets with and without

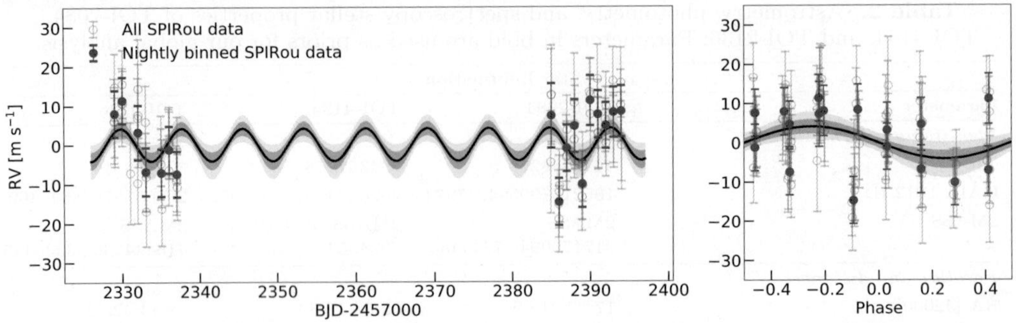

Figure 2. *Left panel*: Time series of the SPIRou RVs after subtracting the best-fit systemic velocity. The blue circles are all original SPIRou data taken every night while the red dots are the nightly binned RVs. The black solid line is the median RV model from the final joint-fit. The grey shaded areas denote the one and two sigma credible intervals of the RV model. *Right panel*: Phase-folded SPIRou RVs. The presented RV error bars in both panels are the the quadrature sum of the fitted instrument jitter and the measurement uncertainties.

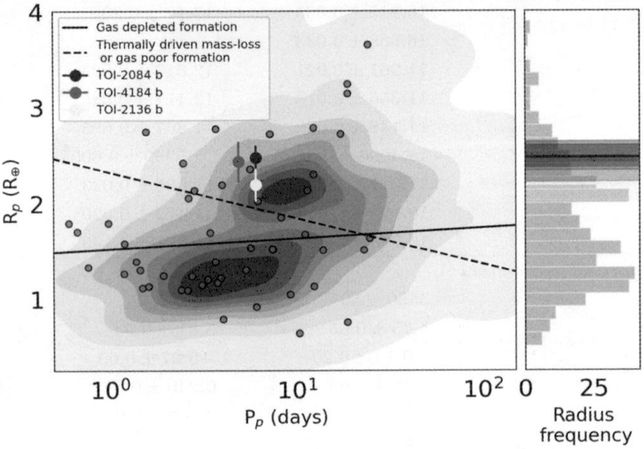

Figure 3. Distribution of planet radii and orbital periods for all confirmed small planets hosted by low-mass stars ($M_* \leq 0.65 M_\odot$). The solid line represents the predicted location of the radius valley based on the gas-depleted formation model (Cloutier & Menou 2020). The dashed line shows the predicted location of the valley for the thermally-driven pho-toevaporation and gas-poor formation models Van Eylen *et al.* (2018). The yellow, blue, and red dots represent the planets TOI-2136 b, TOI-2084 b, and TOI-4184 b, respectively. The 1D radius distribution, along with the locations of the three planets, is shown in the right panel.

Table 3. Derived planetary parameters for TOI-2136 b, TOI-2084 b, and TOI-4184 b.

Parameter	TOI-2136 b	TOI-2084 b	TOI-4184 b
P (days)	7.85193 ± 0.000019	6.0784247 ± 0.0000096	4.9019804 ± 0.000005
T_0 (BJD-2457000)	$2017.7042 \pm +0.0009$	1741.07323 ± 0.00065	2483.65667 ± 0.00023
$(R_p/R_*)^2$ (ppm)	3552 ± 1190	2267 ± 186	9115 ± 937
R_p ($R_\oplus$)	2.20 ± 0.17	2.47 ± 0.13	2.43 ± 0.21
M_P ($M_\oplus$)	$6.37^{+2.45}_{-2.29}$	$6.74^{+5.31}_{-2.81}$	$6.60^{+5.20}_{-2.75}$
b	0.34 ± 0.12	0.336 ± 0.061	0.31 ± 0.08
a/R_*	$35.91^{+1.40}_{-1.87}$	22.66 ± 0.39	30.79 ± 0.96
a (au)	0.057 ± 0.005	0.05006 ± 0.00103	0.0336 ± 0.0015
i (deg)	89.4 ± 0.2	89.15 ± 0.15	89.43 ± 0.16
S_p ($S_\oplus$)	$4.0^{+2.1}_{-1.5}$	12.8 ± 0.8	4.8 ± 0.4
T_{eq} (K)	394 ± 20	527 ± 8	412 ± 8

extended gaseous envelopes (Fulton *et al.* 2017, 2018). Martinez *et al.* (2019) demonstrated that this transition radius is dependent on orbital period, following a power law of $r_{\mathrm{p,valley}} \propto P^{-0.11}$. This aligns with predictions from thermally driven atmospheric mass-loss scenarios, such as photoevaporation and core-powered envelope escape, which suggest $r_{\mathrm{p,valley}} \propto P^{-0.15}$ (Lopez *et al.* 2018). In contrast, Cloutier & Menou (2020) proposed that the transition radius for small planets around M dwarfs adheres to a gas-poor formation model, following $r_{\mathrm{p,valley}} \propto P^{0.11}$.

For early-type M dwarfs with masses around 0.64 $M_{\odot}$, Cloutier *et al.* (2020) indicated that thermally driven atmospheric mass-loss remains influential for sculpting close-in planets. However, Luque *et al.* (2021) suggested a potential shift, proposing that the planetary radius valley for stars between 0.54 and 0.64 $M_{\odot}$ may arise from gas-poor formation. The competition between these physical processes at low stellar masses is still uncertain, underscoring the importance of gathering a larger sample of small planets with known bulk compositions.

As shown in Figure 3, TOI-2136 b, TOI-2084 b, and TOI-4184 b occupy positions near the predicted radius valley for M dwarfs, with TOI-2136 b lying just above it according to the thermally driven atmospheric mass-loss model, thereby contributing to our understanding of this transition.

Regarding the radius valley's behavior as a function of stellar mass, both physical processes predict a positive correlation between the center of the radius valley and stellar mass. However, the slopes differ across mass bins (Lopez *et al.* 2018; Gupta *et al.* 2019; Wu 2019). Cloutier & Menou (2020) noted a similar trend using a subset of Kepler and *K2* planets; however, the limited sample size, particularly for mid-to-late M dwarfs, introduces significant statistical uncertainty. Thus, TOI-2136 b, TOI-2084 b, and TOI-4184 b contribute to the growing sample of planets around M dwarfs, which will enhance our understanding of the transition radius's evolution with stellar mass in the future.

4. Conclusion

We present the discovery and characterization of three sub-Neptune-sized exoplanets: TOI-2136 b (Gan *et al.* 2022), TOI-2084 b, and TOI-4184 b (Barkaoui *et al.* 2023), all identified by the TESS mission and confirmed through a combination of photometric, spectroscopic, and imaging techniques.

TOI-2136 b is a sub-Neptune orbiting an M4.5 dwarf ($V = 14.1$ mag), with a radius of $R_p = 2.20^{+0.17}_{-0.17}\ R\oplus$, a mass of $M_p = 6.37^{+2.45}_{-2.29}\ M\oplus$, and an equilibrium temperature of $T_{\mathrm{eq}} = 395^{+24}_{-22}$ K. Its bulk density suggests a water world or a rocky planet with a moderate atmosphere. TOI-2136 b lies near the radius valley for M dwarfs, making it a compelling target for further studies on planetary formation and evolution. Its host star's small size and quiet nature make this planet suitable for atmospheric characterization by JWST.

TOI-2084 b orbits an M2-type star ($K_{\mathrm{mag}} = 11.15$) every 6.08 days. The planet has a radius of $R_p = 2.47 \pm 0.13\ R_{\oplus}$, an equilibrium temperature of $T_{\mathrm{eq}} = 527 \pm 8$ K, and a predicted mass of $M_p = 6.74^{+5.31}_{-2.81}\ M_{\oplus}$. The host star also has a wide M8 co-moving companion. This discovery contributes to the growing population of sub-Neptune-sized planets orbiting low-mass stars.

TOI-4184 b is a sub-Neptune orbiting an M5.5-type metal-poor star ($K_{\mathrm{mag}} = 11.86$) every 4.9 days. The planet has a radius of $R_p = 2.43 \pm 0.21\ R_{\oplus}$, an equilibrium temperature of $T_{\mathrm{eq}} = 412 \pm 8$ K, and a predicted mass of $M_p = 6.60^{+5.20}_{-2.75}\ M_{\oplus}$. The host star's low metallicity and small size make TOI-4184 b an interesting object for studies of planet formation in metal-poor environments.

These three planetary systems represent valuable laboratories for studying the formation and evolution of small planets around M dwarfs, with particular relevance to understanding the radius valley and atmospheric properties of such planets. Further observations, particularly with JWST, will provide insight into their atmospheric compositions and potential for habitability.

References

Barkaoui, K., Burdanov, A., Hellier, C., et al. 2019, AJ, 157, 43
Burdanov, A. Y., de Wit, J., Gillon, M., et al. 2022, PASP, 134, 105001
Barkaoui, K. et al.2023 *A&A* 677. doi:10.1051/0004-6361/202346838
Bonfils, X., Almenara, J. M., Jocou, L., et al. 2015, in SPIE Conference Series, Vol. 9605, ed. 96051L
Borucki, W. J., Koch, D., & Basri, G. 2010, *Science*, 327, 977, doi:10.1126/science.1185402
Brown, T. M., Baliber, N., Bianco, F. B., et al. 2013, PASP, 125, 1031
Chen, H. & Rogers, L. A. 2016, *ApJ*, 831, 180, doi:10.3847/0004-637X/831/2/180
Chen, J. & Kipping, D. 2017, ApJ, 834, 17
Cloutier, R. & Menou, K. 2020, *AJ*, 159, 211, doi:10.3847/1538-3881/ab8237
Ciardi, D. R., Beichman, C. A., et al. 2015, *ApJ*, 805, 16, doi:10.1088/0004-637X/805/1/16
Cloutier, R., Rodriguez, J. E., Irwin, J., Charbonneau, D., et al. 2020, *AJ*, 160, 22, doi:10.3847/1538-3881/ab9534
Cutri, R. M., Wright, E. L., Conrow, T., Fowler, et al. 2021, *VizieR Online Data Catalog*, II/328
Demory, B. O., Pozuelos, F. J., Gómez Maqueo Chew, Y., et al. 2020, A&A, 642, A49
Dominik, M., Jørgensen, U. G., Rattenbury, N. J., et al. 2010, Astronomische Nachrichten, 331, 671
Fressin, F., Torres, G., & Charbonneau, D. 2013, *ApJ*, 766, 81, doi:10.1088/0004-637X/766/2/81
Fulton, B. J., Petigura, E. A., & Howard, A. W. 2017, *AJ*, 154, 109, doi:10.3847/1538-3881/aa80eb
Fulton, B. J. & Petigura, E. A. 2018, *AJ*, 156, 264, doi:10.3847/1538-3881/aae828
Gaia Collaboration, Brown, A. G. A., Vallenari, et al. 2021, *A&A*, 649, A1, doi:10.1051/0004-6361/202039657
Gan, T. and et al. 2022. *MNRAS* 514, 4120–4139. doi:10.1093/mnras/stac1448
Gillon, M., Demory, B. O., Madhusudhan, N., et al. 2014, A&A, 563, A21
Ginzburg, S., Schlichting, H. E., & Sari, R. 2018, *MNRAS*, 476, 759, doi:10.1093/mnras/sty290
Gupta, A. & Schlichting, H. E. 2019, *MNRAS*, 487, 24, doi:10.1093/mnras/stz1230
Gupta, A. & Schlichting, H. E. 2020, *MNRAS*, 493, 792, doi:10.1093/mnras/staa315
Gupta, A. & Schlichting, H. E. 2021, *MNRAS*, 504, 4634, doi:10.1093/mnras/stab1128
Henry, T. J., Kirkpatrick, J. D., & Simons, D. A. 1994, *AJ*, 108, 1437, doi:10.1086/117164
Howard, A. W., Marcy, G. W., & Bryson, S. T. 2012, *ApJS*, 201, 15, doi:10.1088/0067-0049/201/2/15
Jin, S., Mordasini, C., Parmentier, V.,et al. 2014, *ApJ*, 795, 65, doi:10.1088/0004-637X/795/1/65
Johnson, J. A., Petigura, E. A., & Fulton, B. J. 2017, *AJ*, 154, 108, doi:10.3847/1538-3881/aa80e7
Kaltenegger, L., & Traub, W. A. 2009, *ApJ*, 698, 519, doi:10.1088/0004-637X/698/1/519
Kempton, E. M.-R., *et al.* 2018, *PASP*, 130, 114401, doi:10.1088/1538-3873/aadf6f
Kirkpatrick, J. D., Reid, I. N., Liebert, J., Cutri, R. M., et al. 1999, *ApJ*, 519, 802, doi:10.1086/307414
Kupke, R., Gavel, D., Roskosi, C., et al. 2012, *Proc. SPIE*, 8447, 84473G, doi:10.1117/12.926470
Lamman, C., Baranec, C., Berta-Thompson, et al. 2020, *AJ*, 159, 139, doi:10.3847/1538-3881/ab6ef1
Lopez, E. D. & Fortney, J. J. 2014, *ApJ*, 792, 1, doi:10.1088/0004-637X/792/1/1
Lopez, E. D. & Rice, K. 2018, *MNRAS*, 479, 5303, doi:10.1093/mnras/sty1707
Luque, R., Serrano, L. M., et al. 2021, *A&A*, 645, A41, doi:10.1051/0004-6361/202039455
Martinez, C. F., Cunha, K., Ghezzi, L., & Smith, V. V. 2019, *ApJ*, 875, 29, doi:10.3847/1538-4357/ab0d93
Miller, J. S., & Stone, R. P. S. 1994, *The Kast Double Spectrograph*, University of California Lick Observatory Technical Reports, 66

Narita, N., Fukui, A., Yamamuro, T., et al. 2020, in SPIE Conference Series, Vol. 11447, 114475K

Nutzman, P., & Charbonneau, D. 2008, *PASP*, 120, 317, doi:10.1086/533420

Owen, J. E. & Wu, Y. 2013, *ApJ*, 775, 105, doi:10.1088/0004-637X/775/2/105

Owen, J. E. & Wu, Y. 2017, *ApJ*, 847, 29, doi:10.3847/1538-4357/aa890a

Petigura, E. A., Howard, A. W., & Marcy, G. W. 2013, *PNAS*, 110, 19273, doi:10.1073/pnas.1319909110

Petigura, E. A., Howard, A. W., & Marcy, G. W. 2017, *AJ*, 154, 107, doi:10.3847/1538-3881/aa80de

Ricker, G. R., *et al.* 2015, *JATIS*, 1, 014003, doi:10.1117/1.JATIS.1.1.014003

Scott, N. J., Howell, S. B., et al. 2021, *Front. Astron. Space Sci.*, 8, 138, doi:10.3389/fspas.2021.716560

Simcoe, R. A., Burgasser, A. J., Bernstein, R. A., et al. 2008, in SPIE Conference Series, Vol. 7014, ed. I. S. McLean & M. M. Casali, 70140U

Skrutskie, M. F., Cutri, R. M., Stiening, R., et al. 2006, *AJ*, 131, 1163, doi:10.1086/498708

Stassun, K. G., Oelkers, R. J., Pepper, et al. 2018, *AJ*, 156, 102, doi:10.3847/1538-3881/aad050

Van Eylen, V., Agentoft, C., Lundkvist, M. S., et al. 2018, *MNRAS*, 479, 4786, doi:10.1093/mnras/sty1783

Winters, J. G., Henry, T. J., Lurie, J. C., et al. 2014, *AJ*, 149, 5, doi:10.1088/0004-6256/149/1/5

Wu, Y. 2019, *ApJ*, 874, 91, doi:10.3847/1538-4357/ab06f8

Zacharias, N., Finch, C. T., Girard, T. M., Henden, A., Bartlett, J. L., Monet, D. G., Zacharias, M. I. 2012, *VizieR Online Data Catalog*, I/322A

Ziegler, C., Tokovinin, A., Briceño, et al. 2020, *AJ*, 159, 19, doi:10.3847/1538-3881/ab55e9

Planetary Science and Exoplanets in the Era of JWST
Proceedings IAU Symposium No. 393, 2026
Z. Benkhaldoun, Y. Moulane & A. Szentgyorgyi, eds.
doi:10.1017/S1743921324002564

VLT imaging survey of giant planets around nearby young white dwarfs

Hans Zinnecker[1], **Wolfgang Brandner**[2] **and Matthias Schreiber**[3]

[1]Universidad Autonoma de Chile, Santiago de Chile
email: hzinnecker50@gmail.com

[2]Max-Planck-Institut für Astronomie, Königstuhl 17, 69117, Heidelberg, Germany

[3]Universidad Técnica Federico Santa María, Valparaíso, Chile

Abstract. We are imaging Gaia-selected young massive white dwarfs in the 40 pc solar neighborhood with the ESO-VLT (ERIS LGS AO instrument) to search for >5 Jupiter mass companions. These white dwarfs have 2.5-5 $M_\odot$ Main Sequence (MS) progenitors, and offer the unique possibility to test the formation of giant planets around intermediate mass stars, assuming these planets can survive post-MS evolution of their host stars. White dwarfs feature key advantages over their progenitor MS stars to spatially resolve giant planet companions in terms of (1) contrast and (2) angular separation. We limit ourselves to young white dwarfs with total ages (MS lifetime + white dwarf cooling age) < 1 Gyr which assures that any giant planets would be self-luminous and bright enough to be detectable (H-band < 25 mag). So far, we have obtained high angular resolution data for 10 white dwarfs with VLT/ERIS, with no confirmed detections, which might imply giant planets do not form around intermediate-mass MS stars, due to the rapid photoevaporation of their circumstellar disks caused by the host stars' elevated FUV and X-ray irradiation. We keep searching.

Keywords. white dwarfs, giant planets, VLT, JWST

1. Context

Imaging white dwarfs in search for giant planets around them has as a double advantage over imaging main sequence stars (the progenitors of white dwarfs) in search of the same giant planets. Advantage number 1 is that white dwarfs are faint, much fainter than their progenitor Main Sequence stars, by a factor of the order of 1000, due to their much smaller size. This provides a huge gain in contrast, some 7.5 magnitudes. Advantage number 2 is that giant planets that survive post-Main Sequence evolution migrate outwards due to conservation of specific angular momentum during the adiabatic mass loss phase of the host stars, by a factor of 4 to 5 in orbital radius. Thus, giant planets around white dwarfs are expected to be located 4 to 5 times further away from them than the same giant planets would be around their progenitor Main Sequence stars (Burleigh et al. 2002; Brandner et al. 2021). This translates into a huge gain in angular resolution, greatly facilitaing their detection.

Thirdly, VLT/ERIS LGS AO imaging is now available in the H- and K-band (¡0.1" resolution) overcoming the limitation of using faint white dwarfs as natural guide stars for AO (see Xu et al. (2015), using VLT/SPHERE, ex. GD 50). And finally, a GAIA white dwarf complete sample is now available in the wider (<100 pc) solar neighborhood (Gentile Fusillo et al. 2019) that enables downselecting the best white dwarf candidates (young: <1Gyr, massive: MWD > 0.7 $M_\odot$). Together, VLT/ERIS and GAIA/DR3 allow

Figure 1. Size comparison of an imtermediate mass main-sequence star, a giant planet, and a white dwarf.

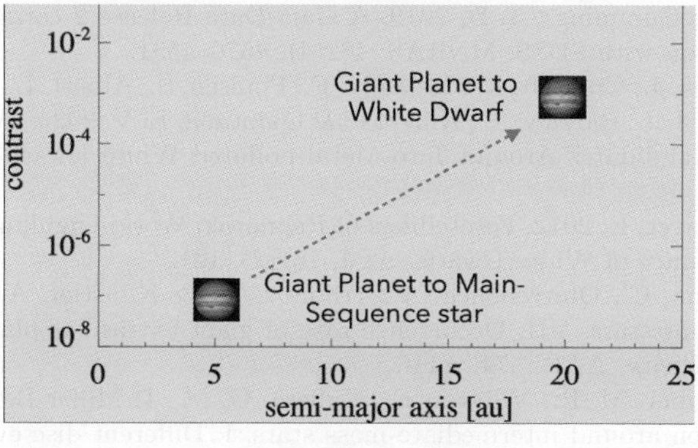

Figure 2. Evolution of planet-star contrast and separation from main-sequence to post-main-sequence stars.

us to study a new parameter space of giant planet formation, namely around intermediate mass stars in the MS mass range of 3-8 M$_\odot$ (here we focus on the progenitor range of 2.5-5 M$_\odot$)

2. Results

We performed angular differential imaging of 10 young, massive field white dwarfs within 32 pc of the Sun, using the VLT/UT4/ERIS instrument with LGS adaptive optics. No clear giant planet candidates were detected (down to a limiting mass of about 5 Jupiter masses). This tends to confirm the Reffert et al. (2015) result (radial velocity survey of 350+ red giants) that giant planets are rare for MS-star masses $>$ 2.5 M$_\odot$. However, our observed sample is still small, and we continue searching.

3. Implications

Why have we not seen any detections in our small sample? Several reasons may explain the non-detections. First, giant planets may exist but at lower masses or closer to the white dwarfs than detectable with our VLT direct imaging survey (see Vanderburg et al. (2020)), who detected a candidate giant planet transiting a white dwarf). To detect lower mass giant planets around white dwarfs we need JWST NIRCAM or MIRI direct imaging (Mullally et al. (2024) have two candidate detections). Secondly, giant planets can be lost being engulfed or tidally swallowed during common envelope evolution, spiralling into the RGB or AGB envelope (e.g. Mustill and Villaver (2012)), especially if originally close to the host star. Thirdly, giant planets do not form around intermediate mass stars due to fast photoevaporation of their circumstellar disks (Ronco et al. 2024).

4. Outlook

The advantage of a possible detection of any spatially resolved giant planet (to be confirmed with common proper motion 2nd epoch data) would be to allow us to do JWST spectroscopy of such a planet with a well-determined age and mass (luminosity) and provide us with a benchmark for comparison with our solar system gas giants.

References

Brandner, W., Zinnecker, H., & Kopytova, T. 2021, Search for giant planets around seven white dwarfs in the Hyades cluster with the Hubble Space Telescope. MNRAS, 500(3), 3920–3925.

Burleigh, M. R., Clarke, F. J., & Hodgkin, S. T. 2002, Imaging planets around nearby white dwarfs. MNRAS, 331(4), L41–L45.

Gentile Fusillo, N. P., Tremblay, P.-E., Gänsicke, B. T., Manser, C. J., Cunningham, T., Cukanovaite, E., Hollands, M., Marsh, T., Raddi, R., Jordan, S., Toonen, S., Geier, S., Barstow, M., & Cummings, J. D. 2019, A Gaia Data Release 2 catalogue of white dwarfs and a comparison with SDSS. MNRAS, 482(4), 4570–4591.

Mullally, S. E., Debes, J., Cracraft, M., Mullally, F., Poulsen, S., Albert, L., Thibault, K., Reach, W. T., Hermes, J. J., Barclay, T., Kilic, M., & Quintana, E. V. 2024, JWST Directly Images Giant Planet Candidates Around Two Metal-polluted White Dwarf Stars. APJL, 962(2), L32.

Mustill, A. J. & Villaver, E. 2012, Foretellings of Ragnarök: World-engulfing Asymptotic Giants and the Inheritance of White Dwarfs. APJ, 761(2), 121.

Reffert, S., Bergmann, C., Quirrenbach, A., Trifonov, T., & Künstler, A. 2015, Precise radial velocities of giant stars. VII. Occurrence rate of giant extrasolar planets as a function of mass and metallicity. AAP, 574, A116.

Ronco, M. P., Schreiber, M. R., Villaver, E., Guilera, O. M., & Miller Bertolami, M. M. 2024, Planet formation around intermediate-mass stars. I. Different disc evolutionary pathways as a function of stellar mass. AAP, 682, A155.

Vanderburg, A., Rappaport, S. A., Xu, S., Crossfield, I. J. M., Becker, J. C., Gary, B., Murgas, F., Blouin, S., Kaye, T. G., Palle, E., Melis, C., Morris, B. M., Kreidberg, L., Gorjian, V., Morley, C. V., Mann, A. W., Parviainen, H., Pearce, L. A., Newton, E. R., Carrillo, A., Zuckerman, B., Nelson, L., Zeimann, G., Brown, W. R., Tronsgaard, R., Klein, B., Ricker, G. R., Vanderspek, R. K., Latham, D. W., Seager, S., Winn, J. N., Jenkins, J. M., Adams, F. C., Benneke, B., Berardo, D., Buchhave, L. A., Caldwell, D. A., Christiansen, J. L., Collins, K. A., Colón, K. D., Daylan, T., Doty, J., Doyle, A. E., Dragomir, D., Dressing, C., Dufour, P., Fukui, A., Glidden, A., Guerrero, N. M., Guo, X., Heng, K., Henriksen, A. I., Huang, C. X., Kaltenegger, L., Kane, S. R., Lewis, J. A., Lissauer, J. J., Morales, F., Narita, N., Pepper, J., Rose, M. E., Smith, J. C., Stassun, K. G., & Yu, L. 2020, A giant planet candidate transiting a white dwarf. NAT, 585(7825), 363–367.

Xu, S., Ertel, S., Wahhaj, Z., Milli, J., Scicluna, P., & Bertrang, G. H. M. 2015, An extreme-AO search for giant planets around a white dwarf. VLT/SPHERE performance on a faint target GD 50. AAP, 579, L8.

Author index

Ábrahám, P. – 126
Affer, L. – 150
Aigrain, S. – 150
Akimkin, V. – 126
Amuda, T. O. – 105
Azagrouze, O. – 138, 171, 176

Bardelli, S. – 87
Barkaoui, K. – 187
Baron, F. – 77
Bellotti, S. – 34
Benkhaldoun, Z. – 14, 53, 138, 171, 176, 187
Bolduc-Duval, J. – 77
Bonomo, A. S. – 150
Bonsall-Pisano, A. – 182
Boskri, A. – 138, 171, 176
Bourrier, V. – 34, 150
Brandner, W. – 196
Buchhave, L. A. – 150

Cameron, A. – 150
Čapek, D. – 14
Cegla, H. – 150
Chafi, J. – 138, 171, 176
Chouqar, J. – 53
Chupina, N. V. – 97
Cortes-Zuleta, P. – 150
Cosentino, R. – 150
Costin, G. – 182
Cretignier, M. – 150

Damasso, M. – 150
Danielski, C. – 34
de Gregorio-Monsalvo, I. – 60
de la Villarmois, E. A. – 60
Di Giacomo, F. – 87
Donati, J. F. – 34
Dumusque, X. – 150
Duras, F. – 87

Ehrenreich, D. – 150
El Azhari, Y. – 138, 171, 176
El Fatimy, A. – 72
Evensberget, D. – 34

Fares, R. – 34
Fateev, A. M. – 97
Fateeva, A. M. – 97
Filippelli, G. – 87
Fiorenzano, A. M. – 150

Gan, T. – 187
Ghedina, A. – 150
Giacomini, L. – 87
Gil, S. – 60
Gómez, J. – 60
Gozaliasl, G. – 156, 161
Guennoun, M. – 14
Guzmán, V. – 60

Hasan, P. – 129
Hasan, S. N. – 129
Hébrard, É. M. – 34
Hojaev, A. S. – 100
Howarth, G. H. – 182
Hubbard-James, H.-S. – 144
Hussain, G. A. J. – 34

Ipatov, S. I. – 9, 29

Jabiri, A. – 53, 138
Jassur, D. M. – 156, 161
John, A. A. – 150

Kapranov, S. – 1
Klein, B. – 150
Kóspál, A. – 126
Koten, P. – 14
Kuznetsov, E. D. – 133, 167

Latham, D. – 150
Lavail, A. – 34
Leke, O. – 105
Leleu, A. – 150
Leng, X. – 114
Linsky, J. L. – 20
Lopez-Morales, M. – 150
Lovis, C. – 150
Lüftinger, T. – 34
Lustig-Yaeger, J. – 53

Malavolta, L. – 150
Maldonado, J. – 150
Mantovani, G. – 87
Micela, G. – 34, 150
Mignone, C. – 87
Molinari, E. – 150
Molyarova, T. – 126
Morin, J. – 34
Mortier, A. – 150
Mosunova, D. A. – 97
Moulane, Y. – 72

Mousavi-Sadr, M. – 156, 161
Moutou, C. – 34

Naud, M.-E. – 77
Negu, S. H. – 97
Nicholson, B. – 150

O'Sullivan, N. – 150
Ouellette, N. N.-Q. – 77

Peel, C. – 182
Pepe, F. – 150
Pérez-Sánchez, A. – 60
Perminov, A. S. – 133, 167
Petit, P. – 34
Pinamonti, M. – 150
Piotto, G. – 150
Plunkett, A. – 60
Postnikova, E. S. – 97

Qazbour, A. – 138, 171, 176

Redfield, S. – 20
Rice, K. – 150

Safronova, V. S. – 133
Saikia, S. B. – 34
Santamaría-Miranda, A. – 60
Sasselov, D. – 150
Scalise, E. – 87
Schreiber, M. – 196
Schreiber, M. R. – 60
Semenov, D. – 126
Shaikh, I. – 129
Silva, A. M. – 150
Sizova, M. D. – 97
Soubkiou, A. – 53, 187
Sozzetti, A. – 150
Stalport, M. – 150
Szentgyorgyi, A. – 53

Tchernyi, V. V. – 1

Udry, S. – 150

Valero, K. V. – 121
van Dishoeck, E. F. – 65
Vanderburg, A. – 150
Vasileva, M. A. – 133
Vaubaillon, J. – 14
Vereshchagin, S. V. – 97
Vidotto, A. A. – 34
Vissapragada, S. – 150

Wakjira, H. Y. – 97
Wandel, A. – 45
Watson, C. – 150
White, H. – 77
Wilson, T. – 150

Zinnecker, H. – 196
Zwicky, L. – 126